The New Solar System

The New Solar System

SECOND EDITION

edited by

J. Kelly Beatty
Brian O'Leary
Andrew Chaikin

introduction by
Carl Sagan

CAMBRIDGE UNIVERSITY PRESS
Cambridge
London New York New Rochelle Melbourne Sydney
&
SKY PUBLISHING CORPORATION
Cambridge, Massachusetts

Published by the Press Syndicate of the University of Cambridge
The Pitt Building, Trumpington Street, Cambridge CB2 1RP
32 East 57th Street, New York, New York 10022, USA
296 Beaconsfield Parade, Middle Park, Melbourne 3206, Australia
and by Sky Publishing Corporation
49 Bay State Road, Cambridge, Massachusetts 02238-1290, USA

First published 1981
Second edition 1982
Printed in the United States of America

Library of Congress Cataloging in Publication Data

The new solar system.

Includes index. Includes bibliographical references.

1. Solar system. I. Beatty, J. Kelly.
II. O'Leary, Brian, 1940- . III. Chaikin, Andrew, 1956- .

QB501.N47 1982   523.2   81.2661   AACR2
ISBN 0-521-24988-0 hard covers
ISBN 0-521-27114-2 paperback

(First edition
ISBN 0-521-23881-1 hard covers)

Table of Contents

Preface

Twenty years ago, Gerard P. Kuiper and Barbara M. Middlehurst published what would become a standard reference in planetary science, *Planets and Satellites.* The third of a four-volume set, this work brought together an enormous body of knowledge about the solar system, using an unorthodox format that dealt with techniques rather than objects. It included chapters on photometry, polarimetry, and radiometry; on planetary temperature measurements made at infrared wavelengths; and on recent photographic surveys conducted with world-class telescopes. Conspicuously absent, however, were chapters devoted to individual planets — the editors thought it more desirable to present the material according to the investigative method used, since that allowed a better interpretation of which information was valid, and which was suspect.

In their preface, the editors pointed out that the growth of planetary astronomy had occurred in three distinct phases. The first spanned nearly three centuries of "classical" revelations by Galileo, Copernicus, Kepler, Laplace, and their contemporaries — a period rich with discovery, when astronomy was almost wholly concerned with the planets, and stars were more suited to philosophical discourse than scientific study. But the situation reversed dramatically in the late 1800's with the advent of astrophysics and reliable astrophotography. In fact, the "market" for planetary science fell so precipitously that only a few astronomers deigned to pursue such studies at all.

"The third phase of planetary astronomy," Kuiper and Middlehurst noted, "is that of the 1960's, during which entirely new data are expected to become available through the application of research with rockets. The first chapter of this volume [entitled "Planet Earth as Seen from Space"] shows at once the power and some of the limitations of this new tool." But they cautioned, "No doubt the planning of these new techniques, which will be several orders of magnitude more expensive than the astronomical observations hitherto made, will continue to depend heavily on ground-based observations, which are the basis of other chapters in this volume."

Kuiper, who died in 1973, is considered the father of modern planetary science. During the 1940's, he began to devote his considerable talent to the planets after already distinguishing himself with work on double stars and stellar evolution. Kuiper showed the power of the telescope to discern aspects of planets and their satellites. He conducted his research with considerable finesse, picking meticulously through observations to glean everything within reason.

Two decades have passed since the publication of *Planets and Satellites.* In that time spacecraft have been flung through space toward (and sometimes onto or into) all of the planets known to the ancients. Much has been learned by these robot messengers; the vastness of their bountiful returns fills entire libraries, with success measured by dozens of missions, tens of thousands of photographs, and billions of bits of information. Yet Kuiper's prophetic insistence that telescopic observation from Earth would remain an important investigative ingredient rings just as true now as then. Without it, much of what you will read about on these pages would still lie waiting to be discovered — the rings of Uranus, Earth-crossing asteroids, and Pluto's satellite Charon are three such examples of recent telescopic "finds."

The New Solar System is not intended to match the exhaustive contents of the Kuiper-Middlehurst series. But one interesting analogy can be drawn. We also have chosen not to organize our chapters in a planet-by-planet rundown the way most of us have learned of these worlds. Our organization is a direct consequence of the way planetary science has evolved since Kuiper's time — today solar system research revolves around what has come to be called comparative planetology. For example, it is easier to discuss the atmospheres of the terrestrial planets (Mercury, Venus, Earth, and Mars) together in one chapter, and their surfaces in another, than to attempt the total characterization of each planet separate from the others.

There are a number of reasons for this comparative approach. First, it reduces the amount of overlap considerably; without it, definitions and ground rules would repeat every few pages. Second, the comparative approach permits a fuller understanding of the total picture. A third reason, which is far from insignificant, is that very few planetary scientists today possess a complete working knowledge of the various scientific disciplines involved in the comprehensive study of "a planet" — these would truly be renaissance scholars, versed in physics, geology, mathematics, fluid dynamics, and biology, among other fields. Finally, the solar system no longer lends itself to convenient partitioning into "planets," "satellites," and "others." For instance, the criteria previously used to distinguish comets from asteroids (comets have tails and pass near the Earth; asteroids do not and stay in the asteroid belt) just aren't applicable any more. In some cases the criteria have changed, while others have simply been discarded entirely.

Of course, we are a long way from knowing all the answers. While the remarkable progress of the last two decades allows us — finally — to piece together reasonable scenarios for the birth, evolution, and present ongoings of our various interplanetary neighbors, we are reminded too often that Nature still retains the upper hand. Each of the authors has endeavored not only to provide the most up-to-date information available, but also to present the gaps in our understanding that beg for further investigation. Nor is the content intended to be entirely self-consistent. Many topics are the subject of disagreement or even outright feuds. Others are presented even though confirmation is lacking. Such is the case with many small outer-planet satellites, glimpsed but once either by spacecraft camera or by telescopic observer.

In assembling this book, we have attempted to bring the fruits of recent planetary exploration to a wide audience. This is really neither a textbook nor a "coffee-table"

volume — it lies somewhere in between. By the same token, we have encouraged the authors to provide material that avoids both sweeping generalizations and incomprehensible details. Above all, we strove to make this enjoyable reading for those with either professional or casual interest.

Our deep gratitude must be extended to those whose time and talent have made this project come to fruition. The groundwork laid by Leif Robinson of Sky Publishing Corporation, and by Paul Wehn, Simon Mitton, and Kathy Aversano of Cambridge University Press, was critical to the overall effort. Our thanks also to David Stryker, Paul Ferragamo, Bob Casoni, Terry Tyler, Chuck Harpel, and William Hoyt for their expertise and unwavering cooperation. The Jet Propulsion Laboratory's Public Information Office, and especially Jurrie van der Woude, were quick to provide numerous "must-have" illustrations. The artistic excellence of Charles Wheeler, Don Davis, and Jon Lomberg contributed greatly to this project. The extraordinary efforts of designer Jennifer Craddock, graphic supervisor Charles Baker, production manager Anita Leek, Connie Springer, Ron Arruda, Richard Backer, Susan Bencuya, and Cheryl Nygaard exceeded all reason; for their tireless work we are truly grateful. And finally, we wish to thank our authors for sharing their insights with a waiting world.

March, 1981

J. Kelly Beatty
Brian O'Leary
Andrew Chaikin

PREFACE TO THE SECOND EDITION

When *The New Solar System* was first published in May, 1981, Voyager 2 was roughly three months away from its encounter with Saturn and closing fast. Before long, Voyager scientists who had barely a chance to digest the tremendous harvest of the first Saturn flyby were faced with a stream of new data on the ringed planet and its family of satellites. As the year closed, two Soviet spacecraft sped not outward but toward the Sun, for a landing on cloud-hidden Venus. Historic color pictures and crucial chemical analyses of its mysterious surface were in store. On Earth, astronomers continued to coax secrets from the dim light of the solar system's most elusive objects — comets, asteroids, and faint satellites — all the while revising and refining their interpretations of recent spacecraft findings.

The solar system of today is in many ways the same as that of spring, 1981. But there are enough differences to warrant a second look, and thus a second edition of this

volume. An assortment of planetary maps, originally envisioned for the first edition, has been added. The result, we hope, will serve as a benchmark until the current hiatus in deep-space exploration has passed.

The production of a second edition proved a sizable undertaking, to the mild surprise of the editors, and additional acknowledgments are thus in order. The efforts of Charles Baker, Ron Arruda, Susan Bencuya, Richard Backer, Richard Hollick, and Anita Leek were again invaluable. Charles Wheeler and Robert Hess contributed fresh artistic input, and we thank Raymond Batson of the U. S. Geological Survey and Marie Tharp for their help in assembling the handsome maps at this volume's end. And special thanks go to the authors, who rushed us their new information and bravely accepted the limitations of available space. Ultimately, it is they and their colleagues who decide when the next reassessment will be warranted.

June, 1982

Introduction

Carl Sagan

A little less than five billion years ago, something happened. A cloud of interstellar gas and dust, perhaps triggered by a nearby exploding star, collapsed and condensed to form the solar system. The central mass in the cloud contracted under its own gravity and heated, until temperatures became so high that thermonuclear reactions were initiated and the early Sun was born. Subsidiary, smaller lumps of matter did not achieve such high temperatures and pressures and did not become stars. Shining by reflected light, they evolved into the planets, satellites, asteroids, and comets that today form the Sun's entourage.

In one of those smaller lumps of matter, the heating of the interior drove out gases, forming an atmosphere. One variety of gas condensed on the surface, forming protective lakes and oceans. Chemical reactions in the air and water produced complex organic molecules, which eventually — about four billion years ago — resulted in a molecular system capable of making identical copies of itself from the surrounding molecular building blocks. The world on which these events happened, third from the Sun, is, of course, the Earth; and we are some of the descendants of that first self-replicating molecular system.

We humans are new to this solar system. We tend to think in short time scales. But the surfaces of many other objects in the solar system date back billions of years to a very different epoch, a time of titanic collisions and catastrophes. We have a tendency to think of other worlds as like our own and other epochs as placid as this one. This is not the case. We inhabit a solar system rich in wonders: a stifling hot world, where heat is trapped in a thick atmosphere surmounted by clouds of sulfuric acid; an extinct volcano three times the height of Mount Everest; delicate circumplanetary rings twisted in astonishing, braided patterns; a world crisscrossed with elaborate straight and curved valleys; hurtling balls of ice, trailing tails of gas and dust as long as the distances between the planets; an underground ocean of molten sulfur. There exists a giant planet with an interior of liquid metallic hydrogen, into which a thousand Earths would fit; a planet-sized moon with an atmosphere as thick as the Earth's, possibly dotted with lakes of liquid methane, and surrounded by an unbroken haze layer made up of complex organic molecules; and a beautiful, small, blue and white world on which organic molecules have evolved to

The subtle yet profound relationships between matter and life provide the theme for "DNA Embraces the Planets," by Jon Lomberg.

produce creatures able to contemplate the diversity of worlds and the passage of ages, slowly becoming aware of the richness that the solar system offers, in space and time.

Voyagers 1 and 2, in their exploration of the lovely system of Saturn, its rings, and its moons, complete a preliminary reconnaissance of all the worlds known to the ancients, of all the planets visible to the naked eye — from Mercury to Saturn. This is an historic moment in the evolution of the solar system: the first time, so far as we know, that one world in the Sun's family has achieved some fair understanding of the others. For human history, this moment is certainly more significant than the discovery of America by Christopher Columbus, and perhaps as significant as the colonization of the land by the first amphibians about 400 million years ago. We are leaving the world of our origins and have begun — first by machines, but later, almost certainly, with spacecraft piloted by human beings — exploring other worlds.

This is an effort where the benefits, while very real, are long range. A consistent commitment to planetary exploration has been difficult to organize. Now many nations are beginning to explore other worlds — not just the United States and the Soviet Union, but also Japan, the constituent members of the European Space Agency, China, and perhaps others. It seems quite probable that if we do not destroy ourselves — thereby ending this promising local experiment in the evolution of matter — we will continue our explorations, garnering more information on the atmospheres, surfaces, interiors, and possible biology within our solar system. We will discover what else is possible. We will better understand our own world by comparing it with others. We will send our little vehicles bravely into the turbulent atmosphere of Jupiter, across the exotic surface of Mars, through the tails of comets, past the asteroids, settling gently down onto the unimaginable landscape of Titan. These are all tasks well within our engineering capability. We need only a steadfast determination, a continuation of that ancient and honorable human exploratory tradition.

But even the complete exploration of our solar system is unlikely to settle the uncertainties surrounding those tumultuous events from which the Sun and planets emerged some five billion years ago. For that, we need to search for other examples of planetary systems. This is possible from the ground, but it will be easier from large telescopes in Earth orbit. There is a real chance that by the middle of the 21st century, we will have performed not only a deep reconnaissance of all of the worlds from Mercury to Pluto and beyond, but also a systematic survey of the planetary systems — if such exist — of hundreds of nearby stars. We will then be able to say, with a fair likelihood of being correct, something of the cosmic generality and origins of our solar system, our planet, and ourselves. This is an enterprise with deep significance for every inhabitant of the planet Earth.

The Golden Age of Solar System Exploration

Noel W. Hinners

The exploration of the solar system by spacecraft has now spanned more than two decades, producing a wealth of basic discovery and new data at a rate unparalleled in history. A host of unmanned spacecraft have transformed our view of the planets from one largely of fuzzy, shimmering telescopic images to one dominated by crisp global perspectives of truly new worlds, simultaneously displaying raw, simple beauty and the awesome end-products of powerful, complex natural forces. Lunar exploration by spacecraft and astronauts, along with all-important laboratory analysis of returned samples, provided us with our first significant glimpse of the history of the Sun and with a chronology of planetary processes in the first billion years of the solar system, a picture unlikely to be updated with knowledge obtained elsewhere in the foreseeable future. The relatively detailed "ground-truth" knowledge we now have of the Moon serves as the primary basis for interpreting the abundance of remotely sensed planetary data.

During this period of unprecedented *in situ* solar system exploration, earth-based telescopic observations led to the discoveries of the rings of Uranus, a satellite of Pluto, methane frost on Pluto's surface, a host of Earth-crossing asteroids, and yielded invaluable compositional information on comets, asteroids, and planetary atmospheres from Venus to Neptune. Finally, intricate terrestrial laboratory studies of meteorites, the only true "free lunch" in the space program, have helped to develop the factual link between pre-solar-system stellar explosions and the birth of the Sun and the planets.

The totality of planetary exploration over the last two decades, summed up in this book, easily justifies, I believe, the appellation The Golden Age of Solar System Exploration. No previous exploratory effort can match it in terms of rate of basic discovery, immediate impact on scientific thinking, eventual significance to society, and rapid communication of the results to the masses of earthly inhabitants. This in no way denigrates previous eras of discovery, the most important of which, in aggregate, was the early exploration of the Earth and the determination of our true place in the solar system and in turn of the solar system's place in the galaxy and universe. Consider, however, that this occurred over an extended period of thousands of years in a combination of tedious, lengthy geographical and astro-

nomical ventures including: Minoan voyages in the Mediterranean (2000-500 B.C.); Phoenician navigations around Africa and the west coast of Europe (1500-500 B.C.); Alexander's conquest of the Persian empire (ca. 330 B.C.); the religiously motivated travels of Hsüan-tsang to India (ca. 600 A.D.); river and ocean voyages of the Vikings (ca. 1000); Marco Polo's expeditions to China (ca. 1300); that great spurt of major ocean voyages by Portuguese, Spanish, French, and English explorers of the 15th and 16th centuries; Cook's Australasian explorations (late 1700's); and the Lewis and Clark expedition (early 1800's). These explorations and others, one by one, were integrated into a partial global picture of the physical attributes of our home planet. But only partial, because even the first "orbit" of the Earth by Magellan's expedition, a venture that took from September, 1519, until September, 1522, did not reveal the whole planet. Only in the last 200 years has the extent of the polar regions become evident and not until the past 40 years has the vast ocean bottom region begun to yield its essential topographic and geologic secrets.

Progress in establishing our astronomical context was similarly slow: more than a century elapsed between Copernicus' development of the heliocentric nature of the solar system in the early 1500's and the confirming observations and analyses of Galileo, Tycho, and Kepler. Not until the early 20th century did our galactic "address" become known and even today our yardstick of cosmic distance (and thus age) has a nasty habit of gaining or losing billions of light years at a clip.

The recent pulse of planetary exploration parallels in many ways that which is also evident in other sciences, especially biology, astronomy, physics, and terrestrial geology; in fact, one might easily wish to reference all of these in aggregate as the Golden Age of Science. This spurt in many sciences is not simply coincidence, as common denominators link their phenomenal growth: a solid foundation in mathematics and physics, new technologies, emphasis on mass education, realization of the long-term importance of basic research, popularization of science, potential near-term economic benefits, "discretionary" economic means, and political competition. These factors interact in often subtle and complex ways to determine the "health" of the scientific process and the potential for future research.

Unlike planetary exploration, however, other sciences are apt to continue a relatively steady pace of activity because of significant differences both in exploitation of the common denominators and in underlying motivations. It is those differences which are now leading to a precipitous decline in the rate of planetary exploration and which may well change its nature in the future. That would leave the exploration to date, along with what little is underway or contemplated for the 1980's, to mark the Golden Age of Solar System Exploration.

One can rationally inquire: with such an impressive and fruitful planetary exploration program spanning two decades, why is it that at this writing, only one American planetary mission has been approved for the 1980's and why that one, the Galileo orbiter and probe of Jupiter, hangs by a thread as various forces conspire to kill it? To understand this dilemma better, and to prepare viable plans for the future, we must examine and understand the origin and evolution of NASA's solar system exploration program.

WHY EXPLORE THE SOLAR SYSTEM?

Solar system exploration must be viewed with two minds, one asking why it exists, the other how it has been accomplished. To confuse the two, as frequently happens, makes it difficult to understand the problems of the future. In this section, we deal with the basic reasons behind the program; later, we examine how it grew and evolved.

There appear to be four dominant reasons why the United States has a solar system exploration program: national prestige, vision, knowledge, and applications. These are listed in order of the priority they held in the late 1950's, when the national space program was formally initiated; subsequently the emphasis has shifted such that today national prestige is scarcely recognized as a significant factor by the "decision makers." Let us now examine each and see where and how they played a compelling role in the past and what their role might be in the future.

National prestige, a product of how we view ourselves and how others view us as a nation, is an elusive commodity and difficult to measure. Surely, it is dependent upon real or imaginary perceptions by the citizenry and its elected representatives. How those perceptions come about is dictated in large part by how the news media perceive and report them and upon how the politician chooses to present them. Few would deny that the Soviet launching of Sputnik in October, 1957, caused an immediate national concern and soul-searching. Since 1945, we had enjoyed the psychological luxury of being "first" in science and technology; we thought our democracy to be the best political system, one in which science and technology could flourish. In contrast, conventional wisdom supported a view that the Soviet political and industrial system, despite its earlier-than-anticipated development of nuclear weaponry, was somehow archaic, backward, and relatively undeveloped. The collective shock and wounded national ego, exacerbated by the realization of potential military applications of orbital and satellite technology, resulted in the near-immediate, politically based decisions to beef up science education programs; establish a public, civil space program; and reinvigorate the military applications of space activity.

The initial, politically inspired thrust into space was reinforced and extended in 1961 when President Kennedy stated his goal of sending American astronauts to the surface of the Moon and safely returning them to Earth by the end of the decade. This decision and the resulting actions which spawned the Apollo program were also politically inspired, driven largely by the loss of face suffered from continuing Soviet space firsts (including Gagarin's orbital flight in 1961) and reinforced by the Bay-of-Pigs fiasco.

It is in the above political context that the planetary program was born and evolved and to which we will return later. For the moment, consider that the first success in the American planetary program — Ranger 7's photo-reconnaissance of the Moon in July, 1964, coming as it did after a long series of failures — was much more of a political success than a scientific one; in showing that the U.S. could accomplish a major technological objective, the achievement regained some of our lost prestige, eased the passage of the NASA budget, and led to a more favorable climate for the conduct of the Apollo project. For 12 years thereafter, the manned programs of Gemini, Apollo, and Skylab overshadowed unmanned planetary programs in public attention and international acclaim. Automated planetary probes regained the spotlight with the Viking landings on Mars in 1976, by which time the entire political climate had changed to one in which the Soviets were no longer viewed as serious competition and the public had become somewhat numbed to space activity. Although news articles frequently highlighted the fact that the U.S. was the first to land on Mars, in contrast to several Soviet failed attempts, editorial comment focused more on the remarkable technological and scientific accomplishment *per se,* and on the exploration ethic and its place in American culture. That pattern continued with the Pioneer mission to Venus and the Voyager missions to Jupiter and Saturn.

Political motivation is not, of course, limited to the United States space program. That of the Soviet Union is well known. Its planetary program in particular has, compared to NASA's, been more preoccupied with attempts to be first, and the Soviets made much ado of the planting of the Soviet flag and Lenin pendants on the Moon and Venus with the implied superiority of the Soviet political system. Aside from their exploration of Venus, a systematic, primarily science-based, long-term Soviet strategy and program plan for solar system exploration does not seem to exist. Both lunar and Martian exploration attempts ceased relatively soon after the spectacular American successes. In a different vein, and a bit surprising to some scientists, countries and organizations which used to depend upon cooperation with the U.S. to conduct planetary missions now have, or soon will, the capability to conduct their own flights. Both the European Space Agency and Japan are desirous of and working toward implementing their own planetary programs, starting with missions to Halley's Comet, which will reach perihelion in 1986. That their first missions are not as sophisticated as those of the United States is beside the point; the political attributes which accrue to a planetary program are viewed as significant and worth the financial investment.

The increased recognition by foreign countries of the political benefits of planetary exploration stands in stark contrast to that of this country, which, unless positive decisions are made soon, will forego the opportunity to conduct a flight to Halley's Comet and indeed, may end up with no viable planetary program in the 1980's. The competitive nature of science and the role of science as a driving force behind technological advances have apparently not been recognized either by the Congress or recent administrations. Rather, the prevailing view (encouraged by some of the scientific community) is that science is conducted for the benefit of all humanity and transcends political boundaries. In that vein, American political leaders have encouraged NASA to seek significant foreign participation in its scientific missions. In view of the changing nature of the foreign capability and attitudes and of the complex managerial aspects (for both sides) of joint endeavors, the policy begs for reexamination.

Vision, a concept frequently invoked as a reason for conducting planetary exploration, carries specific justifications that run the gamut from such simplistic vagaries as the "exploration imperative" in human nature and "climb the mountain because it's there" to more erudite, thoughtful, philosophical discourses. "Vision" is a difficult concept to use as a selling point for new explorations and can even backfire. Its very definition — "something seen other than by ordinary sight: an imaginary, supernatural, or prophetic sight beheld in sleep or ecstasy" — demonstrates the problem: it is easy, in the political world, for one who speaks of vision to be labeled as an idle dreamer or sci-fi nut. For our purposes, a better, more applicable working definition is required: the perception of a challenge to be met and knowledge to be gained, wrapped up in the belief that we possess the will and the means to pursue those objectives, and capped off with the conviction that, in so doing, humanity is well served.

This definition fits well Kennedy's 1961 commitment to Apollo: "If we are to win the battle that is going on around the world between freedom and tyranny, if we are to win the battle for men's minds, the dramatic achievements in space which occurred in recent weeks should have made clear to us all, as did the Sputnik in 1957, the impact of this adventure on the minds of men everywhere who are attempting to make a determination of which road they should take. . . . We go into space because whatever mankind must undertake, free men must fully share." It also fits well what must have driven Robert Goddard through decades of isolated, dedicated research on rockets. Although both cases involved vision, there is a spectrum of talent necessary to enable such endeavors as space exploration of which Goddard and Kennedy represent the extremes: the early researcher who develops the scientific and technological basis, frequently with a particular goal in mind; and the often political leader who, for whatever reasons, can galvanize the necessary resources, resolve, and patience to pull it off. One might cogently argue that an earlier phase is required, one in which someone plants the basic concept in the public consciousness, a seed that stimulates future generations to make once-crazy dreams come true. John Wilkins was such a person, publishing *Discovery of a World in the Moone* in 1638, two centuries before another, Jules Verne, gave his great, albeit unpowered, boost to the concept of lunar exploration.

In the first 15 years of planetary exploration, it was primarily the scientific community who invoked vision as a selling point, with the attendant problem that the potential self-serving aspect sometimes surmounted genuine altruism in the eyes of a wary public or body politic. The recent Viking and Voyager successes, being viewed more and more in the context of a declining exploration program, have triggered nonparticipating observers to comment in the mass media. For example, Robert Cowen states "The U.S. planetary program has arrived at this sad state through lack of vision more than any other single factor." Tony Reichardt, waxing rather eloquently on what may constitute one aspect of vision, writes with regard to Voyager's Saturn pictures: "A momentary elevation of our sight, a halt to routine and preoccupation, a spark of the old, nearly forgotten emotions of wonder all evoked by strange visions sent by a robot from a planet a billion miles away.

"From the great optical distance our own situation comes into focus for a second; we feel refreshed for having seen something 'new' yet enduring and bigger than our own lives. The illusion may be fleeting, but it's a good one — of a common humanity united in a single effort toward something grandly mysterious.

"In a better world this would be reason enough to have an active space program. Never mind the inventory of newly discovered rings and moons, or methane atmospheres or low-density ammonia ice particles. What is involved here is something decidedly more spiritual, something to do with a nation's character and its sense of purpose."

Knowledge. The planetary exploration program's greatest contribution is certainly the eventual addition to the great store of our collective knowledge. In terms of sheer bulk of new data, it is overwhelming; in terms of amount of new information, mind-boggling. Both the data and information are immediate returns whose true value cannot yet be adequately measured. That must await their more complete conversion into the cache of human knowledge, where knowledge implies a sensible understanding of the natural processes involved in creating the particular data. However, even if our current understanding falls short of detailed comprehension, we are already capable of recognizing much of the significance of what is observed, be it for scientific or exploitative purposes. Indeed, the convincing evidence summed up in this book shows that we are making demonstrable progress toward the planetary exploration program's goals of understanding the origin and evolution of the solar system, better understanding Earth through comparative studies, and deciphering the relationship of life and the chemical history of the solar system.

Applications. The large amount of time involved in the assimilation of planetary data and information is typical of many scientific fields; basic research and geologic exploration, in particular, frequently take even longer to yield so-called practical applications.

The belief that practical benefits can be obtained from planetary exploration is largely a matter of faith; in discursive support of such faith, one can do no better than to call upon history. For example, most people are convinced (or can be) that the exploratory expeditions of Lewis and Clark opened up the prospects for, and eventually led to, the development of the West.

Calling upon historical example in support of planetary exploration and asking for an act of faith does not completely satisfy the casual inquisitor, which leads to a modicum of speculation about possible downstream benefits. The possible use of lunar materials for lunar base and other construction projects, the "capture" of Earth-crossing iron-nickel asteroids for terrestrial use as a natural resource, and the colonization of Mars are examples of such speculation. Whether these or others come to eventual fruition is moot; what is not moot is that only by conducting the exploration and research in the first place will we ever have the opportunity to make intelligent choices.

Let us turn from the specific to the general in outlining the practical benefits of planetary exploration. An increasingly popular thesis maintains that by better understanding the origin, evolution, and physical state of the planets, we will better understand the Earth. This concept often surfaces under the rubric of *comparative planetology* and contains the implicit assumption that better understanding the Earth, scientifically, is at least closer to a practical application than, say, better understanding Venus *per se*.

The concept of comparative planetology has demonstrable intrinsic merit: there is no way to develop a decent comprehension of the origin and evolutionary history of a single, highly evolved, complex planet (Earth) by studying it in isolation from the class of objects of which it is but one member. Even though the other planets may also be complex, the fact that they differ significantly in size and composition and have followed different evolutionary paths, enables one to see directly the effects of the various initial conditions. The end result allows the construction of more plausible models of the origin and evolution of planetary bodies both in general and in particular. The same methods are successfully applied toward explaining present-day conditions; for example, global circulation models of the atmosphere (known as GCM's in the trade) are relatively crude but rapidly improving in their ability to reproduce observed behavior in the terrestrial atmosphere. A GCM that takes into account (among other factors) planetary rotation, solar input, atmospheric density, and chemistry can be better tested directly by applying it also to planets other than our own, and comparing the results with observations.

These and other applications of planetary exploration must never be construed as the *raison d'être*, but rather as valuable and essential spin-offs that enhance the enterprise as a whole.

BEGINNINGS

Conceptually, planetary exploration began with the dreamers of yore. They, of course, did not worry much about the practicality of the venture, and it wasn't until the late 1800's and early 1900's that the theoretical and experimental basis for escaping Earth's gravity was laid in the works of Konstantin Tsiolkovsky, Hermann Oberth, and Goddard. But, as with many good ideas and inventions, and as unfortunate as it may be, it took the technological stimulus of war and its aftermath to effect the transition from the mind and laboratory to the field. Thus did space-age planetary exploration have its physical origin in the same root mass as the rest of the space program: the development of rocket technology during World War II, epitomized by von Braun's V-2 missile, and in post-war military development of intermediate and long-range ballistic missiles. It was then but a relatively small step in scale and efficiency to progress from carrying suborbital atomic weapons one-quarter of the way around the Earth to putting a satellite into orbit or leaving for other worlds.

Planetary studies cannot, or at least ought not, be conducted without planetary scientists. As a breed, they came to the space program relatively late (post-Sputnik); atmospheric or "sky" scientists gained a head start through experiments with sounding rockets and captured V-2's in the late 1940's and early 50's. It is thus no great surprise that the first space exploration plans were dominated both by political considerations (primarily to upstage the Soviet Union) and by technologists and atmospheric research scientists. In its proposal for a national space program, dated November 21, 1957, the Rocket and Satellite Research Panel made no mention of the planets; another proposal submitted one month later envisioned only crude unmanned exploration of the Moon followed by manned lunar missions. Even a year later, the newly formed space-science division of NASA did not consider the study of any solid bodies (as included in the NASA charter); not until the fall of 1959 was a formal division formed for lunar and planetary programs.

This gradual evolution finally spawned a five-flight Pioneer "lunar program" sponsored by the Defense Department's Advance Projects Research Agency, and a pair each of Venus flybys and lunar orbiters sponsored by NASA. But the discovery of the Van Allen radiation belts by Explorers 1 and 3 soon altered these plans: Pioneer flights were recast to probe the radiation belts more fully, and the successful Soviet probe Luna 1 provided the political impetus to redirect NASA's program toward Moon-only objectives. (The fact that lunar science would have been but a small part of either program was made moot by a host of launch failures.)

The first coherent plan for planetary exploration was published by the Jet Propulsion Laboratory in April, 1959. Its extremely ambitious program utilized a planetary spacecraft (called Vega), to be launched by Atlas-Vega and Saturn 1 boosters. The plan soon suffered obsolescence, however, as the Atlas-Vega was abandoned for the Atlas-Agena B, and the Vega concept itself became eclipsed by an updated NASA lunar program involving Ranger hard-landing spacecraft followed by more sophisticated, soft-landing Surveyors. The impetus for this new emphasis on lunar programs was both political and scientific: here was an opportunity to impress the world with an early technological success, while leading scientists began to impress upon NASA the importance of lunar science to the secrets of planetary origins.

Of course, the Apollo program, announced in 1961, changed the program's character dramatically. Because of the vast number of engineering and scientific unknowns confronting a manned lunar landing, both Ranger and Surveyor became supporting players to Apollo. They were soon joined by the Lunar Orbiter photo-mapping program. Exploration of the Moon got off to a shaky start with a string of six Ranger failures, then picked up with the three final Ranger flights and the enormously successful Surveyors and Lunar Orbiters. But the momentum from these achievements did not carry over into the Apollo era, scuttled by the scientific promise of manned lunar exploration and the budgetary constraints of the mid-1960's.

The untimely culmination of automated lunar exploration was not as disastrous as it might have been, because the anticipated scientific returns from Apollo were indeed forthcoming. Our extensive present-day knowledge of the Moon stems mostly from the analysis of samples collected by Apollo astronauts, geophysical measurements transmitted by the long-lived experiment packages they emplaced there, and remote-sensing studies conducted from lunar orbit during the last three Apollo flights. "Ground-truth" scientific data from these missions has, in turn, actually increased the largely photographic harvest gathered by earlier unmanned spacecraft and enhanced the interpretive analyses of lunar samples returned by the unmanned Soviet Luna probes. Some have maintained that the same yield of lunar science could have been obtained using unmanned spacecraft for a lot less than Apollo's $23 billion price tag — a lively topic of debate for academic philosophers and historians of science. The simple fact is that no automated program to accomplish the same objectives would have been approved.

PLANETARY EXPLORATION STRATEGY

Planetary exploration planning and mission implementation have followed two divergent paths since the late 1950's. The planning activity evolved by the mid-1960's into an iterative process involving scientists and engineers working in NASA-sponsored groups, through both the agency's field centers and the National Academy of Sciences, to come up with scientifically desirable and technically feasible missions fitting an exploration strategy. The strategy has become highly refined and thoroughly documented by the Committee on Planetary and Lunar Exploration (COMPLEX) of the National Academy's Space Science Board; it should remain valid well into the 1980's.

The extant exploration strategy builds around explicit goals for specific levels of investigation and establishes a hierarchy in which the accomplished goals of one level form the basis for detailed definition of the next higher level. Grossly oversimplified, the progression starts with the *reconnaissance phase,* in which the objective is to characterize a planet. This could also be thought of as the discovery phase because it usually provides the first good look at a planet, often taking us from a blurry telescopic image to one of eye-opening crisp detail, and hinting at what geologic and atmospheric processes have been or are at work. Reconnaissance missions generally make use of broad-range sensors on a flyby spacecraft. Mariner 2 (Venus), Mariner 10 (Mercury and Venus), Ranger (Moon), Mariners 4, 6, and 7 (Mars), and Pioneers 10 and 11 (Jupiter and Saturn) demonstrated the power (and limits — recall that Mariner 4's glimpses led to conclusions that Mars was Moonlike!) of the flyby reconnaissance mission. Reconnaissance must be conducted for all the planets and small bodies (comets, asteroids, and meteoroids) to accomplish the basic characterization of the solar system needed to optimize subsequent exploration (we all know how frequently "the one we're looking for" is at the bottom of the pile).

The second phase, *exploration,* constitutes an effort to determine the present physical state of a planet and to comprehend the physical processes that have shaped its evolution. The workhorses of the exploration phase include remote-sensing, long-life orbiters, soft-landers, and, especially for planets with atmospheres, entry probes. Significant improvement over reconnaissance is accomplished by dint of the global perspective, the ability to detect time-varying phenomena, and greatly increased resolution and sensitivity of instrumentation. Lunar Orbiter, Mariner 9,

Viking (Mars), and Pioneer Venus are striking examples of the exploration mission, as is the Galileo (Jupiter) probe and orbiter. The recent Voyager missions to Jupiter and Saturn were strongly reconnaissance oriented but, being very versatile spacecraft carrying sophisticated instrumentation, they accomplished much that is considered exploration.

The last level of investigation defined by COMPLEX is termed *intensive study* and seeks "to define or refine the remaining scientific questions of the highest order that have been revealed by reconnaissance and exploration and that can be studied in depth." The technical means for conducting intensive study would in most instances involve *in situ* atmospheric and surface measurements from sophisticated entry probes and surface landers, and might well require the return of samples to Earth for analysis. Lunar exploration, by both American and Soviet sample-return missions, comprises the only intensive study program to date.

Occasionally, NASA adds a fourth level or phase of investigation, variously referred to as "comprehensive understanding," "exploitation" or "utilization." It is a level which has been reached so far only for the Earth and clearly transcends scientific considerations. The construction of a permanent base on the Moon (or in space) using lunar materials, the mining of asteroids, and "planetary engineering" might be part of the utilization phase. However far from implementation, they deserve consideration for the simple reason that evaluating the desirability and feasibility of such projects usually requires the existence of a base of comprehensive scientific knowledge. To the degree that the necessary questions for this evaluation can be formulated now, the better can be the design of the precursor scientific missions. This brief excursion into the general exploration strategy is the preamble to mission implementation and to the crucial issues facing continued planetary exploration.

From the point of view of the scientist, it is logical and rational to lay out a sequence of missions that fulfill the strategy. An ideal sequence would carry forth exploration on a balanced front, progressively marching through the inner planets, outer planets, and primitive bodies, with sufficient time between missions to similar bodies to allow for considered analysis and interpretation followed by a review of the specific objectives for the next mission.

It all sounds so neat; so what's the problem? In reality there is a host of problems, some due to the inherent nature of the strategy, others to external forces. It is essential to try to understand these (largely nonscientific) factors in order to formulate the most viable exploration options for the future; a viable option is one with a high probability of being implemented.

Let's look first at the strategy. By our definition, most of the reconnaissance and a fair swatch of the early exploration may be completed within the 1980's. After that, planetary exploration will be ready to move on a broad front into an era of extended exploration and intensive study. At that juncture, public perceptions of the program's overall thrust and structure will change in such a way that the acquisition of scientific knowledge will outweigh national prestige and vision as the *raison d'être* of solar system exploration. This leads to the practical question: Will the total support necessary — public, political, scientific, budgetary, and technological — exist for a continued exploration program?

A casual observer of the planetary program's impressive

successes to date might be inclined to ask, "Why not?" The initiated observer, however, sees trouble lurking at every decision point — be it in the scientific community, NASA, the Office of Management and Budget, or Congress — and anticipates the worst. Indeed, the 20-year history of lunar and planetary exploration is replete with tortuous changes in direction from the outset, with the initial episodes previously described providing good practice for the future. There may have been a momentary lull in 1965, for as Peter Haurlan noted, "...in contrast to the frenzy of program selection seven years ago, there has now evolved a reasonably orderly progression of events — even though still somewhat frenzied — in the process of program selection." One judges that 1965 must have been the Year of the Quiet Frenzy. As we will see later, turmoil hit again during the 1970's (Voyager, later called Viking; and Grand Tour, restructured into Voyager) and is upon us again in 1981 with Galileo delays, cancellation threats, Venus Orbiting Imaging Radar deferral, and no Halley Comet mission. These peaks in frenzy — around 1958, 1970, and 1981 — correlate rather well with the solar cycle as indicated by peak sunspot activity. Those who study solar-terrestrial relationships should take note. Planetary scientists should not anticipate clear sailing in 1986; there are sunspots even at solar minimum.

The reader may be a tad curious now as to just what the nature of "frenzy" is, why it exists, and whether there is a practical (never mind sensible) alternative. Frenzy stems from the constant deviation from, and indeed lack of, a long-range commitment to exploration. The explanation is simple: under ideal circumstances, the time between mission concept and launch is 5-10 years (in practice, more often 10-15 years). This flies directly in the face of the United States' yearly federal budgeting process and four-year Administrations. Add to this the competing considerations of relative science merit, technical feasibility, launch-vehicle availability, absolute and relative cost, and political and popular support — the result is a very inefficient process.

MISSION INITIATION

The formal initiation of space missions comes to a head in the annual budget process, during which a proposed "new start" must wend its way through the NASA Planetary Programs Office, the Office of Space Science, the NASA Administrator, the Office of Management and Budget, and the Congress. In doing so, planetary missions must compete for a share of the available funds. In some years, the size of the total Federal budget constrains planning from the start — there were no new NASA missions of any sort in Fiscal Year 1980 due to the dismal state of the national economy, and the same may hold true in FY 82. At other times, the limiting factor is the total NASA budget. In 1961, for example, both James Webb and Homer Newell had to make overt and strong efforts to continue a space-science program in the face of the Apollo funding commitment. Today, total Federal budget aside, the number-one priority given to funding the Space Shuttle has severely curtailed NASA's ability to initiate additional scientific missions (as well as those with more direct practical applications). In fact, the continued treatment of the Space Shuttle as sacrosanct will likely force the cancellation or indefinite deferral of space-science programs.

Budget constraints at all levels have increased in significance because of a steadily rising cost per planetary mission (not unique to planetary programs). In the early 1960's, Mariner 1 (which failed during launch) and 2 together cost about $44 million. In the mid- to late-1970's, the cost of the Voyager project was over $350 million and Viking about $1 *billion*. The now-defunct Venus Orbiting Imaging Radar proposal would have cost $600-700 million. There are obvious and justifiable, even if unpalatable, reasons for cost escalations: the sophistication of scientific objectives, the operational complexity, and the duration of planetary missions have all increased. Mariner 2's flight to Venus was a simple flyby of about four months duration; the spacecraft weighed 656 kg and carried 18 kg of simple, nonimaging instruments. In contrast, the Voyagers have already been "flying" for years, and Voyager 2 will have spent almost nine years in space when it encounters Uranus in 1986. Moreover, each Voyager weighs about 816 kg, not that much more than Mariner 2, but carries over 113 kg of sophisticated instrumentation — a scientific payload weighing six times that of Mariner 2.

It can be argued that inflation alone has caused a significant increase in the apparent cost and that a valid comparison with earlier missions would take this into account. This is true but rather irrelevant because of a problem in the public's perception: In 1962, $44 million sounded like a lot; in 1981, $500 million still sounds like a lot, even if you do remind people that a loaf of bread cost 15 cents in 1962. This cost perception problem with regard to inflationary growth is worsened by the fact that few people have anything but the foggiest notion of what millions or billions of dollars really mean. Regardless, absolute cost is the major factor in the approval process for planetary missions, and on three occasions a major mission's high cost has led to its restructuring or cancellation.

The first such restructuring occurred early in the planning for the Mars landing. The Voyager project called for two lander-orbiter pairs, to be launched in 1973 on a single Saturn 5. But the mission was restructured in the late 1960's into the less-ambitious Viking project, consisting of two flights launched on separate Titan-Centaurs. The second victim of high cost was the "Grand Tour of the Outer Planets" designed to fly past Jupiter, Saturn, Uranus, Neptune, and Pluto by taking advantage of a rare (once in every 179 years), favorable planetary alignment that would have enabled the tour spacecraft to use the slingshot effect of each planet's gravity to send them on to the next planet. In the early 1970's the Grand Tour "evolved" into the Mariner Jupiter-Saturn mission (later named Voyager). The third example, actually an addendum to the restructured Grand Tour, was the Mariner Jupiter-Uranus (MJU) mission, proposed within NASA as an FY 76 new start for a 1979 launch. Approval of the mission would have necessitated an expensive extension of the Titan-Centaur launch vehicle program for two more years after the 1977 Voyager launches. That cost, the desire in high levels of NASA to phase out expendable launch vehicles (the Space Shuttle was due to take over in 1979), the high cost of the mission itself (approximately $400 million), the reconnaissance nature of the scientific objectives, and the potential for targeting one of the two Voyagers to Uranus combined to bury the MJU concept.

NASA's Office of Space Science oversees the budgetary "divvying up" that includes the planetary program. Other classes of missions — those that deal with solar-terrestrial interactions, astrophysics, and life sciences — compete for funds. These are less expensive missions that orbit the Earth, with simpler spacecraft, fewer experiments, and less-involved operations. Furthermore, most planetary missions are assigned to JPL in Pasadena, California, whose man-power costs are included in each mission's price tag, while other mission categories utilize civil-service personnel whose salaries are absorbed elsewhere in the NASA budget. Not surprisingly, planetary programs have historically con-sumed the lion's share of the agency's space-science funding.

But the competition is growing stronger. Astronomical missions like the 2.4-m Space Telescope (cost: $600 million) will enjoy high priority for the near future, and the Space Telescope's X-ray equivalent may also move into a top priority slot during the 1980's. On a dollar-for-dollar-how-much-science-do-you-buy basis, and on popular appeal, space astronomy continues to pose powerful competition to planetary missions.

In a given fiscal year then, how the Space Science budget is apportioned depends on how much money is available, how much each mission costs, what the year-by-year fund-ing profile looks like, and even whose turn it is. The last factor is more important than one might at first suspect, because intrinsic scientific merit is of little import as a dis-criminator in establishing budget priorities: most mission proposals which have survived the weeding-out process within a particular scientific discipline would yield high-quality scientific results if successfully flown. However, as soon as a mission enters the budgetary process, a highly competitive environment develops, and *relative* scientific quality and mission merit do become issues. Any attempt to avoid making a selection on the rationale that one can't compare apples and oranges (for example, planetary versus astronomical missions) is ill-fated; choices must be made — there simply is not sufficient money to do all the good science desired. The trick, then, from the vantage point of the administrators and mission advocates, is to figure out why an apple may taste better than an orange today. "Well, if you like both but can afford only one," they might say, "why not an apple today, an orange tomorrow?" It is this kind of process which led to a rotational basis for approving major new missions among the space sciences. It isn't a bad way to operate if the rotational cycle isn't too long, for example, one major new mission for each discipline every two or three years. The current problem is that the magni-tudes of the aggregate budget constraints and individual mission costs are resulting in four to five or more years between major new starts. At this level of activity there is a real danger both of losing the engineering and operations talent so essential to the conduct of highly sophisticated planetary missions and of having the scientists drift off to more viable, if not as stimulating, careers.

For 15 years or so, at least part of the space-science com-munity has deluded itself into believing that as each major manned effort winds down, there will be room in the NASA budget to accommodate significantly more new science mis-sions. This hallucination is partly a result of long-term budget projections which indeed, in any given year, show the major programs such as Apollo and Space Shuttle going down in budget demand. The amount of funds in the hoped-for NASA budget over and above the cost projections of approved programs is frequently referred to as the "wedge." If a name for the wedge were needed, it ought to be Casper — no space scientist has ever seen its embodiment. Frivolity aside, there are no recognized, legal, unanimous, *a priori* claims on the wedge; each participant in the process — scientist to space-station buff to Office of Management and Budget examiner — eyes it as his preserve.

The carrot is still there, however. In summing up his term as President Carter's science advisor, Frank Press noted: "In addition, [the Space Shuttle's] completion could release significant funds for new space science and applications projects, thus eliminating a concern, which I share with members of the space-science community, that [Shuttle] cost overruns will lead to a decline in planetary exploration and space research."

As one might suspect, the manned space flight aficiona-dos are not about to yield their "piece of the wedge." Not only will Space Shuttle operations in all likelihood cost much more than was projected in 1972, but the concept of a continuously inhabited space station is reappearing under the guise of the Space Operations Center. It is easy, and occasionally justifiable, to blame space-science budget woes on the costs of manned space flight ventures. However, I believe that the other factors discussed here are of greater significance, and that, in the long run, all of NASA's un-manned space flight programs are accommodated to some extent by the manned program's budgetary umbrella.

The role of launch vehicles cannot be underestimated. The best spacecraft ever made is rather useless sitting in Pasadena, California. In the 10-15-year cradle-to-grave life of a planetary mission, the hour or so of critical launch activity takes on a deserved eminence, and planetary exploration owes a debt of gratitude to the rocketeers. The task of launching a spacecraft has not been and shows no prospects of being without a dollop of pain, however. Consider that the first Mariner missions to Venus were scheduled for launch by an Atlas-Centaur in July and August of 1962. In August of 1961, it became obvious that Centaur would not be developed in time. JPL then pro-posed, and soon received authorization for, a scaled-down mission using the Atlas-Agena B and a smaller spacecraft (dubbed Mariner-R) which combined attributes of the original Mariner design and those of Ranger.

The vulnerability of planetary exploration to snags in launch-vehicle development has been long recognized inside and outside of NASA; the Space Science Board has expres-sed its explicit concern on the subject. Recently, such prob-lems have had a major impact on planetary missions. The Galileo mission to Jupiter is scheduled to be the first planetary spacecraft carried by the Space Shuttle. A booster known as the Inertial Upper Stage (IUS) was to be carried up with the spacecraft (and on other Shuttle-launched mis-sions) to propel it out of Earth orbit. But the schedules of the Shuttle and the IUS slipped so much that planners were forced to postpone Galileo's launch date from 1982 to 1984. In 1981, the increased cost and reduced performance of the IUS led NASA to consider the substitution of a Centaur booster for the IUS. Even that plan was not without prob-lems, as the Centaur would have needed modification to work compatibly with the Shuttle. In the 1983 budget,

Galileo is shown as IUS-launched in 1985, encountering Jupiter in late 1989 or early 1990.

Another hoped-for mission will not take place. The Comet Halley rendezvous mission, designed to take advantage of the celebrated comet's next perihelion passage in 1986, was to make use of solar-electric propulsion (SEP). Such low-thrust propulsion is essential to certain missions where conventional boosters would be impractical — comet or asteroid rendezvous, and planetary sample return. The technological basis for SEP was demonstrated in the 1960's and 70's, but further development did not follow. NASA soon found itself unable to respond to the opportunity to conduct the Comet Halley rendezvous and lost the chance to use SEP on an alternate mission that would have included passing Halley en route to a rendezvous with Comet Encke.

Political and popular support are certainly important and necessary ingredients at all steps in the planetary mission approval process. Given an Administration lacking either a neutral or a positive stance on the space program, NASA would not receive funding approval from the Office of Management and Budget. Once the Administration submits a budget to Congress, it is then required to garner the support of four subcommittees — the House and Senate authorizations and appropriations subcommittees — and their hierarchal parent committees. Despite some interesting tussles and an occasional floor show, the Congress has generally supported NASA budgets largely as proposed by the Executive Branch. When there is serious congressional disagreement concerning planetary missions, it has usually been tied to peripheral but real problems of overall and specific science priorities in the national budgets, the inherent value of basic science and exploration, the relative balance within NASA between science and applications, and the availability of launch vehicles.

A significant portion of the public and news media is favorably disposed toward planetary exploration, a fact that has certainly been a positive influence on the overall political support of, and thus the progress of, the planetary program to date. Members of the executive and legislative branches and their staffs are frequently aware of the widespread editorial support given planetary missions. They could hardly have avoided seeing the extensive media coverage of the Viking and Voyager results, especially the spectacular photography. Such responses to public interest can be viewed as passive advocacy. In addition, over 50 special groups maintain strong or central interest in overall space exploration, frequently carrying their enthusiasm to the point of active advocacy. Their combined influence is difficult to evaluate, but without it much of the civil space-exploration program would fade from existence.

FUTURE PLANETARY EXPLORATION

The results from U.S. planetary exploration have been undeniably spectacular as regards human exploration, basic scientific discovery, and technological achievement. This has been accomplished over two decades for a variety of reasons and under constantly changing circumstances. Through all the complexities one can discern a few basic trends that have led to a decreased rate of planetary explora-

tion over the past decade and lead one to project very lean years ahead. To reverse this trend, a desire of many of us, calls for recognition of the increased mission sophistication and costs, tighter overall budgets, the transition from basic planetary discovery to more detailed scientific exploration, and increased competition from other fields of science.

For now, the wisest approach would be to carry through the Galileo mission to Jupiter and aspire to conduct the basic reconnaissance missions of Venus' surface, comets, and asteroids. What additional missions should be advocated is not yet clear; NASA has recently established a Solar System Exploration Committee to wrestle with that question and the complex economic, political, scientific, technological, and sociological interactions that ultimately determine what happens. Although no answer is yet in hand, it is clear that those involved in devising planetary missions should seriously consider making them less expensive, possibly by assigning more focused and limited objectives, by concentrating exploration on the nearby inner planets, or both. As the Space Science Board has noted, comparative planetology would greatly benefit from the concentrated exploration of Venus and Mars. It may well be that sophisticated flights to orbit and probe the Saturnian system or to fly past Neptune and Pluto are not in the cards for now. If so, increased attention must be given to making clever use of the Space Telescope and ground-based astronomy as planetary exploration tools, to follow up on Voyager, and to build upon prior observations of Neptune and Pluto.

Our more visionary goals are certainly a long way from realization. But even though potential utilization of lunar resources, asteroid mining and planetary engineering, or global atmospheric experimentation on Mars seem fanciful now, it may make sense to focus on missions to provide the scientific basis that will enable us to make utilization decisions farther downstream.

Whatever is concluded in the near-term regarding specific missions, a few obvious facts stand out. No amount of remote sensing will ever yield the kind of information attainable by analysis of planetary materials in terrestrial laboratories. The return of samples to Earth from Mars, Venus, comets, and asteroids is essential to making the desired leap in basic knowledge of the composition and physical state of those bodies. To do so requires the development of low-thrust propulsion. Let us then get on with developing the techniques of sample return and solar-electric propulsion, fully recognizing that this should be done prior to proposing specific missions.

We should not become overly depressed by the current state of affairs regarding planetary exploration. In contemplating the human footprint on the Moon, stream channels on Mars, volcanoes on Io, and braided rings at Saturn, recall the tortuous route to get there. Let us use the current situation to stimulate innovation, focus our resources, plan the options, and be ready to respond to new exploration challenges. Let us not idly await a hoped-for opportunity or for the spirit of the "exploration imperative" to strike; opportunities are more often a result of creation than random occurrence. Most things happen because determined people want them to. The solar system beckons.

The Sun

John A. Eddy

*"Hundreds of dewdrops to greet the dawn,
Hundreds of bees in the purple clover,
Hundreds of butterflies on the lawn,
But only one Mother the whole world over."*

The hundreds of planets and lesser objects that make up the solar system are held together as a family by a single parent: probably, though not certainly, their real mother. Whether or not they were once part of the Sun, the present conditions on planets, planetoids, and planetesimals depend almost entirely on its continued presence and steady warmth. As inhabitants of the Earth we are particularly aware of this dependence since not only our light and heat, but our food and energy stores and the renewal of oxygen all depend in one way or another on the Sun. Thus we study it not only as a representative, nearby star, and a remote laboratory of physics, but for its terrestrial effects, both known and suspected.

ANATOMY OF THE SUN

The Sun is a star: a burning sphere of gas that derives its heat from a nuclear furnace at its very center. There and only there in the solar system are temperatures, densities, and pressures great enough to sustain nuclear fusion — the same process that lights all the stars in the sky. In the *solar core* hydrogen atoms, our star's most abundant fuel, are transformed to helium at the rate of about 5 billion kg per second, with the release of energy that gradually works its way outward to the *photosphere*, or visible surface of the Sun. Beneath the solar surface, and hidden from our view, the outward flow of solar energy becomes turbulent, creating giant, rising bubbles of gas that make up the *convective zone* of the solar interior. The upper surfaces of these boiling bubbles are visible as a pattern of *granulation* in pictures of the white-light surface of the Sun. Subsurface turbulence also interacts with interior solar magnetic fields to produce dark *sunspots* that come and go in the photosphere, and a larger-scale network of supergranulation that characterizes the *chromosphere*.

The chromosphere and more extensive *corona* above it make up the outer, extended atmosphere of the Sun. These dynamic layers are invisible to ordinary view because they are diffuse and so much dimmer than the blinding glare of the white-light photosphere. The temperature of the photosphere is about 6,000°K, roughly that of an iron welding arc; sunspots, in contrast, are nearly 2,000° cooler, but still as hot as an acetylene flame. Sunspots are cooled by the inhibiting effects of the intense magnetic fields that hide within them: awesome magnetic forces that are thousands of times stronger than the feeble magnetic fields of the Earth or other planets. Arched lines of force that connect sunspot magnetic fields are visible in the chromosphere and lower corona as solar *prominences* that soar tens of thousands of kilometers above the solar surface in a variety of shapes and forms (Fig. 1). Magnetic forces dominate the structure of the chromosphere and corona, and provide the prodigious energies that are cataclysmically released from time to time as *solar flares,* seen well in Fig. 2.

After falling steadily outward from the solar core, the temperature of the Sun rises abruptly in the chromosphere and corona, reaching temperatures nearly as hot as at the center of the star. Radiation from the cooler photosphere passes easily outward through these transparent layers to bathe the Earth and the rest of the solar system in a steady flow of light and heat.

The solar diameter is about 1.4 million km — nearly 100 times larger than that of Earth and 10 times larger than giant Jupiter. Still, the Sun is mostly nothing. Like any fire, the solar atmosphere (photosphere, chromosphere, and corona) is made of very little substance. The corona and chromosphere, though extremely hot, are so diffuse that they would be called a vacuum here on Earth. Even the photosphere is as rarified as the uppermost atmosphere of Earth, where spacecraft operate. It would be quite possible to fly through cloudlike prominences arching high into space without ever seeing them, or into the gaping maw of a sunspot, or beneath the white-light surface of the Sun, if we could stand the heat or the force of the Sun's intense magnetic fields. The density of the Sun increases as we go inward, but it is not until we have gone a tenth of the way to the core that we encounter gas as dense as the air we breathe on Earth, and not until halfway to the center before we find anything as solid as water.

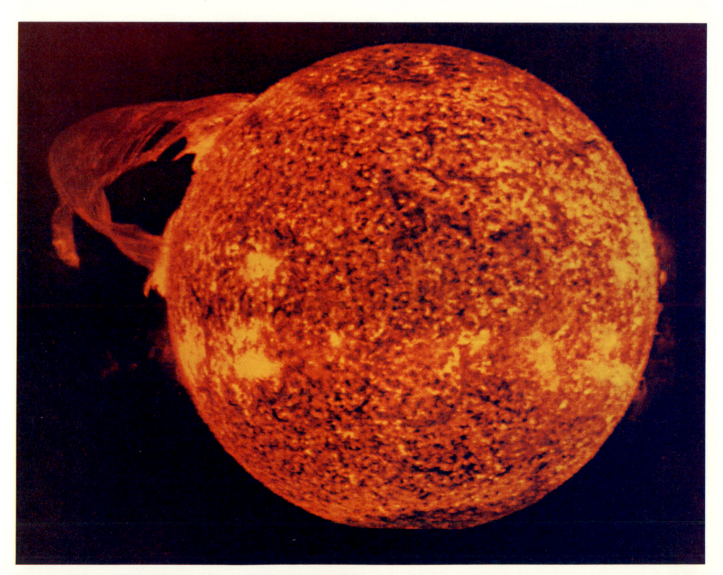

Fig. 1. An arched, eruptive prominence lifts off from the Sun, propelled by magnetic forces. Chromospheric material, at temperatures of 10,000°-20,000° K, is visible in this Skylab ultraviolet photograph made in the 304-angstrom line of singly ionized helium. Footprints of the growing prominence stand in the photosphere and chromosphere; twisted lines of force connect them.

AN HISTORICAL PERSPECTIVE

Physical study of the Sun had progressed a long way before the age of rockets and spacecraft. Quantitative interpretation of the Sun's surface and atmospheric layers in terms of chemical composition, temperature, and pressure began with spectroscopy, more than 100 years ago; soon after the turn of the century almost all the fundamental physical properties of the star were known. "Solar Physics," a new name given a large book by J. Norman Lockyer in 1874, was a mature science when most of us now active in the field were born. Thus the solar discoveries of the space age, though colorful and important and sometimes surprising, were more refined than those that characterized more exploratory voyages to the Moon or the planets or excursions to unseen environs of the Earth.

By 1950 the basic physics of the Sun seemed well in hand: its gaseous composition, interior structure and nuclear source of energy, the high temperatures of the chromosphere and corona, and the dominant role of solar magnetic fields. The Sun's 11-year cycle of activity had been documented to the point of boredom. Though we could not predict them (and still cannot), solar flares had been seen and studied since 1851, the chromosphere, corona, and prominences since the early 1700's, and sunspots for at least a century before that. The total luminosity of the Sun had been known to within a few percent for a quarter of a century and the limits of variability of the total flux of energy from our star were equally well-bracketed.

Still, the possibility of observing the Sun from above the atmosphere held out new hopes: for one, praise be to God, the seeing would improve. But the real promise of spaceborne solar telescopes lay in opening up the short-wavelength spectrum — the last frontier of observational solar physics. Here the hope was not for any breathtaking surprises, as with missions to dimly lit outer planets, but for quantitative measurements of what we knew we would find, when we could at last look at the Sun with X-ray and ultraviolet eyes.

Observations of these radiations, to which even the atmosphere above Mt. Everest is opaque, promised a big step forward in diagnosing chromospheric and coronal conditions, since these high-temperature regions of the solar atmosphere should emit most of their radiation at such wavelengths. Ultraviolet observations of the Sun and other stars were long hoped for: in 1937 the theoretician Meghnad Saha pleaded for a "Stratospheric Solar Observatory" to record the yet unobserved Lyman-α emission at 1216 angstroms; in December of 1929 the observationalist Svein Rosseland had journeyed through winter snow to Honningszåg above the Arctic Circle to test, without success, whether an ultraviolet hole opened up in the sky in the dead of the long polar night.

SOLAR SPACECRAFT

Simple observations of the Sun and solar radiation were tried during 19th-century balloon ascents and from aeroplanes, dirigibles, and unmanned balloons in early years of the present century, in attempts to rise above the ocean of air that cheated earthbound views. On a summer day in 1914, Charles Greeley Abbot, the persistent pioneer of solar constant measurements, sent an automated pyrheliometer (a device to measure incident solar radiation) aloft from Omaha, with hydrogen-filled, rubber balloons that reached an altitude of 24.4 km (80,000 ft.). In 1935, the much heavier Explorer II balloon gondola reached the same altitude in the stratosphere, but with two men and a load of scientific apparatus aboard, including a solar radiometer. None of these attempts pushed the frontiers very far, however, for ozone, molecular oxygen, and nitrogen in the remaining atmosphere overhead still blocked most of the solar ultraviolet shortward of about 3000 angstroms.

A breakthrough came in October of 1946, when a wartime V-2 rocket, with a solar spectrograph tucked in its tailfin, climbed over New Mexico to 55 km and photographed the solar spectrum down to 2400 angstroms, showing patterns of expected lines and a strong continuum. By 1948 — ten years before NASA was born — other rocket payloads had detected solar X-rays and Saha's Lyman-α line. For a dozen years, while solar scientists waited for real spacecraft to come upon the scene, rockets continued to probe and sample the rich, shortwave spectrum of the Sun down to wavelengths of a few angstroms. Rockets fired from balloons captured fleeting X-ray emission from flares in 1956, as the 11-year cycle of solar activity was building toward its highest peak in history. In 1960 the first solar spacecraft, called Solrad, was launched to keep a constant watch on whole-sun X-ray and Lyman-α flux. In the same year the first crude X-ray images of the Sun were secured, using pinhole photography in rockets, with an enticing preview of the wonders that would come later with better resolution (Fig. 3).

Fig. 2. A solar flare sprays upward in this photograph taken in hydrogen-alpha light (6563 angstroms) at Big Bear Solar Observatory in California on May 22, 1970. At this wavelength, bright *plages* mark the hotter, denser regions of the chromosphere that overlie sunspots. Note also the tiny, grasslike chromospheric *spicules.*

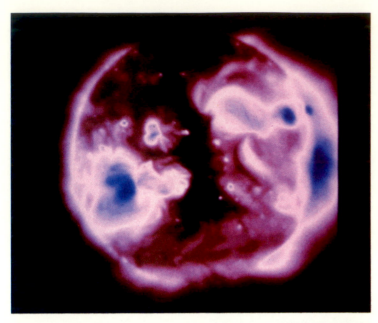

Fig. 3. The Sun, seen with Skylab's X-ray eyes, burns in eerie colors. Blue, white, and violet have been added in processing to emphasize brightness differences, with blue brightest and hottest. The empty, black zone that runs almost vertically down the center of the Sun is a coronal hole.

A vastly more ambitious series of Earth-orbiting, solar spacecraft was initiated in 1962, with the launch of the first Orbiting Solar Observatory, OSO 1. In all, eight were launched, keeping the Sun under nearly continuous and unprecedented short-wave surveillance for 1½ cycles of solar activity. The OSO's were the first astronomical spacecraft designed to point continuously at their target, and the first to try the concept of a sophisticated, shared observatory. Competing experiments packed together in crowded quarters learned to live together. For seventeen years the orbiting solar observatories served as the workhorses of solar physics from space: each small enough to fit in the back of a pickup truck, limited by parsimonious budgets of electrical power and telemetry time, but loaded to the gunwales with whatever deserving experiments could be squeezed aboard.

Clearly the greatest of Earth-orbiting solar observatories was Skylab (Fig. 4), the manned space station that carried a battery of eight observatory-sized telescopes collectively called the Apollo Telescope Mount (ATM) that wanted nothing more than to look out at the Sun, hour after hour. The ATM telescopes were immense compared to the miniaturized solar telescopes flown before: each one was 3 m long and weighed as much as a small automobile. They covered the spectrum from hard X-rays to visible light with instruments that were ahead of their time. They were also versatile, and were pointed with unerring aim by trained astronaut observers from a *Star Wars* control panel in a *Star Trek* setting (Fig. 5). Seven of the eight telescopes utilized photographic film, carried up in bulky canisters that returned to Earth with the astronauts.

For nine months, ending in February of 1974, the Skylab telescopes, astronauts and teams of solar physicists in Houston put together the best organized and most intensive study that has ever been directed at any astronomical object anywhere. The awesome power of that kind of regimented attack, focused on specific, predetermined problems, set an example whose impact on science has probably yet to be fully felt: one can dream, as did Saha with Lyman-α, of similar, intensive manned observing stations of the future, orbiting not the Earth, but Venus or Jupiter or the moons of Saturn, and looking not up, but down.

Fig. 4 (left). Three crews of astronauts occupied Skylab during 1973-74. Seen here at the center of its windmill-shaped solar panels is the Apollo Telescope Mount, a cluster of advanced telescopes and other instruments that are covered by sunshades in this photograph. Fig. 5 (right). Solar physicist Edward Gibson demonstrates the pointing controls of the ATM console during his three-month flight aboard Skylab.

While the world watched Skylab (or waited for it to drop), a new pair of spacecraft, Helios A and B, left the Earth in 1974 and 1976 for the Sun itself, in elliptical orbits that passed inside the orbit of Mercury, two-thirds of the way to the Sun. They carried not telescopes, but instruments designed to measure the atomic particles and magnetic fields that are drawn out from the Sun by the solar wind.

The Helios sentinels continued their lonely vigil of measurement for six years. Other, similar spacecraft had made extensive measurements of solar particles and fields, beginning in 1960; these included observations from early Russian interplanetary probes, plus American Explorer, Apollo, Imp, Mariner, Vela, and Pioneer spacecraft. But none had come so close to the Sun itself.

In early 1980 the latest of the Earth-orbiting solar observatories was launched, to keep a special watch on the Sun during the past maximum in the Sun's 11-year cycle of activity. The Solar Maximum Mission spacecraft was, in a sense, an updated and automated ATM with increased emphasis on the hard X-ray and gamma-ray emissions from the Sun. Complemented by a coordinated set of ground-based studies, the craft's observations were directed primarily toward understanding solar flares, one of the major unsolved problems of solar physics.

THE SUN IN THE ULTRAVIOLET

The color response of the human eye begins to fade at wavelengths between 3000 and 4000 angstroms, in the deep violet of the rainbow of colors that makes up the so-called "visible" spectrum of light. It is no accident that the deep violet is also the rough limit of transmission of the blanket of air above us, for we and other animals of the Earth have evolved through hundreds of millions of years in sunlight that has always been cut off in this way. The total energy of sunlight that is excluded from our view in the ultraviolet is only a few percent of the total radiation that comes to us from the white-light photosphere of the Sun. That is not the case for the thinner, hotter, upper atmosphere. The chromosphere and corona, where much of the action takes place, radiate very little of their energy in the visible spectrum, which is why they are hard to see with ordinary instruments; as mentioned, most of it is in the ultraviolet and X-ray regions (Fig. 6). Thus these normally obstructed short-wavelength regions of the spectrum are of particular value for the study of solar flares, active solar regions, and the physical processes that govern the Sun's outer atmosphere.

Until the second half of this century our understanding of the outer layers of the Sun progressed slowly, paced chiefly by the amount of data we could steal during the fleeting moments of total solar eclipses. At these times it is possible, through a genuine accident of nature, to see the chromosphere and corona, edge-on, without the blinding glare of the photospheric disk. When the opportunity came to observe the Sun at shorter wavelengths, from space, we could see the chromosphere and corona continuously and unobscured. The picture quickly cleared, with an explosion of new knowledge of the physics of the solar atmosphere. Because different ultraviolet wavelengths originate at different heights in the chromosphere and corona, tuning through the ultraviolet spectrum enabled astronomers to focus on specific solar layers, much as adjusting the focus of binoculars lets you see closer or farther away.

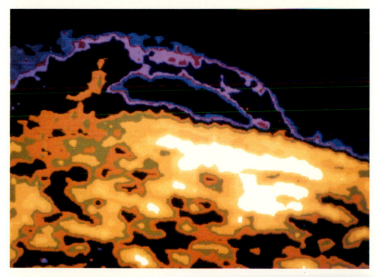

Fig. 6. Solar prominences stand out dramatically when pictured in the unseen ultraviolet. Colors have been added in processing to enhance differences in intensity. At top, a surging, active prominence lashes into the corona; below that, loops and arches in quiescent prominences delineate magnetic lines of force that thread into the corona, confining cooler, chromospheric material at temperatures of 10,000° to 50,000° K.

In the ultraviolet the Sun is a garish star. Variable solar features are enlarged and dramatically enhanced to the point that it seems almost a caricature of the more familiar, stolid, white-light disk. That is because the chromosphere and corona are ever changing and thoroughly dominated by dynamic, nonradiative processes: a Dantean inferno of heaving waves and fiery, twisted arches, ruled by magnetic fields of awesome power that shape and then disrupt the scene. Features that seem subtle and bland in visible light leap out, enlarged and exaggerated, in invisible wavelengths that sample higher, hotter layers.

Ultraviolet pictures of the Sun allowed us to study, for the first time, the thin and elusive *transition region* where temperatures in the solar atmosphere jump abruptly from 50,000° K in the upper chromosphere to 2,000,000° in the corona. The transition region is more than a curiosity: the physical processes that happen there are the keys to conditions in the corona which, in turn, dictate the particle and magnetic-field environment of the Earth. The important jump in temperature between the chromosphere and corona happens in a layer so thin that it is almost a discontinuity; nothing like it occurs in the temperature structure of the atmosphere or oceans of the Earth — or anywhere else in the solar system.

A geographical comparison helps put the thinness of the solar transition region in perspective. The white-light photosphere that we see with the naked eye is itself a very thin layer of the Sun — only 400 km deep, or about the width of Alabama (Fig. 7). The jagged, hotter chromosphere stretches above it, toward the west in our terrestrial comparison, for at least several thousand kilometers, enough to reach from Alabama to Los Angeles. The million-degree corona, above (or to the west), extends too far for our Lilliputian yardstick. Between the two vast regions a dramatic jump of a million degrees takes place in a layer no thicker than the width of the city of Los Angeles — a few tens of kilometers. In that small space is where all the action is.

In the transition region, sunspots and the chromospheric network of spicules disappear to make way for the more diffuse and ethereal forms of the corona. Magnetic forces now rule the eerie world of a near-vacuum at a temperature of more than a million degrees. The spectrum of the transition region, like that of the chromosphere, is one of bright emission lines of highly ionized metals: the last atoms that can hold on to their electrons in the trying conditions of high temperature and low density. In transition-region lines we see the lower legs of the magnetic loops that thread the corona; their feet stand in active-region magnetic fields within the chromosphere and their lofty arches soar sometimes a million kilometers above. Since the region is so thin, spectroheliograms (pictures of the Sun made in transition-region lines) are particularly sharp.

By far the biggest benefit of ultraviolet observations of the Sun is the ability to examine and therefore model the three-dimensional structure of the solar atmosphere: to see what had only been guessed at before. In the ultraviolet we see it all, in quiet or active regions, from equator to pole, all the way from the photosphere up through the corona, in spectral lines that are ideally suited for physical diagnosis. In one use of that examination, time-lapse observations from the OSO 8 spacecraft tested the role of acoustic waves in heating the corona (our standard explanation in the 1960's

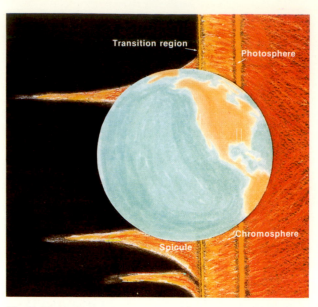

Fig. 7. Planet Earth serves as a yardstick to scale the thickness of solar atmospheric layers. The photosphere (orange) is about as thick as Alabama is wide, or about 400 km. The less dense and more turbulent chromosphere (red-orange), spans several thousand kilometers, stretching on our scale from Alabama to Los Angeles in its thinnest parts; spicules extend its outer boundary even higher into the corona (black). The transition region (yellow) separating the chromosphere and corona is no thicker than the width of metropolitan Los Angeles.

and early 1970's) and found it didn't work at all. Ultraviolet observations of flares revealed the real nature of these explosions on the Sun, filling in their third dimension (height) and high temperature, and their relationship to magnetic fields in the corona.

THE X-RAY SUN

Beyond the ultraviolet lies the X-ray spectrum: shorter-wavelength, high-energy electromagnetic radiation that is of particular interest in astrophysics because it comes from high-temperature plasmas. Strong X-ray emitters include neutron stars, black holes, and the million-degree corona of the Sun (Fig. 8). X-ray photons have extremely high energies compared with visible or ultraviolet light, and for this reason they have long been employed in medicine, as a source of piercing light to make shadow-pictures of our bones. Yet X-rays from celestial sources are unable to get through all the atmosphere of the Earth, and thus X-ray astronomy is a child of the age of rockets and spacecraft.

The first solar X-ray detectors were carried up in rockets in 1948. By 1960, photographic X-ray images had been obtained with resolution sufficient to demonstrate that the Sun was bright where there were active regions and nearly devoid of X-ray emission elsewhere. These coronal X-ray enhancements coincided with bright regions seen at the Sun's limb using optical coronagraphs, and with areas of intense solar radio emission that indicated increased coronal electron density. Through cleverness, such as firing instrument-laden rockets during solar eclipses, scientists further isolated the source of coronal X-rays. By 1968, X-ray telescopes aimed at the Sun from space equaled the resolution of optical instruments. The dramatic pictures that came from them put the solar corona in a clear, new light.

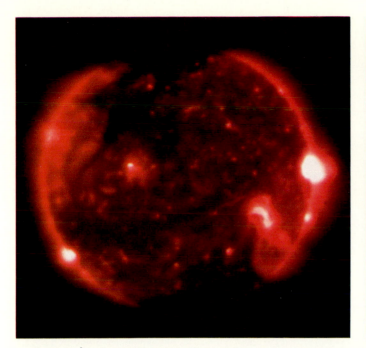 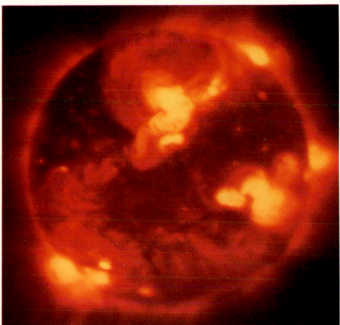

Fig. 8. Two views of the Sun acquired by sounding rockets reveal its appearance at "soft" X-ray wavelengths (here from 7 to 65 angstroms). The image at left, taken in 1976 during a minimum in the Sun's 11-year activity cycle, shows the solar atmosphere to be filled with low-lying, magnetically confined material; large coronal holes present during Skylab missions are absent, although dark areas mark holes in the polar regions. The image at right, made in 1978 near the maximum in the solar cycle, shows bright, dynamic regions with arching loops of material along the limb. Courtesy John Davis, American Science and Engineering, Inc., Cambridge, Massachusetts.

In such images, the photosphere and chromosphere are black; only the corona can be seen. This gives the great advantage of revealing the low corona over the entire solar disk; at times of eclipse, or with coronagraphs, we can see the corona only around the limb of the Sun, edge-on. The difference is that of exploring a forest from the edge of the woods or from a helicopter above.

A major revelation using this new perspective was that the corona is made almost entirely of loops, loosely laced between the regions of opposite magnetic polarity on the surface of the Sun. Coronal continuum X-ray emission comes from atomic particles trapped in these patterns of arched magnetic fields. Solar physicists had long talked of a background, "quiet corona" but in X-rays it wasn't there; where there were no loops, there was little if any corona. This confirmed the pattern of secular changes observed in the white-light corona at times of eclipse: since the 1870's we had known that the corona is nearly blank at times of minima in the 11-year sunspot cycle and filled with arched streamers at times of maxima, when concentrated magnetic fields prevail.

A second major discovery was that of holes in the corona, large patches without trees in our forest analogy, that had escaped detection from the edge of the woods. Coronal holes cannot be seen in visible-light spectroheliograms and appear confused in edge-on pictures from eclipses. In these "bald" regions, there are no magnetic loops to constrain coronal material; from coronal holes the solar wind flows outward freely into space along open magnetic field lines. Thus the discovery of coronal holes paved the way for an important breakthrough in understanding fluctuations in the flow of solar particles at the Earth.

A third discovery was that of tiny, isolated "bright points" of intense X-ray emission that cover the Sun like chicken-pox. (The same bright points also appear in ultraviolet pictures made from space.) They are the mark of concentrated, magnetic areas on the Sun, dipole fields that are even smaller than sunspots. Unlike all other, earlier-known features of solar activity, coronal bright points seem to pay no attention to the well-established rules of solar behavior. They are not restricted to the equatorial bands where other magnetic structures are always found, but exist everywhere, including the solar poles and in coronal holes. There are so many bright points — at least a hundred can be seen on the Sun at any time — that they represent sources of a significant fraction of its total magnetic energy. Perhaps most important, they seem to increase in number as other known signs of solar activity fall, as though to balance the overall magnetic budget of the Sun.

THE OUTER CORONA

High above the loops and arches of the lower corona stretches the Sun's wispy outer corona, the long-tailed streamers seen at eclipse that have historically been likened to the petals of a dahlia, or drawn-out tulip bulbs. Beyond a distance of about half a solar radius above the solar limb, the corona — still at a temperature of a million or more degrees — consists chiefly of widely separated atomic particles that once were parts of atoms in the lower solar atmosphere: electrons, and protons, and a few heavy atomic nuclei. When we look at the outer corona in visible light (Fig. 9), we see sunlight scattered from coronal electrons and hence we can trace the solar forces that control them. Two forces dominate the movements of these hot, charged particles: solar magnetic fields and the expulsive force of the solar wind, whose high temperature carries solar electrons and ions away from the gravitational field of the Sun. The bulbous shape of coronal streamers results from a combina-

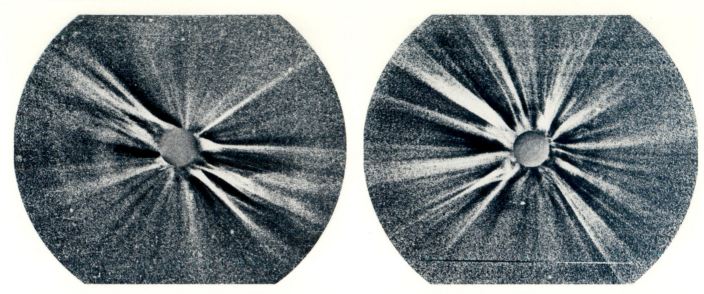

Fig. 9. Charles Keller and his colleagues at the Los Alamos Scientific Laboratories obtained these composite images of the Sun's outer corona from specially equipped high-altitude aircraft during recent solar eclipses. Note the dramatic change in coronal activity between the 1973 eclipse (during solar minimum) and the one in 1979 (near solar maximum). Their technique, which utilizes computer enhancement of several composited images, has recorded the corona out to 11 solar radii from the Sun's limb. (The innermost circles are not the eclipsed solar disk but a computer-produced mask 1.2 solar diameters across.)

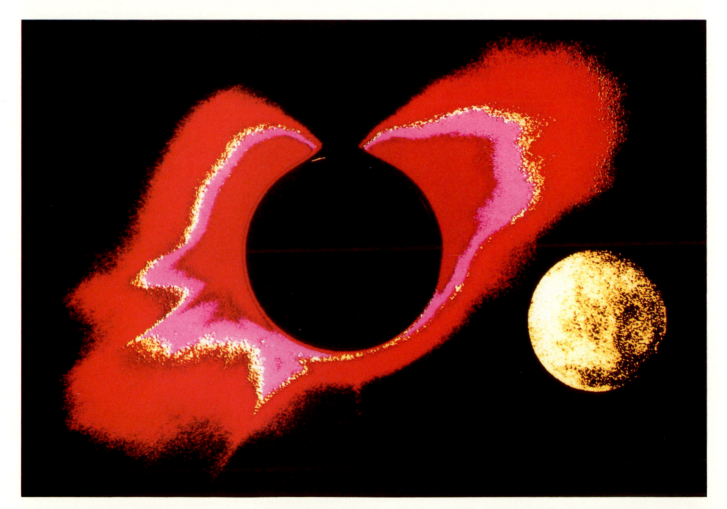

Fig. 10. A psychedelic eclipse of the Sun was created by special color processing of black-and-white coronagraph film from Skylab. The black circle at center is the occulting disk of the coronagraph that hides the (smaller) solar disk. A new Moon, here tinted yellow, moves toward a flaming pink and red corona where, some minutes later, it produced a total eclipse of the Sun on June 30, 1973. Observers on the ground were treated to a rare 7 minutes of totality as the Moon's shadow crossed Africa.

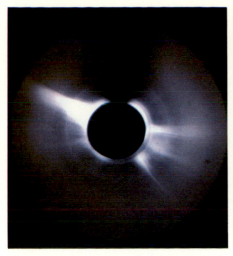  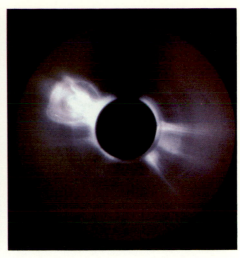

Fig. 11. A transient disturbance in the corona, typical of those seen with spaceborne coronagraphs, pushes its way outward from the Sun into interplanetary space, at a velocity of about 500 km per second. The action seen in these three Skylab pictures took place in less than two hours. In this brief time the blob of coronal material, larger than the Sun itself, traveled nearly a million kilometers. Such gargantuan disturbances are now realized to be commonplace features of the Sun.

tion of these two effects. Their rounded, wider bases near the Sun betray the shape of closed, magnetic arches and give evidence of atomic particles that are still bound to the Sun. Their drawn-out extensions — beginning about half a solar radius above the limb — tell of the eventual victory of the forces of the solar wind in wresting the charged particles away from the grip of solar magnetism. Thus features in the outer corona take the form of stream lines flowing outward into space.

The corona defined in this way reaches to the Earth and well beyond, but the material of the outer corona is so diffuse that we can hardly see it. During a solar eclipse, when the Sun's disk is covered and the sky is dark, we can only trace coronal extensions to perhaps one degree from the limb, even under the best conditions. At other times, coronagraphs at high mountain stations can map the outer corona to about one-third of this distance before the brightness of the sky overwhelms what can be seen. Coronagraphs above the atmosphere, on the other hand, are not so limited, as was demonstrated in the 1960's when such instruments were carried up in balloons and rockets. A coronagraph in space operates in its ideal environment, for it sees the corona against the black background of space (Fig. 10).

Orbiting instruments proved capable of detecting the outer corona much farther from the Sun than could be seen at eclipse; moreover, they could keep the outer corona in view almost continually, for periods of months to years. An orbiting coronagraph easily gathers more information on the outer corona in the first few weeks of its operation than observers have gleaned from natural eclipses in all of time.

As the spatial resolution of these instruments improved it became clear that the outer corona of the Sun was continually changing, on time scales of months, days, and even hours — a fundamental property that had eluded eclipse-chasers for a hundred years. The supernatural, months-long eclipses made possible by coronagraphs in orbit also revealed the three-dimensional nature of outer coronal forms and their association with solar surface activity. A great surprise was the frequent occurrence of transient disturbances, of gargantuan scale, in the outer corona.

Following certain flares and prominence eruptions, solar material ejected from the chromosphere was seen to expand like a balloon, as in Fig. 11, pushing other solar material ahead of it as it shoved its way through the outer corona and disrupted existing rays and streamers. These coronal transients, moving outward at speeds of about 500 km per second, were a new form of solar activity and among the largest objects observed in the solar system — before passing out of the coronagraph's field of view, many coronal transients had grown larger than the Sun itself. Yet they occurred in step with other forms of solar activity, on an average of several times a week. Like comets, coronal transients contain very little matter and are visible only because the outer corona itself is almost a perfect vacuum. Had the Sun blown one of these gigantic bubbles every day since it was born, the material lost would still be an insignificant fraction of its total mass.

THE SOLAR WIND

The solar discovery of greatest impact in the first two decades of the space age dealt not at all with conventional descriptions of the star, but with the intrusion of its outer atmosphere into interplanetary space: the outward flux of atomic particles now known as the *solar wind*. Intermittent streams of "solar corpuscular radiation" had long been suspected, as a way to account for the aurora borealis and australis, geomagnetic disturbances, and fluctuations observed in cosmic rays. In 1931 physicists Sydney Chapman and V.C.A. Ferraro proposed a physical model to explain geomagnetic storms: they suggested that a hypothetical cloud of ionized gas streamed outward from the Sun at times of solar flares. But it was generally thought that the space between the planets was empty and static, disturbed only infrequently by the sort of impulsive solar intrusions that Chapman and Ferraro described. The precipitation of these bunches of charged particles directly onto the polar atmosphere of Earth seemed adequate to trigger the northern lights (Fig. 12), and their compressive effects on the dipole field of Earth served as an explanation of geomagnetic storms.

In 1951, Ludwig Biermann also suggested that a more con-

Fig. 12. Auroras sometimes produce richly colored curtains and streamers that can literally dance in the sky. This variegated display was recorded by amateur astronomer David Huestis on May 4, 1978, from his backyard observatory in Uxbridge, Massachusetts.

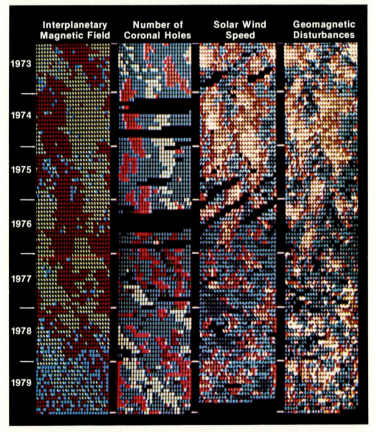

Fig. 13. This comprehensive record of solar-related data demonstrates the connection between the appearance on the Sun of low-latitude coronal holes and geomagnetic disturbances on the Earth, through the connecting link of high-speed streams in the solar wind. Each row in the four broad columns contains color-coded measurements made over a 27-day period (or about one solar rotation). In the first column, red indicates positive polarity, green or white negative polarity; in the last three, brighter colors (like red) correspond to higher values. Long-lasting features recurring over several solar rotations show up as vertical bands within each column, and similar patterns among columns demonstrate the cause-and-effect nature of Sun-Earth interactions. Courtesy Neil R. Sheeley, Jr., and John Harvey.

tinuous flux of atomic particles from the Sun might explain accelerations noted in anti-solar spikes in the tails of certain comets; a steady flow of solar electrons was also invoked to explain properties of the zodiacal light. But it was thoughts and not observations that led to a dramatic theoretical conclusion by Eugene Parker in 1958, that the Sun's atmosphere could not be contained by gravitational forces: that the corona did not terminate near the Sun but stretched out almost forever in steady expansion, filling the solar system with an outward flow of boiled-off bits and pieces of the star. Though immediately challenged on theoretical grounds, Parker predicted that if we could only measure it, the space between the planets should be filled by a continual wind of atomic particles whose force and numbers depended chiefly on coronal temperatures; at the Earth's orbit they should reach velocities from 400 to 800 km per second — faster than a speeding bullet.

Within two years the outlandish prediction was confirmed. Russian interplanetary space probes launched between 1959 and 1961 carried rudimentary particle detectors that measured a high flux of rapidly moving ions of positive charge. Our own Explorer 10, launched also in 1961, found the same result with more sensitive apparatus. Mariner 2, sent to Venus in 1962, compiled a nearly continuous record three months long that left no room for any other explanation: there was, and always had been, a solar wind — and therefore stellar winds around all the other stars. Solar wind flow speeds measured by Mariner were exactly as predicted, ranging from 319 to 771 km per second; wind density ranged from a single particle per cubic centimeter to some 50 times that amount. Parker, four years earlier, had predicted a steady, structureless flow. What was found was something a lot more complicated: recurrent high-speed gusts, or streams, that sometimes reached 1,000 km per second and whose pattern of recurrence fit the Sun's rotational period. Flare-produced shock waves were found and an unexpected "sector structure" in which broad angular divisions of the solar wind, rotating with the Sun, preserved a dominant positive or negative magnetic charge.

The existence of recurrent, high-speed streams in the solar wind provided an important piece in a puzzle that had stalled progress in solar-terrestrial physics for decades. Long-term records of geomagnetic perturbations showed a dominant, recurrent period of 27 days that correlates with solar rotation (Fig. 13). The spacecraft measurements of gusts of solar wind sweeping past the Earth with the same period left no doubt that these were direct, *in situ* measurements of the long-sought link. What remained was to identify their source on the Sun. Some held that it must be regions of solar activity, such as those that produced solar flares; others, looking differently at correlations between the Sun and the activity around Earth, proposed that these unknown, solar "M-regions" might be associated with exactly the opposite: solar regions with no visible activity.

Since it may take several days for particles released at the Sun to reach the Earth, and since the exact route and speed of their passage across the 150 million kilometers is not known, it is difficult to establish a connection between a pattern of observed events on the Sun and a sequence of geomagnetic disturbances at the Earth. With the discovery of the solar wind, and an appreciation for the way that magnetically closed, solar active regions could inhibit the outward flow of coronal particles, the prime suspect for the enigmatic "M-regions" shifted to long-lived, quiet areas on the Sun — regions undefined and hard to see, where magnetic field lines open radially into space, channeling particles out. The final pieces fell into place when rocket and spacecraft pictures made at X-ray wavelengths revealed the existence of precisely such regions: large, long-lived holes in the corona that are free of confining, magnetic loops. Quickly the chain was forged: coronal holes, turning with the Sun, release blasts of atomic particles into space — like a rotating garden sprinkler. These recurrent streams of higher-than-normal velocity in the solar wind can be detected in interplanetary space; when they reach our vicinity, the high-speed solar wind streams produce a common class of geomagnetic disturbances and aurorae.

THE SOLAR CONSTANT

The principal influence of the Sun on the Earth is through its constant, brute-force input of light and heat: the continuous blackbody spectrum of radiation that comes almost entirely from the 6,000° K solar photosphere, with maximum intensity in the yellow-green region of visible light. In astrophysics the integral of this radiated power is known as the total luminosity of a star. For the Sun, however, a more parochial definition is commonly used: the total amount of energy of all wavelengths intercepted per unit of time by a unit surface at the top of the Earth's atmosphere, corrected to the Earth's mean distance from the Sun. The integrated quantity defined in this legalistic way has been known since 1838 as "the solar constant," although few astronomers have ever seriously considered it to be truly invariable.

The mean value of the solar constant has been known to within a few percent for more than 70 years. Its absolute value is very close to 2 calories per square centimeter per minute. An uncertainty of a few percent in the measurement is all-important, however, since it hides any intrinsic variability smaller than this amount. To date, relative measurements of sufficient accuracy have not been made for an adequate length of time to define the limits of variability of the solar constant. Evidence for temporal variation of the total luminosity of the Sun — even of small magnitude — can be extremely important, both for what it tells of energy generation and storage within the Sun, and for the possibility of solar effects on the climate of the Earth.

Modern models of global climate suggest that a change of just 1 percent in the solar constant will bring about a change of 1-2° in the average temperature of the Earth. That is a lot. During the depths of the last Ice Age, about 20,000 years ago, the average temperature of the Earth was probably 5° cooler than now, equivalent to a drop in the solar constant of at most 5 percent. A continued drop of 10 percent or 10° now seems a ticket to cosmic catastrophe, for it would induce, the models say, an ice-covered Earth. Moreover, since ice is so reflective, a subsequent increase of an improbable 50 percent in the solar constant would be needed for a thaw. A drop of 1 percent in the solar constant seems enough to explain the Little Ice Age that chilled Europe and the Americas between the late Renaissance and the start of the Industrial Age, and a change of but a tenth of that, if lasted, would alter meterological conditions enough in marginal areas to bring major economic impacts.

There is an important proviso: the circulation of the lower

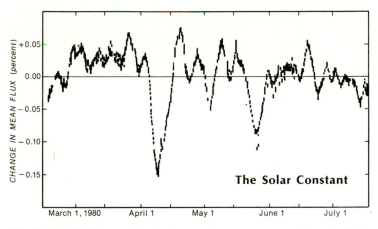

Fig. 14. The first five months of solar-constant measurements from the Solar Maximum Mission spacecraft exhibit deviations from a "steady" solar flux of up to 0.15 percent. From a paper by Richard Willson et al.

atmosphere of Earth is so complex and subject to so many other inputs and prior conditions that almost any change in solar flux, *if brief enough*, will go unnoticed or get lost in the welter of competing factors. Weather and climate seem able to tolerate almost any short-term solar fluctuation. Solar changes that persist, on the other hand, can produce the drastic climatic changes described above. The boundary between ineffective and effective solar changes, we now believe, is about 10 years.

Measurements of the solar constant made from the ground — even on the highest mountains — are severely limited by large corrections for absorption in the air overhead. It was thus a real breakthrough when radiometers to measure the solar constant were lofted above the ocean of air, first on aircraft and balloons, then sounding rockets, and ultimately on spacecraft where continuous monitoring could at last begin. The first spaceborne attempts to monitor the solar constant were made with cavity radiometers carried as low-priority piggybacks on the Mars-bound Mariner 6 and 7 probes, launched in 1969. Useful observations spanned a period of about five months, at the maximum of the 11-year solar cycle. The Mariners' measurements of the solar constant fluctuated about 0.1 percent, at the limit of measurement accuracy. Better radiometers (launched on Nimbus 6 in 1975, Nimbus 7 in 1978, and the SMM) gave more accurate results; these latest spaceborne measurements established, for the first time, unequivocal evidence for real solar constant fluctuations. In the SMM data the maximum excursions are about 0.1 percent (as seen in Fig. 14) and last for periods of several days to a week or more. Early analysis suggests that these small fluctuations are the result of simple blocking by large sunspot groups, leaving many questions as to where the blocked radiation goes or how the Sun can store it very long.

Fluctuations of at most 0.2 percent in the solar constant lasting but a week may be of academic interest only, so far as climate is concerned. Day-to-day cloud cover varies a good deal more than that. The important thing, however, was the demonstration that the total output of the Sun varied at all; and if it changes perceptibly in so short a span of measurement, what might it do in a longer time? For this reason, the Nimbus and SMM measurements of minuscule change in the solar constant were probably as important in the history of solar physics as were Galileo's first observations of dark spots on the surface of the Sun. In each case, a long-held premise of perfect constancy was forever shattered and thrown aside.

A LOOK AHEAD

There are surely as many surprises ahead in the study of the Sun from space as we have found in the first 20 years, yet who can say what they will be? The initial excursion into space, characterized for solar physics by Earth-orbiting, observatory spacecraft such as the OSO's, Skylab, and SMM has run its course, to be supplanted (we hope) by a generation of new and more sophisticated spacecraft. For a time Nimbus will extend its run of measurements of the most practical quantity ever observed in astronomy — the solar constant, the Cinderella of solar physics. The SMM, crippled after only nine months in space by a failure in its pointing system, will do the same once repaired by a crew of Space Shuttle astronauts. The daring rescue is NASA's plan to save the satellite, which already has identified coronal loops as the locus of primary particle acceleration in flares (Fig. 15).

On later missions, Space Shuttles will carry up a number of planned Spacelab payloads, two of which are aimed at studying the Sun for 7-day, intensive missions reminiscent of Skylab, with professional solar physicists aboard as payload specialists to guide specific observations. Also being considered is an advanced solar observatory that may be part of a permanent, Earth-orbiting power station to recharge the Space Shuttle's batteries. On trips to this solar-powered "filling station," the shuttle could also service the attached observatory. Other missions under study include a Coronal Explorer and a Solar Dynamics Explorer to study the cycle of solar activity, a novel Pinhole/Occulter telescope capable of unique visible and hard X-ray observations of the corona, and a manned Solar Terrestrial Observatory to concentrate on physical connections with the Earth.

Of unusual interest among planned Shuttle payloads is the Solar Optical Telescope, featuring a 1.25-m-aperture telescope designed for direct observation of the Sun in the visible region of the spectrum. It will capitalize, for the first time from space, on the advantage of perfect seeing and image stability in the broad spectral region from about 1100 angstroms in the ultraviolet through the near infrared, exploring detailed structures in the lower chromosphere and photosphere. Ground-based solar telescopes, and previous orbiting solar telescopes, have resolved features about one arc second in size on the Sun, roughly 750 km across (the size of France or Spain). Occasionally, moments of unusual seeing permit a slightly better view. The SOT, free of distortions in the atmosphere of Earth, promises to capture detail fully 10 times smaller, and do it day after day, with a resolution that will enable us to see solar features the size of Paris or Lake Geneva — from 150 million kilometers away! We can expect, as in the past, that with better resolution new phenomena will be discovered and new insights gained.

A far more bold departure, the International Solar Polar Mission, is planned for late in the present decade. A shuttle-launched spacecraft will leave the plane of the Earth's orbit to explore the Sun's influence on interplanetary space in the vast domains above and below the plane of the ecliptic. We now know the Sun only in one restricted

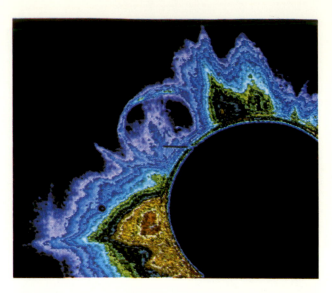

Fig. 15. The High Altitude Observatory coronagraph on the Solar Maximum Mission made this visible light image on April 14, 1980. Coronal streamers at nine and twelve o'clock (red, green, and yellow in this computer enhancement) are the brightest features; the purple loop between them is an outward-moving coronal transient that soon grew larger than the Sun itself.

plane, for all our previous measurements of its flow of particles and fields and radiation have been made largely from the plane of the planets, leaving to theory and conjecture what lies above and below. To conserve energy the spacecraft will first follow a long detour to Jupiter — five times farther from the Sun — to use the mass of that giant planet to sling it out of the ecliptic plane. It will spend the rest of its life in a slow-moving, solar polar orbit, spending about six years above the Sun and then six years below, probing the planetless *terra incognita* of the solar system. Solar-imaging experiments originally planned for a two-spacecraft ISPM (reduced to one craft in response to budgetary pressures) are sadly absent in the present design, leaving for another generation of solar astronomers the chance to look down at the poles of the Sun.

Even more visionary are spacecraft now under consideration that would plunge directly into the solar corona, like moths into the flame, providing the first direct, physical observations of conditions within the near corona of the sun. These "Starprobe" missions could also make near-Sun gravitational measurements that would indirectly sense the structure of the unseen solar interior. Even deeper, instrumented probes flung into the chromosphere and photosphere seem technically possible, to sample, first hand, the raging inferno of the lower solar atmosphere and convective zone. High temperatures are not an insurmountable problem for such probes, since the local density is like that of a vacuum. Once inside the corona, however, signals telemetered by radio cannot escape, and the last, momentous words of a plunging craft might be whispered out from beyond the veil in gasps of laser light.

Like the 1946 launch of the first solar V-2 rocket, these pioneering probes out of the ecliptic and into the Sun will open another new era in the study of the most important star. And if past experience is a guide, their results will open up as well a whole new box of unanswered questions.

Magnetospheres and the Interplanetary Medium

James A. Van Allen

One of the most engaging detective stories of modern science has been recognition of the existence of solar "corpuscular radiation" and of its effects on the Earth, on the other planets, and on comets. The term corpuscular radiation, now supplanted by the term *solar wind,* was adopted to distinguish gaseous material from light and other forms of electromagnetic radiation. This gaseous material and its associated magnetic field constitute the interplanetary medium. In their great 1940 treatise, *Geomagnetism,* Sydney Chapman and Julius Bartels document the early history of the subject. Direct observations of the solar wind by instrumented spacecraft have confirmed some, but not all, of the early inferences concerning solar corpuscular streams and have provided a wealth of detailed knowledge.

THE INTERPLANETARY MEDIUM

The solar wind consists of hot (about 100,000° K) ionized gas, an electrically neutral mixture of ions (principally protons with minor but very significant numbers of heavier ions) and electrons, emitted from the Sun's corona and continuously present in interplanetary space to a greater or lesser extent. After escaping from the gravitational field of the Sun, this gas streams radially outwards to distances at least as great as 27 astronomical units (AU) or 4.0 billion

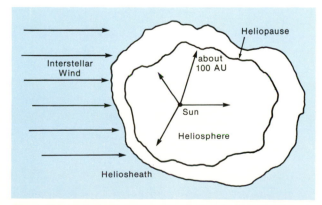

Fig. 1. This conjectural cross-section depicts the interplanetary environment out to more than twice the distance to Pluto. Within the heliosphere the solar wind flows radially outward until it encounters the irregular and fluctuating heliopause. The heliosheath marks the transition to the interstellar world beyond.

km (the current distance of Pioneer 10). At a theoretically estimated distance on the order of 100 AU the solar wind merges with the nearby interstellar medium and loses its identity. The boundary at which this merging begins is called the heliopause, and the region inside it the heliosphere (Fig. 1). Observation of the heliopause is an important objective of the four current deep-space missions: Pioneer 10, Pioneer 11, Voyager 1, and Voyager 2. A spacecraft passing into the nearby interstellar medium will probably experience several effects as it escapes the influence of the Sun.

First, the solar wind flowing radially outward will be replaced by an interstellar wind of different speed, ionic composition, temperature, and direction of flow. Second, the intensity and composition of galactic cosmic radiation will no longer be modulated by the irregular magnetic structure of the solar wind; we expect the interstellar wind to induce much weaker effects. Specifically, the characteristic, quasi-periodic modulation associated with the Sun's 25-day rotation period should disappear. Third, the intensity, ionic composition, and energy spectrum of the cosmic radiation will stop fluctuating and become essentially constant. This last prediction stems from our belief that the interstellar medium contains an average of contributions from many extremely remote sources, an idea supported by the nearly constant character of very high-energy cosmic rays observed from the Earth.

Until now, we have relied solely on our conceptual understanding of the heliopause as a guide to estimating its location. Designing an instrument sensitive enough to detect the dwindling wisps of solar wind at such great distances has proven a formidable challenge. On the other hand, the intensity of relatively low-energy cosmic rays actually increases as we move farther from the Sun. For this reason, I anticipate that observations of galactic radiation have the best potential for revealing the location of the heliopause. Even so, the changeover may occur not at a specific distance but over perhaps many AU.

At the Earth's orbital distance (1 AU) the number density of ions (and electrons) in the interplanetary gas is typically five particles per cubic centimeter under quiet conditions, but sporadic, order-of-magnitude fluctuations occur in response to varying solar activity. The average number density diminishes as the inverse square of the heliocentric distance. Atomic collisions in this exceedingly dilute gas are rare, but complex interactions cause energy exchanges among the constituents of the gas; these give rise to a host of wave-particle phenomena such as Alfvén waves in the magnetic field (generated by the oscillations of the ions around their equilibrium positions), electrostatic oscillations, ion-acoustic waves, and electromagnetic radiation.

Even though each individual parcel of gas moves radially outward from the Sun, a continuous stream of gas flowing from a localized area of the Sun's corona takes on, by virtue of the Sun's rotation, the approximate form of an Archimedean spiral in interplanetary space as viewed by an observer over the solar north pole (Fig. 2). The average speed of the flowing gas, about 400 km per second, is remarkably independent of distance from the Sun, but marked fluctuations in this speed induce a variety of collisionless shock phenomena as fast streams overtake slow ones. The ionized, electrically conducting gas carries with it an entrained magnetic field caused by a system of currents surviving from their origin in the corona of the Sun. The observed magnetic field fluctuates in magnitude and more especially in direction from point to point but generally parallels the theoretical Achimedean spiral, as predicted by Eugene Parker in the late 1950's. Concurrent observations by spacecraft at various heliocentric distances and longitudes have confirmed this spiral form by noting differences in the arrival times of distinctive features of the solar wind and of energetic particle beams. At the orbit of the Earth, the spiral makes an angle of about 45° with a radial line from the Sun but becomes nearly perpendicular at the orbit of Saturn (9.5 AU) and beyond. At 1 AU the magnetic field strength is typically 5 gammas (5×10^{-5} gauss), whereas it is many orders of magnitude greater at localized points in the solar chromosphere. Because the field lines assume a tightly wrapped spiral (and almost cylindrical) form far from the Sun, the field strength decreases approximately as the inverse first power of increasing heliocentric distance.

Direct observations of the solar wind have been confined, thus far, to a thin pancake-shaped region near the plane of the Earth's orbit (the ecliptic plane) which is also approximately the equatorial plane of the Sun. A ballistically difficult space mission to high solar latitudes has not yet been achieved but is planned for the mid-1980's with the International Solar Polar Mission. The velocity necessary to leave the ecliptic plane will come from a Jupiter flyby which makes use of the giant planet as a gravitational slingshot. In the meantime, the detailed properties of the solar wind at high solar latitudes remain conjectural.

COMETARY EFFECTS OF THE SOLAR WIND

The Sun exerts both a solar wind pressure and a radiation pressure (the latter from electromagnetic radiation or photons) on all objects in the solar system. Both of these decrease as the inverse square of the distance from the Sun — moving twice as far away decreases the pressure experienced by four times. On large objects, the pressure of the impinging solar wind is about 3,000 times less than solar radiation (or light) pressure. This means that electromagnetic radiation from the Sun pushes much harder on inert obstacles than does the solar wind, although both are overwhelmed by gravity in most cases. One exception concerns

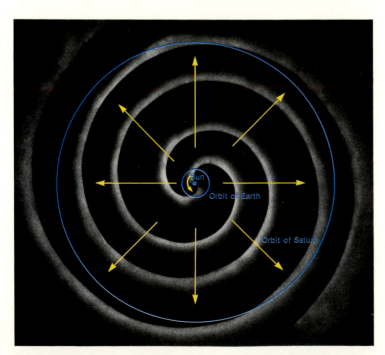

Fig. 2. This bird's-eye view of the solar equatorial plane shows the radially moving solar wind and the Archimedean spiral of the interplanetary magnetic field that results from the Sun's rotation.

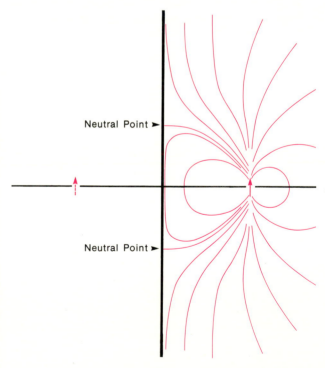

Fig. 3. James Clerk Maxwell's idealization of the interaction of the solar wind with a planet's magnetic field. The heavy black line represents a rigid conducting plate in motion toward the magnetic dipole at right (red arrow). Note that two of the schematic lines of force end in neutral points on the plate. His "mirror dipole" is represented by the light arrow.

cometary dust grains, which have high area-to-mass ratios. Here the effect of radiation pressure becomes more significant, as it would for spacecraft with large arrays of photovoltaic cells or even "solar sails."

The solar wind has a much greater effect on gaseous material in its path; in this case it is more important than radiation pressure. Cometary gas encountered by the ionized, magnetized solar wind gets swept into the flow, undergoing charge exchanges and plasma-physical interactions that are important in various comet phenomena (see Chapter 17). The gas is transported radially outward as an added constituent of the slightly slowed solar wind. In fact, early observations of comets' gas tails by Ludwig Biermann provided one of the first lines of evidence for the continuous nature of solar wind flow (whereas observations of geomagnetic storms had favored the sporadic occurrence of solar corpuscular streams). Ionized comet tails are easily distinguished from the more common dust and neutral-gas tails by their form, optical spectra, and other detailed features.

THE SOLAR WIND AND PLANETARY BODIES

The solar wind has a negligible effect on the movements of large bodies such as planets, even though it is fundamental to the creation of ion tails in comets. Yet the solar wind has profound effects near planetary bodies, for there it creates the phenomena to which we now turn: *magnetospheres*. A magnetosphere is that region surrounding a planet within which its own magnetic field dominates the behavior of electrically charged particles. (The term does not imply spherical shape but is used in a looser sense as in the term sphere-of-influence.)

Because the solar wind is an ionized gas or plasma, it is an electrically conducting fluid. It is also magnetized; that is, it contains electrical currents that were frozen in during its origin in the solar corona. These two properties of the solar wind, when combined with its bulk flow, are the essential causes of magnetospheric phenomena.

James Clerk Maxwell first analyzed and described theoretically the essence of the interaction of the solar wind with a magnetized planet in his 1873 treatise, *Electricity and Magnetism*. He idealized the situation by envisioning a conducting "plate" of infinite area moving toward a magnetic dipole whose axis is parallel to the plate. As part of his calculations, Maxwell described the system of magnetic eddy currents which would be generated on the plate. The strength of the magnetic field associated with these currents would be equal to that of a second magnetic dipole, exactly like the real one, located at the same distance from the plate on the opposite side (Fig. 3). The effect of this "mirror dipole" is to limit the magnetic field of the real dipole to the half-space containing it; to produce two neutral points in that field; and to otherwise modify the form of the field. In addition, the plate is repelled by a calculable force.

The real situation is somewhat similar, but the solar wind, being a fluid rather than a rigid plate, flows around the sides of the magnetic barrier. Like the plate, the solar wind is repelled by the magnetic field; the point at which this repulsive force exactly balances the solar wind pressure is called the stagnation point, and is the point of closest approach by the solar wind to the center of the planet. A planet may possess a magnetosphere containing electrically charged particles in durably trapped orbits only if the stagnation point is located well above the sensible upper limit of the planet's atmosphere. Otherwise certain plasma-physical and electromagnetic effects (like the acceleration of charged particles) occur, but a well-developed magnetosphere does not exist. The distance from the stagnation point to the center of the planet (the standoff distance) is considerably more sensitive to the planet's magnetic moment than to the solar wind pressure. For the Earth, this value is about 64,000 km (about 10 Earth radii) on our sunward or "upwind" side, well above our atmosphere.

The solar wind's magnetic field merges with that of the planet and stretches it out to produce a long, turbulent magnetotail, or wake, on the downwind side of the planet. Spacecraft have determined that the Earth's magnetotail is several million kilometers long. Moreover, the solar wind induces a transverse electric field across the planet's magnetic field, which injects a small fraction of the surrounding interplanetary plasma into the outer magnetic field of the planet.

As the flow of the solar wind and its magnetic field fluctuate, they induce electric and magnetic fields which randomly accelerate and decelerate particles. The effect of this process diffuses some particles inward where they become trapped in the stronger regime of the inner magnetosphere. Some particles ultimately diffuse down to the planet's atmosphere; others manage to work their way back out and are lost to space. The entire process is a dynamic one. The residence time of a magnetically trapped particle varies widely — from hours to years. Additional low-energy ions and electrons come from the ionosphere of the planet and the ionospheres and atmospheres of the planet's satellites. These particles are also subject to the dynamical processes sketched above.

One "internal" source of high-energy protons and electrons is found in the neutrons produced due to bombardment of the atmospheric gases of the planet by galactic and solar cosmic rays. A small fraction of such neutrons undergo radioactive decay as the fly outward and, in the

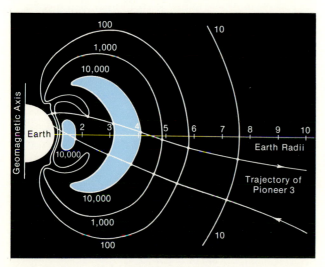

Fig. 4. A cross-section (through the meridian) of the distribution of energetic particles in the Earth's inner and outer radiation belts. The author made this sketch in 1959 based on observations by Explorers 1, 3, and 4, and by Pioneer 3 (whose trajectory is shown). The intensity levels are in relative units.

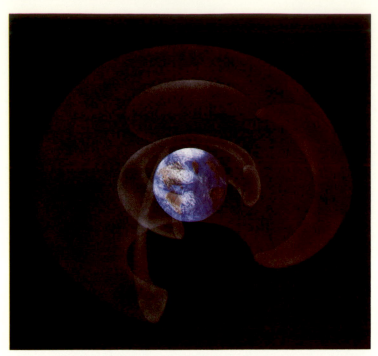

Fig. 5. The doughnut-shaped radiation belts that encircle the Earth have been portrayed in an isometric perspective by artist Charles Wheeler.

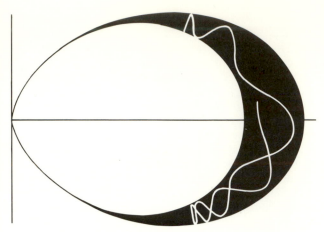

Fig. 6. In this adaptation of a 1907 diagram by Carl Störmer, an electrically charged particle is shown trapped in the magnetic field of a dipole (at the origin at left).

process, inject energetic protons and electrons into the magnetosphere. The residence times of such particles in trapped orbits depend in part on the field strength within the magnetosphere; in the strong magnetic field of Earth's inner magnetosphere, residence times for protons having energies of tens of millions of electron volts are something like a decade, so the accumulated populations from this weak atmospheric source are correspondingly great.

The entire physical process of producing and populating a magnetosphere with energetic particles is quite complex and the attempt to fully understand it is an ongoing effort by many, many researchers. This has been only a sketch of what are currently thought to be the main features.

PLANETARY MAGNETISM

One of the most fundamental characteristics of a planetary body is its state of magnetization, comprised of remanent ferromagnetism and of electromagnetism caused by internal electrical currents. It was shown many years ago that the magnetic field of the Earth must be attributed primarily to a system of electrical currents flowing in its deep interior — the Earth is a huge electromagnet. The necessary electrical currents are maintained by what is called a self-excited dynamo. Theories of such a dynamo remain conjectural and their predictive value is quite limited. But all of the plausible theories require two basic features: a rotating body and a hot, fluid, electrically conducting interior within which convective motion occurs.

Firm knowledge of the state of magnetization of planetary bodies is still acquired only through observation and is one of the central objectives of missions to the planets. The Earth's external magnetic field has been the subject of massive investigation for many years and is extremely well known on a descriptive level, as is its short-term and long-term variability. Spacecraft have also measured the mag-

netic properties of the other six planetary bodies known to the ancients: the Moon, Mercury, Venus, Mars, Jupiter, and Saturn. In addition, some crude upper limits have been determined for the Galilean satellites of Jupiter and for Titan, the largest satellite of Saturn.

The simplest general characterization of a planet's magnetism is its *equivalent dipole magnetic moment*. This is a quantity that takes in the intrinsic strength of the field, its tilt with respect to the planet's rotation axis, and its offset from the planet's geometric center. By dividing the dipole moment by the cube of the planet's radius, one obtains the average value for the magnetic field at its equatorial surface; the actual value can be stronger or weaker depending on longitude. At the Earth's equator, the magnetic field measures 0.308 gauss, or 30,800 gammas.

MAGNETOSPHERE OF THE EARTH

Using the simple radiation detectors aboard the American satellites Explorers 1 and 3, in 1958 my students and I discovered the existence of a huge population of energetic charged particles, durably trapped in the external magnetic field of the Earth. No one had predicted such an effect; indeed, we designed the early instruments for a comprehensive survey of cosmic-ray intensities above the appreciable atmosphere. Soviet investigators S. N. Vernov, V. I. Krassovsky, their colleagues, and I confirmed the discovery later in 1958 and during 1959 by follow-on satellite and space probe investigations using instruments designed with knowledge of actual particle intensities.

The Explorer spacecraft identified two distinctively different "radiation belts" in this early work. They encircle the Earth, with the central plane of each toroid approximately coincident with the Earth's magnetic equatorial plane. Fig. 4 shows in cross-section the inner and outer radiation belts of energetic particles and Fig. 5 is an artist's three-dimensional view.

Carl Störmer showed in his celebrated paper of 1907 that an energetic, electrically charged particle can be permanently trapped, or confined, within the magnetic field of a dipole. In any static magnetic field, the force on a moving charged particle is directed at right angles to both the direction of motion and the magnetic vector. This is called the Lorentz force; its magnitude is proportional to the particle's velocity, electrical charge, and the magnetic field strength.

Because the Lorentz force acts perpendicularly to the direction of motion, it bends the particle's trajectory but does not change its energy. Under this effect, particles move in helical paths (Fig. 6) between the poles of the magnetic dipole. The helices drift slowly in longitude and eventually sweep out a toroidal volume encircling the dipole. In the Earth's field, the helices of electrons drift eastward and those of positively charged ions westward; each specific trajectory depends on a particle's energy. Störmer showed that particles have no access to trapped orbits, but if they are somehow injected onto such orbits they remain so trapped forever. This theory gives us a rough basis for understanding radiation belts, but the really interesting physics comes from the departures from this simple case.

It is possible to trap radioactive nuclei in a dipole field but nearly all of the particles actually present in natural radiation belts (such as hydrogen and other common atoms) are stable, not radioactive. Moreover, these belts are optically transparent; the particle distribution itself does not shield the Earth in any significant way.

In the years following 1958, scientists discovered an immense variety of physical phenomena near the Earth. Thomas Gold first suggested the now-accepted term magnetosphere to describe the entire region containing trapped radiation (Fig. 7). The entire magnetosphere, but especially its outer reaches, is dynamic and constantly changing. The distributions of relatively low-energy particles (less than about 10,000 electron volts, which we collectively call

plasma) have a much greater influence on gross physical phenomena than do those of higher energy. But the latter pose hazards to living things, and their distribution places practical limits on the regions of space around the Earth where organisms (like us) are safe from excessive radiation exposure. The most readily accessible region of safe flight lies at altitudes below 400 km (Fig. 8). The radiation dosage within the equatorial region of the inner radiation belt at an altitude of about 2,000 km is especially severe — even electronic instrumentation has a limited useful lifetime there.

Our general understanding of the Earth's magnetosphere has risen to a fairly good level, with the help of dozens of instrumented spacecraft. Magnetospheric physics is employed in the analysis of such phenomena as auroras, geomagnetic storms, heating of the upper atmosphere by particle precipitation, anomalous propagation of whistler waves from lightning discharges, and the generation and propagation of a rich variety of plasma waves and electromagnetic waves originating in plasma instabilities. Such work continues to be one of the most active areas of space physics; but even with so much attention being applied in this field, many detailed puzzles remain. Our knowledge of the Earth's magnetosphere provides the basis for the interpretation of the magnetospheres of other planets, but each case exhibits distinctive and unique features.

THE MOON

The first extraterrestrial body to be investigated by space-

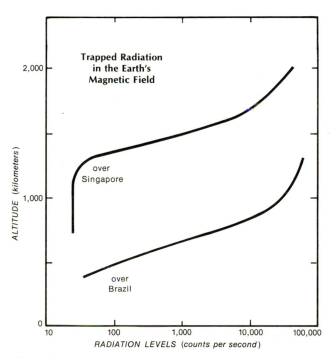

Fig. 8. The increase in radiation intensity with higher altitude is evident in this diagram, derived from Explorer 1 data. The measurements were made within 20° of the magnetic equator, over Brazil (lower curve) and Singapore. The lowest counts, 30 per second, are due to ambient cosmic rays; higher counts result from trapped particles. Such observations were used to determine the offset of the magnetic center of the Earth from its geometrical center (some 400 km). Adapted from a diagram by S. Yoshida, G. Ludwig, and the author.

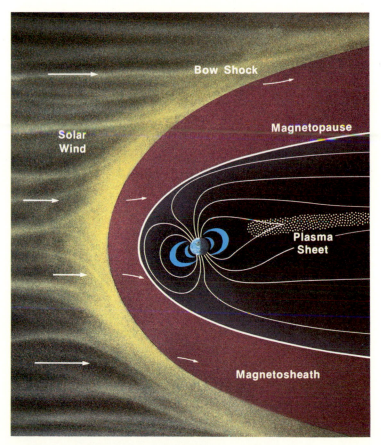

Fig. 7. The general shape and characteristic features of the Earth's magnetosphere. Compare the magnetic field lines here with the lines of force depicted in Fig. 3. The magnetotail extends to the right.

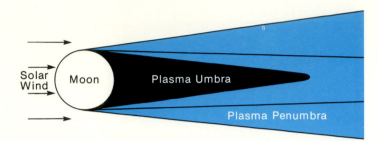

Fig. 9. The interaction of the Moon and the solar wind is shown in this diagram adapted from Y. C. Whang.

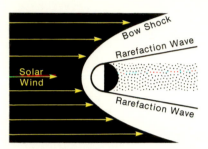

Fig. 10. Venus has no appreciable magnetic field, but the interaction of its electrically conductive ionosphere with the solar wind produces a well-developed bow shock nonetheless. The rarefaction waves shown in this diagram, adapted from one by Herbert Bridge, are a common aerodynamic effect observed when a high-speed flow is interrupted by an inert, spherical body.

age techniques was the Moon, an object studied intensively by American and Soviet flybys, orbiters, and landers. In the context of this chapter, their principal finding is that the Moon has no general magnetic field — its magnetic moment is less than one ten-millionth that of the Earth. The Moon must therefore lack an internal dynamo, most likely due to its slow rotation (sidereal period: 27.3 days) and the probable absence of a hot fluid core. Experimenters have discovered localized magnetic regions on the Moon with surface magnetic fields of 6-300 gammas, but these appear to be geologic anomalies exhibiting residual magnetization that has survived from an earlier epoch.

In its orbit about the Earth, the Moon passes through the Earth's magnetotail for a few days near the time of full moon; for the remainder it is outside the magnetosphere and exposed to the solar wind. Because our satellite lacks a general magnetic field, the solar wind strikes the lunar surface and accumulates there. On the antisolar side of the Moon there is a long, plasma void (or plasma umbra) shaped like an ice-cream cone, with its apex downwind and with the Moon itself as the scoop of ice cream (Fig. 9). Although this plasma void gradually fills in because of lateral motion of the particles in the solar wind, it still may be the most nearly complete vacuum in the solar system. The mere existence of such a plasma void requires a system of weak electrical currents; this has been observed most clearly by the lunar orbiting spacecraft Explorer 35. In addition, the varying magnetic field within the solar wind induces a system of transient electrical currents within the body of the Moon, as observed by the long-lived Apollo magnetometers on its surface.

Overall, the plasma-physical phenomena at the Moon are prototypical of the interaction of the solar wind with an inert, though slightly conductive, object.

VENUS

A succession of U.S. and Soviet spacecraft have observed the magnetic properties of Venus at an increasing level of detail. These include the first-ever approach to another

planet by Mariner 2 in 1962, Mariner 5 (1967), Mariner 10 (1974), Veneras 4, 6, 9, and 10 (1965-75), and, most important of all, Pioneer Venus orbiter (1978–present).

The magnetic moment of Venus is less than 0.00005 that of the Earth despite our "sister" planet's comparable size and probable fluid core. But it rotates extremely slowly (sidereal period: 243 days), and it appears that the system of internal currents is correspondingly weak or nonexistent on a large scale. Nonetheless, the solar wind is prevented from reaching the surface of the planet by its dense atmosphere and, more specifically, by magnetic eddy currents induced in its conducting ionosphere. Thus, Venus has a well-developed bow shock (Figs. 10 and 11), which exhibits many interesting plasma effects but contains no population of durably trapped particles. It is a case intermediate to those of the Earth and the Moon, but more similar to the lunar case.

MARS

In 1965 Mariner 4 made the first observations of the magnetic properties of Mars. The subsequent Soviet missions Mars 2, 3, and 5 (1971-74) have made further contributions.

The magnetic moment of Mars is less than or about four-thousandths that of the Earth. As with Venus, the solar wind interacts with the conducting ionosphere and a weak bow shock is produced (Fig. 11). Mars' sidereal rotation period is 24 hours 37 minutes, similar to that of the Earth, but the planet's radius is only 53 percent of Earth's. The magnetic evidence suggests that Mars lacks a hot, fluid core.

MERCURY

Between 1973 and 1974, the Mariner 10 spacecraft made three successive flybys of Mercury, performing the first and thus far only *in situ* observations of this innermost and smallest of the terrestrial planets. The sidereal rotation period of Mercury is a slow one, nearly 59 days. Nonetheless, the planet has a well-determined although weak general magnetic field (Fig. 11), with a planetary magnetic moment similar in magnitude to the upper limit cited above for Venus. A bow shock has been observed (despite the absence of an atmosphere and ionosphere) as has the transient acceleration of charged particles and other plasma phenomena. But the Mercurian magnetic field is too weak to permit the development of a radiation belt of trapped particles.

JUPITER

In the decades before spacecraft visited the largest planet, radio astronomers obtained approximate but reliable data on its magnetosphere. The results set Jupiter apart from all other planets and provided an impetus for thorough, *in situ* observations of its radiation belts.

As early as 1955, Bernard Burke and Kenneth Franklin reported the first persuasive evidence that Jupiter is a nonthermal source of sporadic bursts of radio noise at a frequency of 22.2 MHz (decametric wavelengths). The term nonthermal distinguishes these emissions from the thermal radiations that are a universal property of all objects whose temperatures are above absolute zero. Nonthermal emissions arise from physical processes that are not contemplated in Planck's theory of the black body radiation of heated objects.

A few years later another type of nonthermal radiation from Jupiter was discovered in a quite different portion of the spectrum, the *decimetric* wavelength range (300-3,000 MHz). The decimetric radiation has an intensity and spectral form that are nearly constant with time. It originates not at the planet's surface, as does the decametric radiation, but within the toroidal region encircling the planet (Fig. 12); the central plane of this torus tilts about 10° to the planet's equatorial plane. In 1959, Frank Drake and S. Hvatum interpreted the decimetric emission as synchrotron radiation by electrons trapped in the external magnetic field of the planet (as in our Van Allen radiation belts of the Earth) moving at relativistic velocities.

The first two spacecraft to make *in situ* observations of the Jovian magnetosphere, Pioneers 10 and 11, were successful far beyond the expectations of their original advocates and even of those who participated in the missions. Pioneer 10 was launched in March, 1972, encountered Jupiter in early December, 1973, and is still operating well and providing good interplanetary data as it continues to escape the solar system. By mid-1982, the spacecraft attained a heliocentric distance of 27 AU (far beyond the orbit of Uranus); it is the most distant of man-made objects and the one most likely to discover the heliopause. Pioneer 11 was launched in April, 1973, encountered Jupiter 20 months later, and continued onward to make the first encounter with Saturn in the late summer of 1979. It also continues to operate properly.

Instruments on Pioneers 10 and 11 provided a massive body of definitive knowledge on Jupiter's magnetosphere. Among other achievements, the Pioneers showed that the pattern of decimetric radio emissions from Jupiter corresponds to the distribution of relativistic electrons there, confirming Drake and Hvatum's interpretation. They measured the configuration and structure of the bow shock and outer reaches of the Jovian magnetosphere, as well as the energy spectra, angular distributions, and positional distributions of energetic electrons and protons within its interior. The magnetosphere of Jupiter generally resembles that of the Earth but its dimensions are over 1,200 times as great. In fact, if we could see the Jovian magnetosphere from here, it would appear several times larger than the full Moon. The magnetotail of Jupiter extends several AU in the anti-solar direction.

These enormous dimensions result from a magnetic moment 19,000 times greater than the Earth's and from the fact that the solar wind pressure there is only about 4 percent of its value at the Earth's orbit. The magnetic moment vector of Jupiter is tilted about 9½° to its rotational axis, in agreement with evidence from radio observations, and has the same sense as the rotational angular momentum vector. This latter relationship is the opposite of that for the Earth. But in the context of our incomplete understanding and in light of the evidence for many reversals of the Earth's magnetic moment over geologic time, this is not especially significant.

Beyond the basic characteristics sketched above, the magnetosphere of Jupiter exhibits a rich variety of special and unique features. The outer magnetosphere occupies an enormous disk-shaped region (Fig. 13), the result of two principal mechanisms. First, a large mass of low-energy plasma trapped within the magnetic field exerts a kinetic pressure on the magnetic field, inflating it in the same

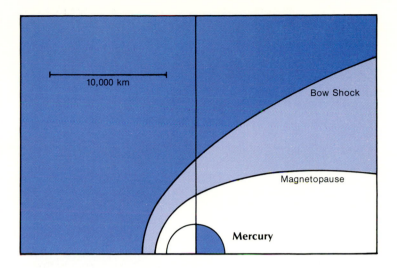

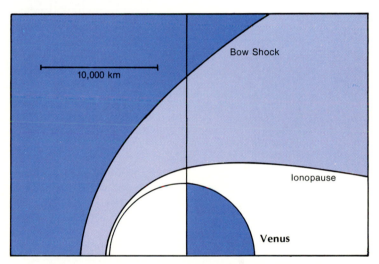

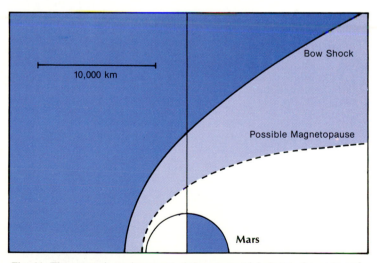

Fig. 11. The magnetic environment of three terrestrial planets shown to scale. The magnetopause is defined as the boundary between the solar wind and a region close to the planet shielded from its flow; the bow shock is a characteristic front that forms upwind from the magnetopause where the solar wind senses the presence of the planet's own magnetic environment. Mercury and perhaps Mars have magnetic fields that deflect the solar wind around them; Venus depends on its electrically conductive ionosphere to divert the flow (this may also apply to Mars). Earth's is not included here, because its magnetic envelope is some seven times the size of Venus'.

manner as an air-filled balloon. Because the field is weakest in its equatorial plane, the distention is most prominent there. Second, because of its interaction with the rotating magnetic field of the planet, the plasma co-rotates in a period of 9 hours 55.5 minutes. Centrifugal force on the rotating plasma adds to the outward pressure.

The first of these effects dominates out to about 50 planetary radii and the second at greater distances. Improved plasma analyzers on the two Voyager spacecraft showed that this plasma originates principally as sulfur dioxide, hydrogen sulfide, and other gases vented from the large, tidally heated satellite Io during its volcanic eruptions. These gases contribute a unique feature to Jupiter's magnetosphere called the Io torus.

Pioneers 10 and 11 provided an excellent survey of the distributions of energetic electrons and protons within the inner magnetosphere of Jupiter and demonstrated that Io and the other inner satellites play an essential role in limiting the total population of trapped particles by absorbing those that venture into the inner magnetosphere. (The Moon does not have the same effect on the Earth's magnetosphere because it is too far away.) Nonetheless, the characteristic energies of such particles at Jupiter are about an order of magnitude greater than such energies in the Earth's inner magnetosphere and the intensity of the trapped radiation is several orders of magnitude greater. Even during the brief encounters of both Pioneers 10 and 11, the integrated dose of ionizing radiation was enough to threaten the survival of transistor electronics and to darken exposed optical glass.

Pioneer and Voyager observations of Jovian magnetospheric properties have spawned an intensive burst of interpretative work and have gone far toward enriching our understanding of the origin of energetic charged particles in a larger astrophysical context. Indeed, we now recognize Jupiter itself to be a copious emitter of previously enigmatic energetic electrons into interplanetary space.

SATURN

In contrast to the situation at Jupiter, previous to late summer 1979 we knew nothing of the state of magnetization of Saturn or of the presence or absence of radiation belts there. Speculative considerations favored a strongly magnetized Saturn whose ring system prevented energetic electrons from becoming trapped within the inner and strongest part of its presumed magnetic field. Moreover, the telltale synchrotron radiation (emanating from outside this region)

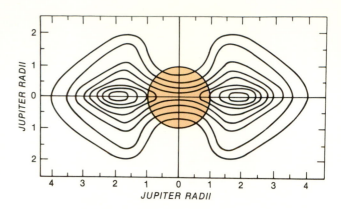

Fig. 12. Glenn Berge's diagram reveals the intensity contours of Jupiter's decimetric (10.4-cm wavelength) radio emission. The circle represents the planet's disk; note two "hot spots" at about 1.9 Jovian radii from its center.

would be attenuated by a factor of about four relative to that from Jupiter, because Saturn is nearly twice as far from us. On the basis of these two considerations, scientists predicted that an intense radiation belt could exist outside of the outer edge of Saturn's ring A without violating the negative evidence from Earth-based radio astronomy.

The first affirmation of Saturn's magnetosphere came in 1979 as various instruments aboard Pioneer 11 detected the presence of a bow shock 24 Saturnian radii (1.44 million km) from the center of Saturn on its sunward side. Later, as the spacecraft passed beneath the outer edge of the planet's A ring, the instruments recorded a guillotinelike cutoff of the charged-particle flux — just as expected. All the data returned during that period were new, rich in interpretable significance, and intensely interesting.

Pioneer found the magnetosphere of Saturn to be intermediate between those of the Earth and Jupiter both in extent and in the population of trapped energetic particles. The magnetic moment of the planet is 550 times greater than Earth's but 35 times less than Jupiter's.

As at Jupiter, the magnetic moment vector and the angular momentum vector have the same sense; but in contrast to the situation at the Earth and Jupiter, the two Saturnian vectors are parallel to each other within an uncertainty of about 1°. Moreover, Saturn's magnetic center lies slightly north of its geometric center by only 2,400 km — just 0.04 of the planet's radius. The unexpected absence of a

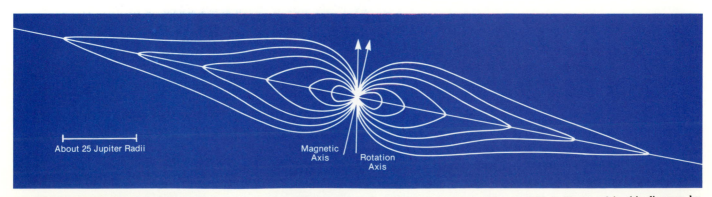

Fig. 13. The magnetodisk model of Jupiter's outer magnetosphere, inferred from Pioneer 10 observations in 1973, is illustrated in this diagram by the author and colleagues.

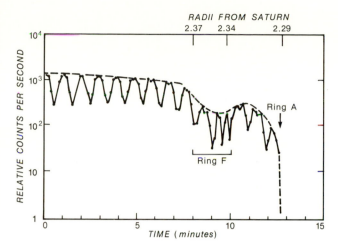

Fig. 14. The abrupt particle absorption cutoff at the outer edge of Saturn's ring A is indicated by these Pioneer 11 observations made September 1, 1979. Also shown is the absorption signature of ring F. This diagram was adapted from one by the author and colleagues.

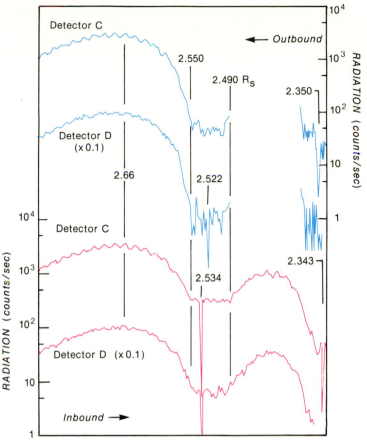

Fig. 15. Charged-particle fluxes, as recorded by two detectors aboard Pioneer 11, are shown for the inbound (red) and outbound (blue) legs of the spacecraft's encounter with Saturn in 1979. The data have been stacked here to show particle absorptions more clearly; one of these, at 2.534 planetary radii from Saturn's center, is attributed to an object some 170 km in diameter (one of Saturn's co-orbital satellites). Other absorption signatures occur at the indicated distances from Saturn's center.

tilt in the magnetic axis tends to exclude certain theories of planetary magnetism.

Beyond about 7 planetary radii, the outer magnetosphere is markedly distended on the dawn side but is much less so on the sunward side. Thus, Saturn's outer magnetosphere has some disklike properties which are again intermediate between those of Jupiter (where it is a prominent feature) and those of the Earth (where it is nearly absent).

Saturn's largest satellite Titan, at an orbital radius of 20.4 planetary radii, moves through the outer fringes of the magnetosphere. According to evidence from the close brush made by Voyager 1 in November of 1980, Titan produces substantial plasma and magnetic effects in its wake and apparently loses nitrogen gas from its upper atmosphere into the magnetosphere (see Chapter 15). The escaping gas is either ionized initially or soon becomes so, thus providing a source for the plasma trapped in Saturn's inner magnetosphere. Other apparent sources of gas are the inner satellites Enceladus, Tethys, and Dione.

During their approaches to Saturn, both of the Voyager spacecraft detected sporadic emission of nonthermal radio noise in the kilometer-wavelength range. The occurrence of such bursts is modulated with a period of 10 hours 39.4 minutes, and some evidence suggests that Dione may also modulate this emission. This period has been adopted provisionally as the magnetospheric, or interior, rotational period of the planet, although the origin of the emissions has not been identified.

The large inner satellites Rhea, Dione, Tethys, Enceladus, and Mimas, like their counterparts around Jupiter, have profound effects on the absorption of inward-diffusing electrons and protons. One of the most remarkable aspects of this influence is that the satellites selectively permit the inward diffusion of electrons of specific energies. At the orbit of each satellite, those electrons drifting in longitude at the same rate as the moving satellite diffuse across the orbit as though the satellite were not present, whereas all non-synchronous electrons have a calculable probability of encounter and absorption by the satellite. All of this alters the energy spectrum of the electrons in a manner analogous

to the effect on white light passing through a succession of overlapping colored filters. At the orbit of Mimas, for example, the surviving population of electrons has a nearly monoenergetic spectrum at about 1.6 million electron volts. No similar effect occurs in the magnetospheres of either the Earth or Jupiter.

But in contrast to the situation for electrons, inward-diffusing protons are strongly absorbed by Dione, Tethys, and Enceladus and perhaps by the tenuous, particulate E ring; there is a nearly total absence of low- and intermediate-energy protons inward of 4 planetary radii. The protons that *are* found near the planet have energies approaching or exceeding 100 million electron volts. Investigators believe these originate as the decay products of neutrons released during the passage of cosmic rays through Saturn's upper atmosphere and ring particles. The high-energy protons are most common 2.7 radii from the planet's center.

The electron density drops dramatically at the outer edge of ring A because of the electrons' absorption by the ring particles (Fig. 14). In addition, the general magnetic field of the planet excludes cosmic rays from the inner magnetosphere. As a result, the region interior to the ring A cutoff is nearly free of energetic particles — the most completely shielded region within the solar system, excluding the atmospheres and solid bodies of planets, large asteroids, and the Sun.

The imaging instrument on Pioneer 11 gave clear evidence for a previously suspected inner satellite (designated 1979 S1) and discovered a previously unknown ring (F). Both of these were identified independently by their electron absorption signatures (see Fig. 15); the satellite was designated 1979 S2 in this case. Other distinctive absorption signatures were designated 1979 S4, S5, and S3; the last was identified by Voyager 1 as another new ring (G). A number of additional small satellites were also found by Voyagers 1 and 2 (see Chapter 19).

Our investigations have been fruitful; Saturn's large and well-developed magnetosphere now joins those of Jupiter and Earth within our limited but growing knowledge. All three appear to be based on a common skeleton of physical principles, yet each possesses unique features.

BEYOND SATURN

Voyager 2 is now targeted to fly by both Uranus and Neptune during the late 1980's. No credible theoretical estimate

of the magnetic moment of either planet exists, but, considering their large sizes and rapid (roughly day-long) rotation rates, there is a widespread expectation that both planets possess extensive magnetospheres. That of Uranus, if it exists, will be of special interest because the planet's rotational axis lies nearly in the plane of its orbit and in 1986 will point approximately toward the Sun (whereas the rotational axes of other planets are approximately perpendicular to their orbital planes).

If Uranus' magnetic axis lies close to its rotational one, the surrounding magnetosphere may respond in surprising ways to the rush of solar wind down its magnetic poles. Pioneer 10, as mentioned, has confirmed the presence of the solar wind to that distance — we have only to wait for Voyager 2's encounter to satisfy our curiosity (Fig. 16).

The solar wind is an important element in the physical processes occurring in all of the planetary magnetospheres investigated so far. But it is not essential to the formation of radiation belts, as amply illustrated by the inner radiation belts of the Earth and Saturn. An important source of trapped energetic particles in both cases is the cosmic-ray-produced neutron flux from the planet's upper atmosphere and in Saturn's case, from the rings and satellites as well. Theorists consider this mechanism to be ubiquitous throughout our galaxy and presumably throughout the universe.

One necessary requirement for the existence of a magnetosphere is that any hypothetical body must have a magnetic moment behaving more like a simple dipole than a more complex quadrupole, octupole, and so on. This allows charged particles to assume helical trajectories between opposite hemispheres and to drift in longitude around the planet. Second, a particle's energy must be small enough to remain bound to the prevailing magnetic field. There are other considerations, too, which involve the total number of particles that can be retained by a given magnetic field; too many of them would tend to break down the magnetosphere partially or even completely.

Scientists expect the gas that pervades the universe to be at least partially ionized and moving some kilometers per second with respect to any planetary body. For example, even in the unexpected event that a reasonably magnetic Neptune lies *outside* the heliopause (and thus out of contact with the solar wind), the magnetospheric particles derived from its atmosphere would not simply accumulate without limit, but would be forced to interact and strike a balance with the interstellar wind. Classical astronomy contains scarcely any mention of electrical and magnetic fields, electrical currents, or energetic charged particles. Yet the solar system is replete with phenomena exhibiting their importance. The magnetospheres of the planets provide striking examples of this point and suggest that the natural acceleration of charged particles is probably a universal phenomenon. The application of knowledge from magnetospheric physics to the study of the radio, X-ray, and gamma-ray emissions from distant objects is already an active field that will likely play a major role in modern astrophysics.

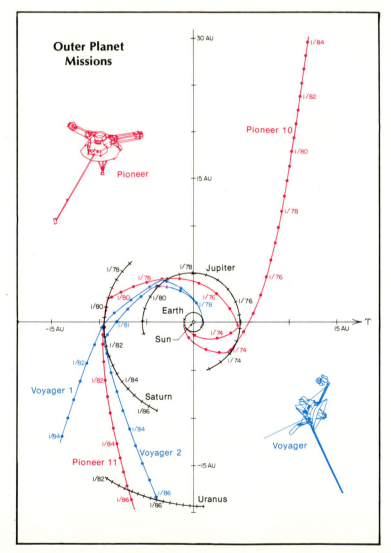

Fig. 16. Trajectories of four American missions to the outer planets are projected on the ecliptic plane (the Sun is indicated by the central dot). The locations of spacecraft and planets are marked at half-year intervals.

The Collision of Solid Bodies

Eugene M. Shoemaker

One of the most striking results from the past two decades of exploration by spacecraft has been the discovery of heavily cratered surfaces on many planets and satellites. These cratered surfaces have been found on bodies ranging from Mercury, on the inner margin of the planetary system, out to the satellites of Saturn. Even as late as 1960, the significance of impact in the sculpturing of the Moon was a subject of lively debate. Today there is no longer any serious doubt that impacts have played a major role in the evolution of the Moon and most solid planetary surfaces. Indeed, one can reasonably argue that impact is the most fundamental process to have taken place on the terrestrial planets — a group of bodies whose very formation probably depended on the collision of smaller objects. Impact craters are by far the dominant land forms on most of the rocky bodies yet surveyed in the solar system.

An examination of the impact history of the planets should start with the present state of the solar system and work backward in time. We can look out into interplanetary space with telescopes and try to estimate the populations of small bodies on planet-crossing orbits, the "stray bodies" as they have been called by Ernst Öpik. From these observations the present collision rates of small bodies with the planets and satellites can be estimated by fairly straight-forward theory which Öpik initially developed. For the Earth, astronomical observations can be checked against the geologic record of recent impact craters, which provides control on assumptions of density and other characteristics of the small bodies. This allows us to estimate the present cratering rates on the other terrestrial planets. Extrapolating from our knowledge of stray bodies' present distribution, we can also infer the cratering history of other terrestrial planets from the impact record of the Earth-Moon system.

PLANET-CROSSING BODIES TODAY

Of the small bodies that cross the orbits of the planets, the ones which can be observed with telescopes include asteroids and comet nuclei. Asteroidlike objects dominate the swarm of objects found at the orbits of the Earth and Mars, while comet nuclei predominate at and beyond the orbit of Jupiter. The distinction between a comet and asteroid, however, relies on our ability to detect an envelope of gas and dust (called a coma), an intrinsic feature of comets. It is pos-

sible that most Earth-crossing asteroids are actually extinct or degassed comets.

Roughly 100 of the numbered asteroids, nearly 5 percent of the ones with well-determined orbits, are planet-crossing. One of these, 944 Hidalgo, crosses the orbits of both Jupiter and Saturn. Its orbit is similar to those of a few short-period comets, which characteristically remain in such orbits for less than one million years before being ejected from the solar system after a close pass by Jupiter. Hidalgo is probably an extinct or inactive comet. Interestingly, a former Earth- or Mars-crossing asteroid of that size (about 30 km across) has only one chance in a thousand of being perturbed into an orbit like Hidalgo's. Another "asteroid," 2060 Chiron, crosses the orbits of Saturn and Uranus. This object is very probably a large comet nucleus that remains too far from the Sun to exhibit any detectable cometary activity (at least as far as we can tell telescopically).

All other known planet-crossing asteroids make close approaches to one or more of the terrestrial planets. About half of these objects do not overlap any planet's orbit at this time. But the celestial dynamicist J. G. Williams has shown that a combination of changes in the orbits of these bodies and that of Mars leads to orbital intersections and possible collisions or close passes with the planet.

Most of these shallow Mars-crossers, as they are called, can be distinguished from main-belt asteroids only by elaborate investigation of their motion. An estimate of their population can be derived from an analysis of well-observed asteroids (those with numbered designations) and of fainter ones surveyed photographically by C. J. Van Houten and his colleagues. For ex-

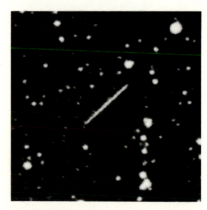

Fig. 1. The Earth-crossing asteroid 2062 Aten, discovered by E. F. Helin in 1976, is 0.9 km in diameter. Moving along an orbit slightly smaller but more eccentric than the Earth's, the asteroid left a trailed image at the center of this 20-minute exposure taken with the 46-cm Schmidt telescope at Palomar Observatory in California.

	Population	Mean probability of one collision per million years	Total frequency of collision per million years
Aten asteroids	≈ 100	0.009	≈ 0.9
Apollo asteroids	700 ±300	0.0026	1.8 ±0.8
Earth-crossing Amor asteroids	≈ 500	≈ 0.001	≈ 0.5
Total	≈ 1,300		≈ 3.2

Table 1. Populations of Earth-crossing asteroids and their predicted collision rates with the Earth. Values listed under *population* are the total number of objects brighter than absolute visual magnitude 18; diameters at this limiting magnitude range from 0.9 to 1.7 km.

ample, I calculate that the total number of shallow Mars-crossers brighter than absolute visual magnitude 18 (which corresponds generally to bodies 0.9-1.7 km in diameter) is about 20,000. Each of these objects stands about a 50-50 chance of hitting Mars or being thrown out entirely in a period comparable to the age of the solar system. Most of them are probably Mars *planetesimals* left over from the accretion of that planet billions of years ago.

All but two of the remaining numbered, planet-crossing asteroids currently overlap the orbit of Mars, and many of these also venture inside or come close to the orbit of the Earth. For a given size, these objects are picked up by telescopic observers much more easily than are main-belt asteroids. Very small and intrinsically faint, they are observable only when relatively close to the Earth. Several are less than 2 km in diameter. Our sampling of the population of objects so small is, to say the least, spotty. The objects that currently come closer than 1.3 astronomical units (AU) to the Sun but do not pass inside the orbit of the Earth have been named Amor asteroids, and those on orbits overlapping the Earth's are the Apollo asteroids. In addition, three numbered and one unnumbered asteroids have been found on orbits with semimajor axes smaller than 1 AU (Fig. 1). The orbits of these objects, termed Aten asteroids, overlap Earth's orbit but lie entirely inside the orbit of Mars.

About 40 numbered and unnumbered Earth-crossing asteroids have been discovered to date; these include about half of the known Amors, as well as nearly all the Apollo and Aten asteroids. The population of these objects can be estimated from their rate of discovery in systematic surveys of the sky, and the total number of Earth-crossing asteroids to absolute visual magnitude 18 is estimated at about 1,300 (Table 1). Taking the orbits of the discovered Earth-crossers as a representative sample of the orbits of the population, about 3 of these bodies, on the average, collide with the Earth every million years. A comparable number collide with Venus, some collide with Mars, Mercury, or the Moon, and about half are ultimately ejected from the solar system. Small-body expert George Wetherill calculated that the typical dynamic lifetime of these bodies is a few tens of millions of years, although some can be expected to survive hundreds of millions or even billions of years.

Clearly, the large majority of Earth-crossing asteroids cannot have occupied their present orbits over most of geologic time. The cratering record of the Earth suggests that the population has been in approximate equilibrium, however, over the last half billion years — apparently these objects have been supplied from other regions of space at

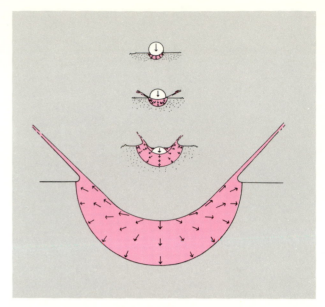

Fig. 2. A schematic representation of the stages in an impact event like the one that formed Meteor Crater, discussed in the text. A compressional wave spreads outward from the impacting meteoroid (arrows indicate the flow of material). A wave of decompression moving behind the shock front allows material mobilized in the event to be ejected in a conical sheet. Engulfed by the shock wave, the meteoroid melts and becomes partially vaporized. Laboratory experiments indicate that for all but very low angles of impact (less than about 20 degrees) the crater produced is circular. Diagram based on studies by Donald Gault.

about the same rate at which they have been swept up or ejected. A few percent may have been supplied by deflection of shallow Mars-crossers during close encounters with that planet. The remainder evidently have come from the main asteroid belt and possibly from more distant sources. Wetherill, Williams, and other investigators have shown how fragments from collisions between the main-belt asteroids can become Earth-crossing if injected into orbits subject to strong resonant perturbations. Such perturbations arise, for example, when the orbital periods are simple multiples or simple fractions of the orbital period of Jupiter (see Chapter 10). Another source may be comets that have been driven into very short-period orbits by a combination of close encounters with Jupiter and the so-called nongravitational accelerations produced by the "thrust" of escaping gases. Comet Encke is an example of a body that has been driven into an orbit like those of some Earth-crossing asteroids. If only a small percentage of such comets yield a core or residue of rocky material when the ices have all sublimed away, they may provide an important source of Earth-crossing asteroids.

Active comets also play a direct role in the impact cratering of the Moon and the terrestrial planets. On the average, each year astronomers find about three comets on nearly parabolic orbits that approach within 1 AU of the Sun. Every comet has a small bright nucleus which, as the comet is followed to large distances from the Sun, becomes nearly stellar or asteroidal in appearance. Many lines of evidence show that the nucleus is a rotating solid body, from which the coma is derived by sublimation of its constituent ices and ejection of dust (see Chapter 17). The form of fresh ray craters on the Galilean satellites of Jupiter, which are probably produced almost entirely by impact of comet nuclei, shows that collision of comets produces craters very similar in form to fresh craters on the Moon.

I estimate that about one-third of Earth's recent impact craters at least 10 km across are due to impact of comet nuclei, about 95 percent of these having been in orbits with periods greater than 20 years. Bodies with shorter periods seldom strike the Earth, because only a small percentage cross our orbit. At the orbit of Jupiter, however, these short-period comets dominate the present population of crater-forming objects.

IMPACT RECORD OF THE EARTH

If a projectile is large enough, it can survive passage through the Earth's atmosphere, more or less intact, and strike the ground or the ocean at high velocity. The threshold size for survival depends on the material strength and density of the body and on its velocity at the time of encounter; for a stony body, this size appears to be about 150 m, close to the diameter of the smallest Earth-crossing asteroid for which we have an accurate orbit. Most smaller objects are sheared apart, and the individual fragments are nearly stopped by aerodynamic drag. A strong shock wave of compressed air forms in front of some fragments, and some of its energy, in turn, escapes as visual and infrared radiation from atmospheric gases shocked into incandes-

cence. Fragments from disrupted iron meteoroids, on the other hand, sometimes reach the ground with substantial remnants of their cosmic velocity and produce swarms of small impact craters. Half a dozen such swarms are known, including a cluster of craters produced by an observed fall in the Sikhote-Alin region of the eastern U.S.S.R. in 1947.

Except for small, very young craters associated with iron or stony-iron meteorites, the impact record of the Earth generally consists of craters or eroded impact structures about 3 km in diameter and larger. A crater 3 km in diameter, formed by a stony body traveling at the average encounter velocity of Earth-crossing asteroids (20 km per second), is roughly 20 times the diameter of the impacting body. An iron asteroid about 50 m in diameter can survive atmospheric entry and form a crater, as at Meteor Crater, Arizona (known officially as Barringer Crater). An assortment of relatively unshocked meteorites surrounding Meteor Crater indicates that numerous fragments were detached and decelerated during atmospheric passage of the principal body, which may have broken apart just before impact. The resulting crater could have been produced by a tight cluster of fragments rather than a single body.

When a solid body strikes the ground at high speed (Fig. 2), a shock wave propagates into the target rocks and another shock into the impacting body. At collision speeds of tens of kilometers per second, the initial pressure on the

Fig. 3. Meteor Crater, near Flagstaff, Arizona, is one of the youngest impact craters found on the Earth. The crater formed about 25,000 years ago when an iron mass (or perhaps several) struck flat-lying sedimentary rocks at a speed of more than 11 km per second. Between 5 and 20 megatons of kinetic energy were released during the impact, which left a bowl-shaped crater roughly 1.2 km in diameter and 200 m deep, surrounded by an extensive blanket of ejecta. The iron meteoroid(s), fragmented and dispersed in the event, was the object of an unsuccessful search by D. M. Barringer in the early 1900's. Barringer made several excavations and drillings which provided data used in the structural and geologic analysis by the author. The crater now has a somewhat square outline, lake sediments blanket the floor, and erosion has removed 15-25 percent of the ejecta. This photograph was taken by U. S. Geological Survey scientists D. J. Roddy and K. Zeller.

rocky material engulfed by the expanding shocks is millions of times the Earth's normal atmospheric pressure. This can squeeze dense rock into one-third of its usual volume. Stress so overwhelms the target material that the rock flows essentially like a fluid. A rarefaction (decompression) wave follows the advancing shock fronts into the compressed rock, allowing the material to move sideways. As more and more of the target rock becomes engulfed by the shock wave, which expands more or less radially away from the point of impact, the flow of target material behind the shock front is diverted out along the wall of a rapidly expanding cavity created by the rarefaction wave. The impacting body, now melted or even partly vaporized, moves outward with this divergent flow and lines the cavity walls. Decompressed material sprays out of the cavity as an expanding conical sheet, and rocky material continues to flow outward until stresses in the shock wave drop below the strength of the target rocks. This, then, is the process by which a crater forms.

Meteor Crater (Fig. 3) is 1.2 km in diameter, 200 m deep, and about 25,000 years old. The surrounding raised rim consists partly of rocky fragments ejected from the crater and partly of bedrock that has been lifted up and shoved radially outward. Roughly half of the volume of the crater stems from the ejection of material and half from the outward displacement of rocks in the crater walls. Ejected debris lies stacked on the crater rim in inverse stratigraphic order: the sequence of layers of the bedrock is preserved upside down in the debris. The eroded surface on the rim deposit has a gently rolling, hummocky character that reflects the irregular distribution of large lumps in the ejected debris.

Beneath the crater floor is a lens-shaped body of breccia — rock that has been broken and smashed by the shock wave. Abundant glass, produced by shock melting of the rocks, occurs near the base of the breccia. It contains microscopic spheres of meteoritic iron — the metamorphosed remains of the impacting body. The distribution of this material shows that the initial cavity excavated by the shock extended to a depth of nearly 400 m. The breccia, which was at first smeared along the cavity walls, collapsed toward the center, producing a much shallower final crater. Relatively fine-grained ejecta that had been arrested in the atmosphere then showered down, leaving a layer of mixed debris about 10 m thick on the floor of the crater. Subsequently, the rim and upper walls of the crater have been eroded, and lakebed sediments 30 m thick lie in the center.

The products of shock metamorphism found in the rocks at Meteor Crater provide important clues for the identification of impact craters and structures elsewhere on Earth. In addition to shock-melted glass and distinctive macroscopic and microscopic deformation of unmelted rocks, two high-pressure forms of crystalline silica have been discovered. One of these shock-formed minerals, coesite, has now been found at many other impact localities around the world.

Crater-hunters have recognized other impact sites on Earth primarily by their general structure and by evidence of shock metamorphism in the rocks. But the Earth's craters are ephemeral; they tend to fill in or erode away. If erosion has not encroached too deeply, part of the impact breccia may be preserved. Impact structures larger than about 30 km in diameter commonly have a fairly thick layer of re-

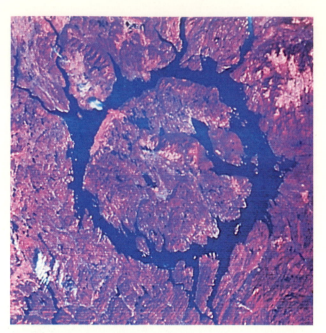

Fig. 4. The Manicouagan impact structure, Quebec, as seen from a Landsat satellite. About 200 million years old and 70 km in diameter, this structure has a central peak of shock-metamorphosed rock surrounded by a thick layer of frozen impact melt that pooled on the original crater floor. The ring-shaped lake formed after continental glaciers carved out soft sedimentary rocks that slumped inward along the original walls of the crater.

frozen, strongly shocked ejecta, which fell back into the crater in a molten state and pooled on the floor (Fig. 4). As shown by Edward Anders and his colleagues, the contamination of the ejecta melt by material from the impacting body can be recognized by analysis for trace elements such as the noble metals (including platinum, iridium and gold) that are relatively abundant in meteorites but greatly depleted in the crust of the Earth.

In most impact craters significantly larger than Meteor Crater, the rock walls slumped inward soon after excavation of the initial cavity. This has occurred in all craters at least 3 km in diameter that formed in soft sedimentary rocks and in most craters larger than about 4 km across that formed in strong crystalline rocks. The converging flow of slumping material produced a pronounced central hill or peak in most of these. A few large craters exhibit a central ring-shaped ridge or more complex central structure. Evidence of uplifted rocks at the center and of subsidence and inward flow from the sides are important clues for the recognition of the deeply eroded cratering sites.

Approximately 70 impact structures on Earth, from 3 to 140 km in diameter, have been recognized by the two basic criteria of structure and shock-induced metamorphism. Most of these are younger than 500 million years, but some of the largest are more ancient. (In general, our geologic record is more complete for recent times than for older ones.) Some 50 additional structures have suggestive but inconclusive evidence for impact origin.

Most of the recognized impact structures occur on the continental shields — areas where the most ancient rocks of the Earth are preserved and which have been structurally stable for the longest time. The shields most thoroughly mapped and investigated by geologists are in North America and western Europe, and these are where most of the well-documented impact structures have been recognized. An analysis by R. A. F. Grieve and M. R. Dence suggests that, on these two shields, craters larger than 10 km in diameter have been produced at an average rate of about 1.4 craters per million square kilometers per 100 million years; where the geologic record is most complete, I find that the frequency may be up to 50 percent higher.

These estimates, which pertain to the last several hundred million years of Earth history, are close to the rate of 1.9 ± 0.8 (for the same area and time span) calculated from observations of Earth-crossing asteroids. The assumption here is that half of the Earth-crossing asteroids are dark, C-type objects with densities comparable to carbonaceous meteorites, and the other half bright, S-type asteroids with densities similar to the most common kinds of stony meteorites (see Chapter 10 for a discussion of asteroid types). On this basis, asteroids should create an average of about 3 craters 10 km in diameter and larger on the land areas of the Earth every million years. Interestingly, geologists have in fact found three 10-km impact craters probably no older than 1.2 million years. Two of these are in the U.S.S.R. and the third is occupied by Lake Bosumtwi, the sacred lake of Ghana's Ashanti tribe. If we include the number of craters expected from collisions with comets, the predicted cratering rate increases by about 50 percent. The fit of recognized impact structures to the predicted cratering rate then becomes poorer, but considering the statistical and observational uncertainties, it is still satisfactory (see Table 2).

IMPACT RECORD OF THE MOON

Unlike the Earth, where ongoing deformation and renewal of the crust tend to obscure the effects of impact, the surface of the Moon preserves a pristine record of bombardment by solid bodies that extends back several billion years. Because of the absence of an atmosphere, even microscopic particles strike the Moon at high speed and produce enormous numbers of tiny craters on the surfaces of exposed rocks. Larger particles pound and churn the Moon's surface, forming a layer of ground-up rocky debris, called the regolith, that blankets nearly all of the Moon. This layer averages about 3 m thick where it covers lava flows some 3 billion years old. Most of the Moon's surface has been darkened by accumulation in the regolith of black, impact-formed glass.

Asteroids and comet nuclei striking the Moon have produced conspicuous craters long familiar to telescopic observers. The youngest of these craters are surrounded by bright deposits of freshly excavated rock that extend away from the craters in discontinuous bright streaks called *rays*. One of the youngest, and perhaps the most spectacular, of the Moon's large ray craters is Tycho (Figs. 5 and 6), 85 km in diameter. A continuous deposit of ejecta surrounds Tycho out to an average distance from the rim crest about equal to the crater diameter. The rays form a great asymmetric splash whose fingers can be traced to distances of up to about one quarter of the lunar circumference.

Within the rays and the outer part of the continuous ejecta deposit are abundant small *secondary craters,* formed by chunks of rock and clots of debris flung out of Tycho. Just outside the crater rimcrest, which rises 1 km or so above the average level of the surrounding terrain, the ejected

	Mercury	Earth	Moon	Mars
Comets	1.3	1.0	0.9	0.3
Earth-crossing asteroids	1.0	1.9	1.0	0.4
Mars-crossing asteroids	0	0	0	≈ 2.0
Total	2.3	2.9	1.9	≈ 2.7
Cratering rate estimated from geologic record of impact structures		1.4 ± 0.4 (Grieve and Dence) 2.2 ± 1.1 (Shoemaker)		

Table 2. The present production of craters on the terrestrial planets by the impact of comets and asteroids. Each entry is the average production rate of craters larger than 10 km in diameter over a surface area of one million square kilometers in a period of 100 million years. Absolute cratering rates for asteroidal impact are uncertain by about 50 percent, while those for comets can be estimated only to the correct order of magnitude. However, for each class of impacting object, the relationships from column to column are roughly correct to the precision shown.

debris flowed down portions of the outer slopes in viscous, lobate masses. Some of the most fluid, shock-melted material accumulated in small depressions to form smooth, dark deposits resembling frozen ponds or lakes. The terraced crater walls are a consequence of the slumping of great slabs of rock during collapse of the initial crater. Shock-melted rock (termed *impact melt*) formed local pools on the terraces and extends in frozen rivulets several kilometers down to the main crater floor, itself filled with a deep layer of once-molten material. A prominent central peak, formed by the inward-flow of material as the crater walls slumped, rises about 2 km from this now-solid sea.

All of the events which led to Tycho's present appearance took place within a relatively short period. Some of these events — the formation of impact melt and the excavation of a crater — happened within a few minutes. Slumping of the crater walls and formation of the central peak may have taken an hour or so. The rain of ejecta — accompanied by the formation of secondary craters — and the emplacement of impact melt probably occurred within a few hours after the impact. The times involved for some of these events at other impact sites varies in proportion to the crater size.

Other large ray craters on the Moon closely resemble Tycho. However, craters smaller than 20 km in diameter usually have smooth walls devoid of terraces. Evidently, at this size, the bedrock walls have enough strength to prevent collapse in the weak lunar gravity (which at the surface is about one-sixth that at the surface of the Earth). Where terraces are missing, central peaks are usually absent as well. Craters 7–20 km across generally have a fairly smooth, level floor that probably formed by settling of the impact breccia from the walls of the initial cavity, as at Meteor Crater. Pools of frozen impact melt generally are not recognizable in or around these smaller craters. Below diameters of about 7 km, lunar ray craters usually have a simple bowl shape.

Certain features are characteristic of all ray craters. The exterior slopes of each rim are marked by a pattern of small hills or hummocks and by lower-lying ridges that are roughly radial in orientation. Abundant small secondary craters invariably occur near the outer limit of the ejecta blanket and in the rays. The hummocky topography high on the rim slopes commonly resembles the hummocky rim of Meteor Crater. As the lunar regolith slowly developed under prolonged pounding by small meteoroids, the rim deposits

Fig. 5. The ray pattern of the crater Tycho, 85 km across and situated in the Moon's southern hemisphere, is strikingly distinct in this Lick Observatory photo.

and floors of old craters gradually darkened and the rays disappeared. About half of the large craters found on the great lava plains (the lunar maria) exhibit the characteristic topographic features of ray craters but lack rays. Hummocky rim deposits and surrounding swarms of secondary craters confirm their impact origin.

Our inventory of lunar samples comes from only a small number of landing sites. But it has nevertheless been possible to use the isotopic dating of these samples as a basis for assigning ages to many extensive geologic units on the lunar surface (see Chapter 7). Geologic units at the landing sites can be correlated with similar units in other areas using orbital photography. Without the benefit of separate samples, the correlation of two distant areas based on photographic evidence can be somewhat uncertain. As a second method, correlations can also be made on the basis of similar densities and size distributions of craters (the two statistics can be combined using a size-frequency diagram). This relies on the assumption that any point on the Moon has an equal chance of being struck by an impacting body. The older a given surface, the more craters it will have of any specific size (but it is important to recognize craters that predate the surface under study). Reflecting the size distribution of impacting bodies, small craters are much more common than large ones.

The accuracy of statistics gathered for numerous small craters on a given geologic unit is higher than for the smaller number of large craters. Statistical and geologic uncertainties limit the confidence with which we can compare the crater densities of a large region with the isotopic ages of returned samples.

Using these methods, I have estimated that over the 3.3 billion years since many of the Moon's lava "seas" (the maria) were formed, the average production rate of craters larger than 10 km has been 0.6 ± 0.3 craters per million square kilometers per 100 million years. This rate can be

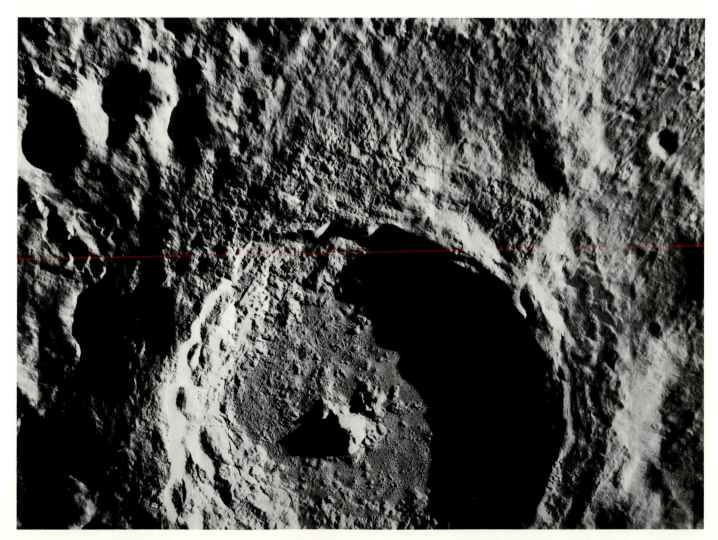

Fig. 6. The Lunar Orbiter 5 spacecraft photographed Tycho under oblique lighting that emphasized topographic detail. Clearly visible are the hummocky debris flows outside the rim, terraced walls, the prominent central peak, and the frozen pool of shock-melted rock on the crater floor.

compared with that for the Earth, but one must take into account that the same size distribution of impacting bodies will produce somewhat larger craters on the Earth than on the Moon, owing to the lower threshold diameter for wall slumping (which widens the crater) in the Earth's stronger gravity. Additionally, Earth's gravity attracts a higher number of bodies, at higher velocities, than the Moon's does; this yields higher crater densities for a surface on our planet compared with a lunar area of the same size and age.

When appropriate corrections for these effects are made, the equivalent rate for Earth is 0.9 ± 0.5. This long-term average just overlaps the estimate based on the study by Grieve and Dence for the last several hundred million years, but it falls short of the present cratering rate expected from the combined impacts of asteroids and comets. Gerhard Neukum and his colleagues, after examining the long-term production of craters on the Moon, conclude that Earth's present rate may be more than three times what it was billions of years ago.

Are these discrepancies real, or is a consensus lurking in all the uncertainties of our calculations? Quite possibly, the difference *is* real. Collisions on the Earth may have occurred two or three times more frequently during the last several 100 million years than during the previous 3 aeons. Perhaps the number of comets passing near the Earth has increased, or maybe several large bodies in the asteroid belt collided and sent fragments careening toward the inner solar system.

In tracing the geologic record of the Moon prior to the

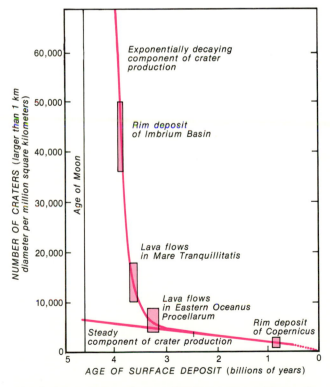

Fig. 7. The variation of crater density on lunar surfaces with different ages. Widths of the small rectangles, which correspond to Apollo landing sites, indicate the uncertainty or possible range in age of each dated surface, and the heights indicate the statistical uncertainty in crater density. The high cratering rate of the late heavy bombardment dropped rapidly between 3.9 and 3.3 billion years ago, giving way to a lower, steady rate of crater production.

eruption of those mare lavas, we find that collisions occurred very frequently at earlier times but dropped off rapidly between 4.0 and 3.3 billion years ago (Fig. 7). For example, the crater density found on a great sheet of ejecta surrounding the Imbrium Basin, formed 3.9 billion years ago, is six times greater than on lavas 600 million years younger. This abrupt decrease in impacts can be modeled mathematically by combining one cratering rate that decayed exponentially with another that remained steady with time. A good fit for the exponential component would be a 50-percent decrease in the cratering rate every 100 million years. According to this model, the formation of craters dropped off by a factor of 35 between 3.9 and 3.3 billion years ago.

These calculations can be translated physically into two distinct populations of Earth-crossing bodies. One group became depleted with time, possibly by collisions with the Earth, Moon, and other planets, or by ejection from the solar system. The second group corresponds to what we observe today as Earth-crossing objects — a population which is apparently renewed from other regions of the solar system at approximately the same rate that it is lost. Without this renewal, the "steady" population would become quickly depleted. The residence time for its members (about 30 million years) is actually short compared to the 100-million-year half-life estimated for the long-exhausted first population.

At least three plausible suggestions have been advanced for the origin of the first, exponentially decaying population: (1) the impacting bodies were long-lived Earth planetesimals — a remnant of the batch of small objects from which most of the Earth accreted; (2) the impacting bodies were injected into Earth-crossing orbits from the asteroid belt, perhaps following major collisions there; (3) the bodies were fragments of a large planetesimal of Uranus or Neptune that was perturbed into an Earth-crossing orbit and became tidally disrupted by close approach to one of the terrestrial planets. Bodies derived from each of these suggested sources could have had a set of orbits necessary to account for the 100-million-year half-life of the population. A fourth hypothesis, which I consider the most probable, will be introduced later.

The documented episode of rapid cratering between 3.3 and 3.9 billion years ago has been referred to as the *late heavy bombardment* of the Moon. The impact events which produced the giant Imbrium and Orientale basins occurred during this period, and some or most of the other large lunar basins probably formed about this time as well. Moreover, calculations suggest that the basins and craters of the densely cratered lunar highlands may all be younger than 4.2 billion years old. Fouad Tera, Dimitri Papanastassiou, and Gerald Wasserburg contend that the late heavy bombardment took the form of a distinct pulse of cratering that peaked about 4 billion years ago. An alternate view holds that the heaviest pummeling occurred as the Moon formed 4.6 billion years ago and that the bombardment and cratering has simply declined in a generally steady way ever since. The late heavy bombardment, in other words, may simply represent the end stage of planetary accretion — the evidence for which has been preserved on the lunar surface. Manifestations of the earlier, more intense bombardment, in this view, have been lost or obscured by the large basins and regional deposits of ejecta formed in the last stage.

MERCURY AND MARS

The production of craters on Mercury from a combination of comets and Earth-crossing asteroids appears to be comparable to the rate here on Earth (Table 2), although the proportions are different. Only one-fifth of the asteroids traversing Earth's orbit pass inside Mercury's as well. But the confluence of their orbits occurs in the relatively small volume of space between Mercury and the Sun, and this increases the probability of impact. Moreover, perhaps 5–10 percent of the asteroids that hit Mercury occupy orbits that lie entirely inside the Earth's. Finally, an object nearing Mercury travels faster because of the Sun's gravitational acceleration; when it strikes Mercury, the impact energy is greater and a proportionately larger crater is formed (Mercury's relatively weak surface gravity is also a factor). Taken together, these effects offset the markedly reduced numbers of asteroids that pass in the vicinity of Mercury and yield a cratering rate per unit area of surface which is about half of that on the Earth.

Roughly one-third of the comets passing inside the Earth's orbit also penetrate inside Mercury's average aphelion distance (the point where it is farthest from the Sun in its orbit). But the same orbital congestion near the Sun that increases the chance of asteroidal impact also holds for comets, resulting in a cratering rate from cometary objects about 30 percent higher for Mercury than for the Earth.

Since the same basic classes of objects strike the Moon, Earth, and Mercury, some idea of Mercury's cratering history can be deduced from the impact records of the other two. For example, one would expect that if the flux of crater-forming objects near the Earth changed, at least during more recent eras, a corresponding change should have occurred for Mercury. Calculations indicate that the present cratering rate on this innermost planet is about 20 percent higher than on the Moon, a proportion that may have remained nearly the same for several billion years. Those who make such estimates were not surprised, therefore, to learn from Mariner 10 in 1974 that Mercury has a heavily cratered surface very similar in appearance to the Moon's.

The surface of Mercury includes smooth plains, somewhat similar to the lava plains of the Moon, and heavily cratered highlands closely resembling the lunar highlands. Fresh-looking craters found on Mercury are remarkably similar to their counterparts on the lunar maria: on both worlds, the youngest of these craters have rays, but older craters do not. The only obvious difference exists in the patterns of secondary craters, which cluster more tightly around their primary craters on Mercury than on the Moon. This difference arises from Mercury's higher surface gravity, which confines crater ejecta to shorter ballistic ranges.

On the smooth plains of Mercury, craters 10 km across are about three times more abundant than the average number on the 3.3-billion-year-old lunar maria. This suggests that the smooth plains formed about 3.7 billion years ago, near the end of the period of late heavy bombardment on the Moon. These plains are associated with an enormous impact structure, the Caloris Basin, which evidently formed about that time as well; it is probably the last giant basin created on Mercury.

The density and size distribution of highland craters on Mercury closely match these characteristics of the lunar

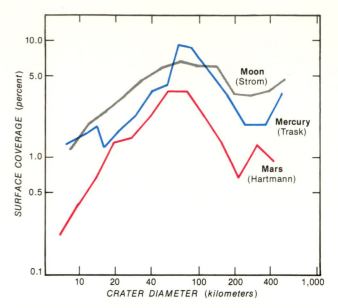

Fig. 8. Crater size distributions on the highlands of the Moon, Mercury, and Mars. This diagram is from the work of Robert Strom; authors whose data have been used are shown by each curve. The ordinate shows the percentage area of the surface covered by craters of a given diameter. The distribution on all three bodies has a peak in the percentage area covered between crater diameters of 40 and 100 km. The similarity in the form of the distributions suggests that the same population of impacting projectiles may have produced most of the craters on the highlands of all three planets.

highlands (Fig. 8). Geologist Robert Strom has interpreted the similarity as an indication that the same population of bodies was responsible for most of the craters in both regions. Moreover, if the orbital distribution of those bodies resembled what we now observe among the Earth-crossing asteroids, collision rates on the Moon and Mercury would have been comparable at any time during the late heavy bombardment. Mercury's surface, like the Moon's, may bear a cratering record as old as 4.1–4.2 billion years.

Impact craters now being created on Mars result mostly from a group of asteroids that do not cross the Earth's orbit. This means that the cratering history of Mars is not as closely coupled as Mercury's is to the history of the Earth-Moon system. Even so, the present formation rate of 10-km craters on the red planet appears to match the Earth's value fairly well. Comets contribute about 10 percent of the impacting bodies and Earth-crossing asteroids another 15 percent. Craters 10 km across now form about as often as they do on the Earth.

Among the abundant impact craters found on Mars, those that have not been significantly modified (by erosion, deposition of sediments, or volcanism) resemble, in most respects, the impact craters of the Moon and Mercury. One rather remarkable difference is that the ejecta around most large, fresh Martian craters flowed radially outward, rather than being deposited by ballistic ejection (Fig. 9). Michael Carr has suggested that the ejected debris' fluidlike behavior may be due to the melting and vaporization of ice mixed in with the target material, or to the trapping of atmospheric gases beneath the ejecta blanket.

The geology of Mars is much more diverse than the geology of the Moon or Mercury, and the surface of the

Fig. 9. Part of the rim deposit of the Martian crater Arandas (28 km in diameter) as seen from a Viking orbiter. The rim deposit has apparently flowed as a fluid or fluidized system.

planet varies greatly in age from place to place. A significant fraction of the Martian plains are only sparsely cratered and perhaps formed by the extrusion of lava or by the deposition of wind-blown sediment (see Chapter 8). Other Martian plains are cratered to about the same extent as the lunar maria (Fig. 10). This diversity implies that the plains have a correspondingly broad range in age, extending nearly from the present back several billion years. If we assume that the flux of Mars-crossing asteroids has been steady with time, the calculated age for some of the older plains would be about 3 billion years. Yet this age is about the same as the typical dynamic lifetime for Mars-crossing asteroids. The collision of these bodies with Mars and their loss due to ejection from the solar system has reduced the population of Mars-crossers during the past 3 billion years. So the cratering rate probably was higher in the past, and these more ancient plains may actually be no more than about 2 billion years old.

Heavily cratered highland regions of Mars (particularly in the southern hemisphere) are similar in both crater density and crater size distribution to the highlands of the Moon and Mercury. And just as in the case of Mercury, the similarities among size distributions suggest that the same population of impacting bodies was responsible for both the late heavy bombardment of the Moon and most of the highland craters of Mars. If these bodies had orbits like those of modern-day Earth-crossing asteroids, their cratering rate on Mars would have been only 40 percent of their rate on the Moon at any given time during the late heavy bombardment. On the other hand, if enough of them originated in the outer solar system, Mars and the Moon could have been subjected to equally intense battering at the same time. Regardless, the cratered highlands of Mars probably record nearly the same episode of late heavy bombardment preserved in the highlands of the Moon and Mercury, although the most ancient Martian craters and basins may have formed slightly before the oldest features on the Moon.

THE SATELLITES OF JUPITER AND SATURN

The Voyager program extended our direct knowledge of collisions in the solar system from the orbit of Mars out to the orbit of Saturn, about 9.5 AU from the Sun. The record preserved for our examination in the vicinity of the giant planets is written in ice. It is the history of icy comet nuclei and icy planetesimals slamming at high speed into the icy crusts of the satellites, dominated in recent times almost entirely by comets traversing this region of the solar system. Close to Jupiter, the cast of impacting bodies is dominated by the so-called Jupiter family of short-period comets.

The Jupiter family of comets has been captured by Jupiter (with important assistance from the other major planets) from a swarm of long-period comets through a succession of close encounters. If a comet revolves around the Sun in the same direction the planets do, its orbital period shortens when an encounter occurs ahead of Jupiter; encounters behind Jupiter lengthen the period. An orbital period of less than 20 years (typical of the Jupiter family) comes about only after a statistically unlikely series of passes ahead of Jupiter or an even less likely single pass at extremely close range. Understandably, this happens to only a very few long-period comets, and dynamicist Edgar Everhart has shown that the capture takes, on average, about 200,000 years. The population of Jupiter-family comets, at any one time, depends on how many long-period source objects are available, their rate of capture, and rate of loss.

Usually, once trapped by Jupiter, a comet is doomed. In no more than a few thousand revolutions its ices are lost during repeated close approaches to the Sun. The comet may simply disappear, or there may be a core or residue of nonvolatile rocky material — an extinct body that resembles an asteroid. Hidalgo may once have been a Jupiter-family comet. Because most of these objects lose all their ices long before they collide with planets or become ejected from the solar system, there may be many more extinct comets besides Hidalgo — totaling perhaps twice the num-

	Satellites of Jupiter				Satellites of Saturn			
	Io	Europa	Ganymede	Callisto	Mimas	Tethys	Dione	Rhea
Long-period comets	0.6	0.7	0.5	0.38	0.5	0.3	0.2	0.12
Jupiter-family comets	1.6	1.3	0.6	0.30	} 0.04	0.01	0.007	0.004
Extinct comets	3.0	2.5	1.2	0.55				
Saturn-family comets	0	0	0	0	≈1.5	≈0.6	≈0.3	≈0.2
Total	5.2	4.5	2.3	1.2	≈2	≈1	≈0.5	≈0.3

Table 3. The present production of craters on satellites of Jupiter and Saturn by cometary impact, with values computed according to the same conditions as in Table 2. Asteroids perturbed outward from the inner solar system reach Jupiter only infrequently; their contribution to the cratering of outer-planet satellites is no more than about 1 percent and so is not included here.

ber of active comets thought to be in the Jupiter family.

Table 3 lists the estimated cratering rates due to the impact of active and extinct comet nuclei on the Galilean satellites of Jupiter. The greatest uncertainty here lies in the size of the nuclei. We can measure the *brightness* of a comet's nucleus far from the Sun (when it is least obscured by its halo of gas and dust), but to compute size we also need a good value for surface reflectivity — something beyond our grasp just now. Recent observations by Johan Degewij suggest that most nuclei are coated with very dark particles, so the cratering rates listed in Table 3 assume that the average comet nucleus is about as dark as a lump of coal. Despite the uncertainties, the estimates of relative rates of cratering among the different satellites should be fairly accurate.

The flux of comets near Jupiter, particularly the short-period ones, is strongly focused by the planet's gravitational field. When comets are pulled in by Jupiter's gravity, their velocities increase. As a consequence, the present cratering rate on Io, the innermost Galilean satellite, is about four times that on Callisto, the outermost one. For craters larger than 10 km across, Callisto's present rate is a little less than the Moon's.

Voyagers 1 and 2 revealed that Ganymede and Callisto, the two largest Galilean satellites, are densely cratered bodies, somewhat like the Moon and Mercury. But no impact craters have been found at all on Io and very few on Europa; volcanism and other processes have erased most of their impact record (see Chapter 14). Ray craters formed on the icy crusts of Ganymede and Callisto look enough like ray craters on the Moon to show that they are of impact origin. As on the Moon, these satellites have developed regoliths that have gradually darkened with time; around older craters the rays have disappeared. An even more interesting effect is the slow disappearance of the craters themselves by plastic flow of the ice. Many large craters have vanished, leaving only faint discolored patches on the surface, called *palimpsests,* and telltale swarms of secondary craters. The shapes of most craters remaining on Ganymede and Callisto have been severely degraded by flow of the icy crusts. Extrapolating the present cratering rates on Ganymede and Callisto back over geologic time, we find that the ray craters represent a record varying from 500 million to nearly 4 billion years old. Rays persist longer on Callisto than on Ganymede, although the retention time varies over the surface of each satellite.

Callisto has far more craters than can be accounted for by a steady rate of cometary impacts over geologic time; the same is true of dark, heavily cratered areas on Ganymede, implying that a period of heavy bombardment took place at some time in the history of these bodies. This heavy bombardment probably did not occur within the last 3.3 billion years; we would have noticed its effects in the cratering record of the Moon. On the other hand, the excess collisions could easily have taken place earlier, coincident with the late heavy bombardment of the Moon. Unfortunately, plastic flow in the icy crusts has prevented us from learning whether the original size distribution of ancient craters on Ganymede and Callisto resembles that of the lunar highlands.

Callisto also has several enormous, targetlike patterns of concentric ridges called multi-ring systems. At their centers are palimpsests that probably mark the sites of former impact basins. Remnants of another multi-ring system are

found on Ganymede. If the past variations of impact rates on Ganymede and Callisto mimic those observed in the Earth-Moon system, then the great multi-ring systems are probably 3.8–4.0 billion years old — comparable to the ages of some large lunar impact basins formed during the late heavy bombardment.

At Saturn, collision-bound objects are now almost entirely comets, just as at Jupiter. Long-period comets account for about one-third to one-half of recent craters on the satellites of Saturn, and Jupiter-family comets only about 1 percent (Table 3). The remaining craters are probably produced by a family of periodic comets captured by Saturn from the long-period population. Most of the Saturn-family comets remain too far from the Sun to generate observable tails or comas, so theorists have inferred their existence almost entirely from theoretical arguments. Chiron is the only apparent member discovered so far, but the total number with nuclei larger than 1 km across may be as high as 10,000.

Television images recorded by Voyager 1 during its encounter with Saturn disclosed another set of icy satellites with heavily cratered surfaces. Calculation of the rate of impact by the long-period and Saturn-family comets shows, once again, that only a small fraction of the observed craters can be accounted for by a steady collision rate maintained over geologic time — one or more episodes of heavy bombardment are indicated. In contrast with the icy satellites of Jupiter, there is little evidence that plastic flow of ice has erased the craters on these very cold bodies. Parts of the cratered surface of the satellite Rhea bear a striking resemblance to the highlands of the Moon (Fig. 18), a match borne out by comparing the crater density and size distribution found there to those of the lunar highlands. Following suit, Rhea may also preserve a record of the late heavy bombardment. Yet large impact basins are absent. It may take some time to study all of the Voyager images of Rhea completely, but the basins have probably been lost by plastic flow or creep in the ice.

Parts of the satellites Mimas and Dione have crater populations quite distinct from those found elsewhere in the solar system. Abundant craters 5–20 km across pepper these

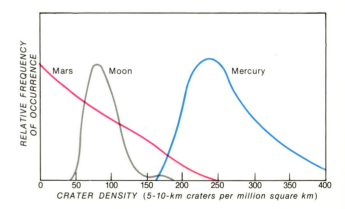

Fig. 10. **The densities of craters between 5 and 10 km in diameter on the plains of the Moon, Mars, and Mercury are compared. The diagram, part of a study by Laurence Soderblom, shows how often different crater densities occur. The plains on Mercury generally have the highest crater densities, while the Martian plains have a broad range of crater densities that indicates diverse surface ages.**

surfaces, but they are practically unaccompanied by craters larger than about 40 km. The implication is that most collisions occurred at low velocities, perhaps involving fragments from the satellites themselves or from other small satellites sharing the same orbits. Very large numbers of such fragments must have been thrown into independent orbits about Saturn after high-energy impacts on Mimas. The result of one such collision is marked by a crater about one-third the satellite's diameter. The scene probably repeated many times during heavy bombardment.

THE OORT CLOUD OF COMETS

In 1950 the Dutch astronomer Jan Oort showed, by careful analysis of the most accurate orbits, that the source of long-period comets is a gigantic, diffuse cloud of comets surrounding the Sun at an average distance of about 40,000 AU. Since the cloud's outer limit extends about halfway to the nearest known star, the comets are only weakly bound by gravity to the Sun. Passing stars perturb some of the comets in the cloud, allowing them to fall toward the Sun; a few eventually pass through the inner solar system, where they can be observed. For stellar perturbations to dispatch the three or so long-period comets discovered annually near the Earth, Oort estimated that there must be about 100 billion comets in the cloud. He also suggested that the comets originated as planetesimals of Jupiter, thrown into the cloud long ago by a succession of perturbations with the young, massive planet. But more detailed study indicates that Jupiter's perturbations were actually *too* strong, slinging most nearby planetesimals completely out of the solar system.

Although some Jupiter planetesimals probably reside in the Oort cloud, likely ultimate sources of the majority of comets are the planetesimals which originated near Uranus and Neptune. If true, their emplacement in the Oort cloud occurred over a protracted period of early solar system history. First, the bodies' orbital eccentricities were amplified by close encounters with Uranus and Neptune. After a while, many Uranus planetesimals were perturbed into Saturn-crossing orbits, and the further evolution of these orbits became controlled by encounters with Saturn.

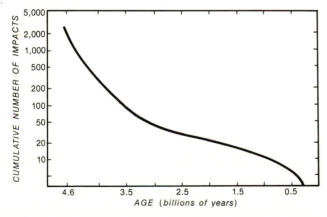

Fig. 11. Time dependence of the impact rate of Uranus and Neptune planetesimals on bodies in the inner solar system, as calculated by G. W. Wetherill. In making the calculation, he assumed that initially an equal number of Uranus and Neptune planetesimals existed. The cumulative number, indicating how many impacts have occurred since a given time, is a relative quantity, as the population of planetesimals was not determined.

Similarly, Neptune planetesimals were occasionally perturbed into Uranus-crossing orbits and some of these, in turn, passed to the control of Saturn. Many also fell under the influence of Jupiter.

The combined action of the major planets forced part of the planetesimal swarm into very long-period orbits characteristic of the Oort cloud. Many planetesimals, of course, collided with the planets (and their satellites) or were ejected from the solar system, and the swarm became gradually depleted with time. Wetherill estimates that the half-life of the Uranus planetesimals was about 100 million years and those of Neptune about 200 million years. Roughly half of the comets may have found their way into the Oort cloud between about 4.4 and 4.6 billion years ago and all but a few percent of the rest in the following billion years.

Using Oort's estimate of the number of comets in the cloud and Wetherill's estimate of the decay rate of the Uranus and Neptune planetesimal swarm, I calculate that the number of planetesimals crossing the orbits of Jupiter and Saturn about 4.0 billion years ago was probably high enough to account for the heavy bombardment recorded on the satellites of these planets. Moreover, the decay rate compares well with the decrease in lunar cratering during the late heavy bombardment (Fig. 11). Jupiter must have captured a fraction of the planetesimals into orbits like those of the current short-period comets. Earth-crossing bodies may well have evolved from this group. If so, the heavy bombardment of the terrestrial planets, the Moon, and the satellites of Jupiter and Saturn all happened at the same time — a consequence of the gravitational shuffling of the accretionary leftovers out to the Oort comet cloud and beyond.

The mass of volatile-rich, cometary material that collided with the Earth during this period may have been about 10 times the present mass of our oceans. Even if all the free water in the impacting bodies had been driven off by sublimation, about one-tenth of their mass could still be water-bound in the crystal lattices of hydrated silicates (certain carbonaceous meteorites provide good examples of this). Therefore, the Uranus and Neptune planetesimal swarm could have supplied the Earth with roughly its entire inventory of water. Other volatile substances may have also reached our planet in this way, and a similar batch delivered to Venus. Of course, the volatile-rich material must also have struck the Moon, Mercury, and Mars, so why has no water been found in the rocks brought back from the Moon? If the high impact velocities of current Earth-crossing asteroids are any indication, volatile substances borne by the Uranus and Neptune planetesimals simply didn't stick to the Moon. They were converted to high-temperature gases during impact and sprayed off at velocities fast enough to escape the lunar gravity field.

Quite possibly, the traces of complex organic molecules found in certain carbonaceous meteorites (see Chapter 18) may also have been incorporated into the Uranus and Neptune planetesimals (and are probably present now in comet nuclei as well). In the larger bodies that struck the Earth at high speed, high temperatures produced by shock of the impact destroyed most of the organic molecules. But a significant amount of organic chemicals probably survived within small fragments that were more gently decelerated in the Earth's atmosphere. These organic molecules, born in interstellar space, may have been the principal flavoring of

Fig. 12. One day, 65 million years ago, a large asteroid perhaps 10 km across may have collided with the Earth. Luis and Walter Alvarez and their colleagues believe the effects of such an impact, described in the text, could have caused the extinction of numerous life forms (including the dinosaurs) at the end of the Cretaceous period. Don Davis has rendered that catastrophic moment, with the flora and fauna shown representative of that time.

the oceanic broth from which life arose. Emplacement of the Oort cloud, in other words, may have supplied not only the aqueous abode for life on the Earth, but also a fairly rich mixture of complex molecules to start the process of biological evolution.

RECENT COLLISIONS OF LARGE BODIES WITH THE EARTH

Over the past several hundred million years, an asteroid or comet nucleus about 15–20 km in diameter probably has collided with the Earth, on the average, about every hundred million years. Each collision produced a crater perhaps more than 200 km in diameter and injected an enormous amount of material into the atmosphere (Fig. 12). The odds are that about two-thirds of these giant craters formed in the ocean and rocks of the ocean floor. In an oceanic impact, giant tsunamis must have been created, and huge quantities of water vapor driven into the upper atmosphere. Öpik first pointed out some of the possible effects of large impact events on the biosphere about 20 years ago. It remained, however, for Luis and Walter Alvarez and their colleagues to discover evidence of a major impact apparently linked to a global biological catastrophe that occurred 65 million years ago.

They found that, at several localities around the world, the boundary between the rocks of the Cretaceous and Tertiary periods in geologic history is marked by a thin layer of claystone. This boundary marker contains anomalously abundant noble metals (especially iridium) and other elements that are normally very rare in the Earth's surface rocks but relatively common in meteorites. They estimated that the amount of these elements found in the boundary

claystone could have been supplied by an asteroid with the composition of ordinary stony meteorites and a diameter of about 10 km. The record of fossils found in the rocks below and above the boundary shows that many species and genera of living organisms disappeared abruptly at the time the claystone was deposited. The scientists suggest that the claystone was deposited from dust thrown up from a great impact crater and distributed worldwide by atmospheric circulation, concluding that the amount of suspended dust was sufficient to blot out the Sun and arrest photosynthesis for a period of months or even years. Accordingly, they reason that the food chain was nipped in the bud and many species simply starved to death.

Another possible explanation for the extinctions at the end of the Cretaceous has been advanced by Cesare Emiliani, Eric Kraus, and myself. If the impact occurred in the ocean, the amount of water thrown into the stratosphere probably led to a strong greenhouse effect. Global temperatures in the lower atmosphere and the surface layer of the ocean could have been raised more than 10° K. Emiliani found that such a pulse of heat could explain why certain species were exterminated, and why others living in protected environments (deep in the oceans or at high latitudes) survived. Similar impacts may have produced global traumas and extinctions at earlier times in geologic history. The ecological niches vacated at these times may have been filled by the very rapid evolution of new species. Thus, bombardment of the Earth, besides possibly providing the starting conditions for life, may also have delivered an occasional (but violent) biological winnowing that spurred subsequent evolution.

Surfaces of the Terrestrial Planets

James W. Head, III

In the last two decades two parallel revolutions in the Earth sciences have resulted in a radically different perception of planets and how they work. The first was the development of the theory of terrestrial *plate tectonics*. This involved the realization that the geologic record of the Earth was not a group of isolated, independent puzzles to be solved individually, but rather a record of the movement and interaction of a small number of large lithospheric (or crustal) plates operating in an integrated, global manner. The second revolution resulted from the unfolding view of planetary surface features derived from spacecraft exploration. In this evolving perspective, the Moon and planets changed from astronomical objects to geologic ones, and in doing so began to provide a framework within which the Earth would be viewed not as a single data point, but as one of a family of planets.

The decade of the 1960's was one of revelation that saw the rapid development and testing of ideas for both revolutions. We began to comprehend the geology and geophysics of the Earth's ocean basins, and the extreme contrasts in the nature of the continents and ocean floors required unorthodox interpretations. Old and new ideas on continental drift and sea-floor spreading finally came together late in the decade in a new theory of global plate tectonics. The extreme complexities of the Earth and of its evolutionary processes had, until that time, precluded geologic studies from being much more than description and data accumulation. The predictive nature of the theory of plate tectonics gave renewed vigor to the geologic sciences. In the 1970's our attention turned to testing and refining plate-tectonic theory and processes and to understanding the last several hundred million years of Earth history in this context.

Planetary exploration in the 1960's saw Mariner spacecraft reveal the surface of Mars and Apollo astronauts begin intensive exploration of the Moon (see Chapter 1). A flood of discoveries, including rocks dating to the planets' formative years and volcanoes on Mars more immense than could be imagined, unleashed us from our Earth-bound view of geology and ushered in an era of comparative planetology. In the decade of the 1970's, data collection continued (with missions such as Mariner Venus-Mercury, Pioneer Venus, and Viking to Mars), concomitant with the analysis and synthesis that provided the broad framework of basic knowledge necessary to understand the important factors in planetary origin and evolution.

Prior to the 1960's, geologic thought focused on specific areas of the Earth's surface and, in a rather passive sense, was analogous to the pre-Copernican, Earth-centered view of the solar system. At present we are indeed abandoning our Earth chauvinism and working and thinking in terms of a new solar system. We now view our own planet as a globe. We integrate our knowledge of geologic and geophysical processes, causes, and effects, then compare the nature and evolution of Earth, the Moon, and the other terrestrial planets. Planetary scientists now think in terms of solar system history, with the Earth providing a record of dynamic processes and the Moon, Mars, and Mercury a record of the first 2½ billion years of planetary evolution.

In this chapter the emphasis is on the nature of the terrain types observed on the terrestrial planets and on what this information tells us about the geologic processes forming and modifying planetary surfaces, their distribution with time, their relation to planetary interiors, and the significant factors in the evolution of planets.

THE MOON

As our closest neighbor in space, the Moon was first to occupy the attention of geologists studying other planetary surfaces. Eugene Shoemaker and his co-workers applied the basic principles of terrestrial stratigraphy to the surface features and geologic units on the Moon. This enabled them to produce geologic maps and thus delineate the major surface processes and sequence of events in lunar history. Similar mapping techniques have been applied to each successive planet explored over the last 15 years, and the collective maps (Figs. 1-4) provide the basis for understanding the history of each planet and comparisons of planetary evolution. For the Moon, the early formation of global crust was accompanied and succeeded for several hundred million years by a massive influx of projectiles impacting the newly formed surface at several kilometers a second, producing impact craters of many sizes (see Chapters 4 and 7).

This *late heavy bombardment* had three major effects: (1) the fragmentation, fracturing, and *brecciation* (or shock-induced cohesion) of the upper few kilometers of the Moon's crust to form a massive soil layer called the megaregolith; (2) the production of global geologic units representing the first few hundred million years of lunar history; and (3) the creation of extremely rough surface topography. The late heavy bombardment ended about 3.8 billion years ago, but not before the largest projectiles had excavated huge depressions (perhaps as large as 2,000 km in diameter) and spread ejecta over immense areas that sometimes affected an entire lunar hemisphere.

Volcanic flooding of the surface of the Moon began during the waning stages of heavy bombardment. By about 2.5-3.0 billion years ago, basaltic lavas had covered approximately 17 percent of the lunar surface, preferentially filling in the low-lying basin interiors to form the lunar seas or *maria*. Tectonic activity on the Moon stands in stark contrast to that of our own planet: instead of the Earth's

multiple, colliding lithospheric plates, the Moon has but a single lithospheric plate — it is a "one-plate" planet." Its limited array of tectonic features occurs only in and near the maria: linear rilles and graben formed by crustal expansion and sinuous ridges formed by compression. The maria bear these geologic undulations because outpourings of basaltic lava have loaded the lithosphere to the point of flexure. Virtually no major geologic episodes have occured on the lunar surface in the last 2.5 billion years.

The Apollo expeditions returned samples from which a chronology of lunar events could be constructed. The Moon thus provides a picture of the first half of solar system history (characterized by impact bombardment and early volcanism), which serves as a cornerstone for the interpretation of the records preserved on other terrestrial planets.

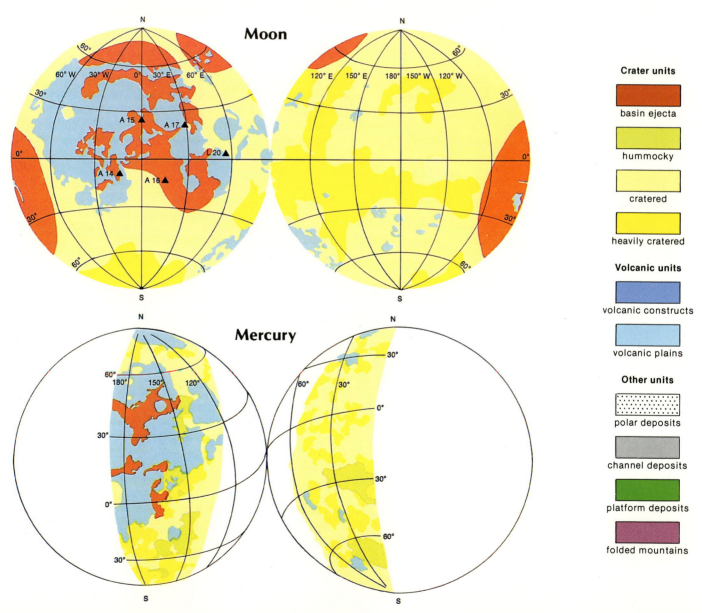

Fig. 1. Reproduced here are maps showing the geologic terrain units on the Moon. Craters and near-side "seas" of relatively thin, solidified lava dominate the lunar landscape; absent are more complex landforms found on the Earth and Mars. Landing sites of Apollo (*A*) and unmanned Soviet Luna (*L*) spacecraft in or near highland areas are shown.

Fig. 2. Mariner 10 photographed only about 35 percent of Mercury's surface but showed enough to demonstrate that the planet looks remarkably like the Moon. Cratered terrain is common, and extensive areas of smooth plains surround the Caloris Basin (Fig. 16), centered at 30° N, 190° W.

MERCURY

The Mariner 10 spacecraft returned images of about 35 percent of the planet's surface and revealed a lunarlike terrain, somewhat surprising since the planet is intermediate in size to the Moon and Mars and has a density comparable to Earth's. Detailed geologic mapping of the surface, however, has shown that Mercury differs from the Moon in several respects. Large areas of relatively ancient *intercrater plains* (Fig. 5) may indicate that more extensive volcanism accompanied the period of heavy cratering on Mercury than on the Moon. Large, extensive scarps on Mercury (Fig. 6) attest to episodes of regional compression and perhaps even the global compression that would result from a modest decrease in the planet's circumference during solidification.

Areas of smooth plains have nearly the same reflectivity as that of heavily cratered regions, which has led to controversy over the origin (volcanic or otherwise?) of the smooth regions; Mariner 10's low-resolution images make it difficult to resolve the uncertainty. With these exceptions, Mercury generally resembles the Moon on its surface, but its high density (5.4 g per cubic centimeter) suggests that it bears more of a resemblance to the Earth in the interior.

MARS

Mariner and Viking spacecraft images revealed that Mars is more geologically diverse and complex than the Moon and Mercury and that it shows a distinct hemispheric asymmetry in the distribution of geologic units (Fig. 3). The often densely cratered southern hemisphere stands 1-3 km above

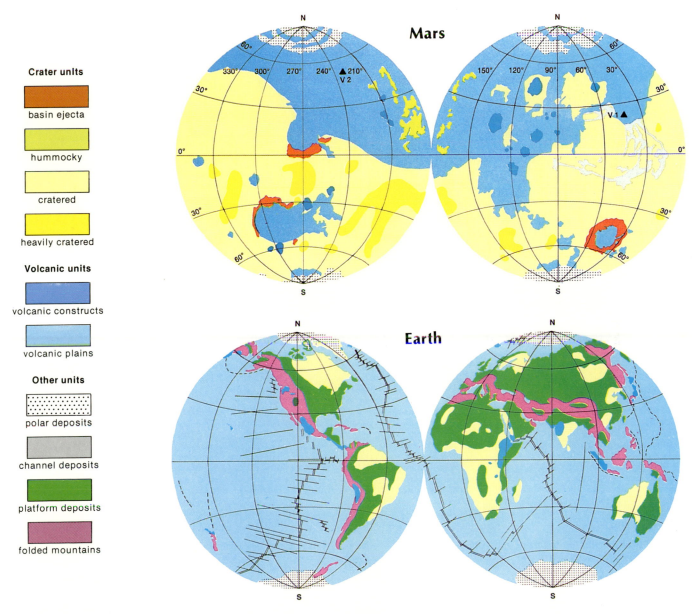

Crater units

basin ejecta

hummocky

cratered

heavily cratered

Volcanic units

volcanic constructs

volcanic plains

Other units

polar deposits

channel deposits

platform deposits

folded mountains

Fig. 3. Diverse volcanic, erosional, and tectonic features distinguish Mars from the less-complex surfaces of Mercury and the Moon. The dichotomy of northern lowlands and southern cratered uplands results in a hemispherical asymmetry which is a significant unanswered puzzle. *V* indicates the Viking landing sites.

Fig. 4. Earth's surface appears to be the most dynamic among the terrestrial planets. Only small portions of our planet are covered with aeons-old rocks; the remainder is relatively young, occurring mostly in the ocean basins. Like others in this series, the maps bring together data from numerous spacecraft investigations and a variety of other sources.

the topographic "sea level" while the northern hemisphere lies generally below that level and is sparsely cratered. The southern hemisphere has two main components; a very ancient crust nearly saturated with large craters and often fractured and cut by abundant small channels; and younger intercrater plains which appear to be ancient as well but less modified. A variety of cratered plains covers the northern hemisphere. Volcanic flows characterize the plains surrounding the large volcanoes discovered by Mariner 9 but elsewhere the plains are featureless except for craters and marelike ridges. At intermediate to high northern latitudes the plains contain various kinds of patterned and striped ground, scarps, and irregularly shaped mesas. Their complexity may represent the influence of volatiles and changing temperature conditions. The boundary between northern and southern hemispheres is extremely complex as well. The older, higher terrain of the southern hemisphere contains numerous channels, tens of kilometers wide and hundreds of kilometers long, that are reminiscent of those formed on Earth by catastrophic flooding.

A major departure from Mars' general surface geology is the Tharsis ridge (Fig. 7), a broad topographic bulge exhibiting ancient heavily cratered units and young shield

Fig. 6. **More than 500 km long and as much as 3 km high, Mercury's Discovery Scarp probably formed when the crust contracted as the planet cooled and solidified early in its history. The scarp is younger than the craters it transects.**

Fig. 5. **In this Mariner 10 image of Mercury's north polar region, the lower half is dominated by numerous impact craters superimposed on an** *intercrater plain,* **while the upper half shows a** *smooth plain* **with relatively few craters.**

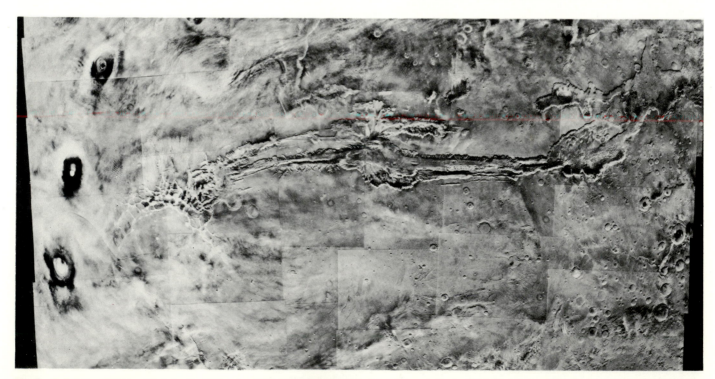

Fig. 7. **A mosaic of images taken by the Viking 1 orbiter in 1980 shows the planet Mars as it appeared on an exceptionally clear day. At far left are the three immense shield volcanoes that sit atop the Tharsis ridge, just southeast of Olympus Mons, while the enormous equatorial canyon system called Valles Marineris runs about 4,000 km across the center of the picture.**

volcanoes. The 8,000-km-wide region, centered at 14° S latitude and 101° W longitude, stands about 10 km above surrounding terrain; some of its volcanic shields extend another 20 km in altitude. The vast majority of the linear rilles and fractures seen on Mars surround the Tharsis region. Valles Marineris, an enormous equatorial canyon system, extends radially away from Tharsis and is probably related to faulting that accompanied the evolution of Tharsis since its formation. Valles Marineris has been extensively modified by the collapse of its walls and also by channel formation.

The origin of Tharsis is uncertain; some believe that it is predominantly a massive uplift of the crust caused by some dynamic process deep within the planet's mantle, while others propose that the lithosphere under Tharsis is thin, rendering it more susceptible than elsewhere to volcanic outpouring and related stresses.

Additional features unique to Mars include craters with odd ejecta patterns that suggest a layer of ice in parts of the upper crust. Much of Mars is covered with wind-blown deposits, which may mix with volatiles at the poles. The surface units thus reveal an early chronology for Mars much

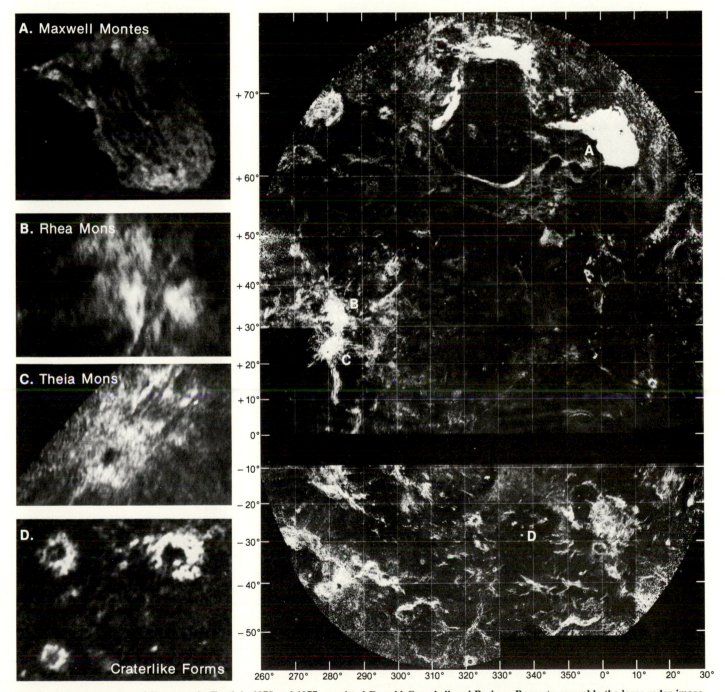

Fig. 8. Close approaches of Venus to the Earth in 1975 and 1977 permitted Donald Campbell and Barbara Burns to assemble the large radar image at right, which covers about one-fourth of the planet at 20-km resolution. Compare this and the enlarged insets at left with the global data from Pioneer Venus (Figs. 10 and 11). Generally, light areas are considered rough and dark areas smooth. Adapted from an article in the *Journal of Geophysical Research,* **85, 1980, 8271-81.**

ЛИНИЯ ГОРИЗОНТА

Fig. 9a. From 1961 to 1980, 23 American and Soviet spacecraft visited neighboring Venus. Yet during that time only two photographs were taken of its surface, by twin spacecraft in 1975. The panorama returned by Venera 9 (*top*) reveals a surface littered with rocks. The sharp corners on many of these show that erosion at the surface is much less intense than had been expected. The Venera 10 image (*bottom*) has been geometrically corrected to remove distortions introduced by the imaging system; the horizon is labeled in Russian. Flat, slablike rocks, showing dark markings which may be mineral inclusions, cover the site. Photographs from Novosti Press Agency.

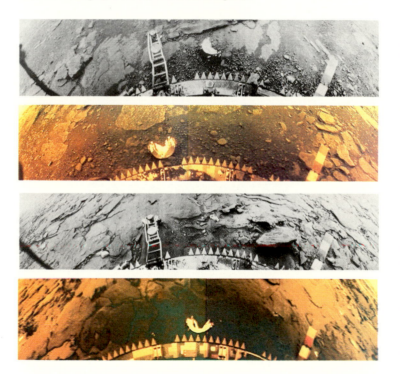

Fig. 9b. In March of 1982, twin Soviet landers obtained these four panoramas of Venus' surface. The top pair, from Venera 13's landing site at longitude 304° and latitude 7° south, detail rocky outcrops and accumulations of small particles that extend to the horizon (upper corners). The spacecraft's shock-absorbing ring has a sawtooth lip to stabilize its descent and touchdown; also visible are calibration standards at right, crescent-shaped lens covers on the ground, and a spring-loaded arm to measure the soil's mechanical strength. The unprecedented color view has an orange cast because the cloudy atmosphere scatters and absorbs the blue component of sunlight before it reaches the ground. Venera 14, at longitude 313°, latitude 13° south, also returned two panoramas, showing flat expanses of rock that apparently lack the fine-grain material seen near the craft's twin. Samples collected by both spacecraft have compositions similar to basalts found on Earth, and the slabs seen by Venera 14 may be thin layers of lava that fractured after solidifying. Tass photos from Sovfoto.

like the Moon's and Mercury's, but volcanism (particularly in the Tharsis region) extended well into the last half of solar system history, perhaps even up to the present.

VENUS

The geology of Venus is of extreme interest because of Venus' similarity to Earth in size, density, and position in the solar system, and its dissimilarity from smaller terrestrial planets. The dense carbon dioxide atmosphere of Venus has served, however, to obscure the surface of the planet from ground-based and spacecraft cameras. Mariner 10 revealed only the clouds of the planet's upper atmosphere (see Chapter 6) and until recently our view of the surface consisted of radar images of small areas of the surface obtained from Earth and six surface panoramas (Fig. 9) from Soviet Venera spacecraft. Since 1978, however, the Pioneer Venus spacecraft has returned radar soundings from hundreds of orbits around Venus. Pioneer's radar altimetry data, covering 93 percent of the surface, has allowed characterization of Venus' global topographic provinces (Fig. 10). About 60 percent of the surface lies within 500 m of the modal (most common) radius and only 5 percent lies more than 2 km above it. Despite this strongly unimodal distribution of altitudes, the range of elevations (about 13 km) is comparable to that of the Earth.

The terrains that cover Venus can be subdivided into lowland plains (20 percent), rolling uplands (70 percent), and highlands (10 percent). Venus' highlands are unlike any topography seen on the smaller terrestrial planets. For example, Ishtar Terra (Fig. 11) is larger than the continental United States and stands several kilometers above the average planetary radius. It is separated from the surrounding rolling uplands by relatively steep flanks, and the western portion is a vast plateau (Lakshmi Planum) some 2,500 km across. Eastern Ishtar contains Maxwell Montes, the single most dramatic topographic feature on Venus. These mountains rise up to 12 km above the mean radius, a height exceeding that of Mt. Everest above Earth's sea level. Ishtar Terra has an elevation comparable to that of the Tibetan Plateau but covers approximately twice the area.

The largest highland region on Venus, Aphrodite Terra covers an area over one-half the size of Africa. Aphrodite extends along the equator for more than 10,000 km and appears topographically rougher and more complex than Ishtar. Several dramatic linear depressions transect east-central Aphrodite; some are up to 3 km deep, hundreds of kilometers wide, and trace across the surface for 1,000 km.

The lowland areas of Venus occupy roughly circular areas (such as the large depression east of Ishtar) and broad but linear depressions (such as that extending along the southern edge of Ishtar). The extensive rolling uplands contain a diversity of topographic features revealed by both Pioneer Venus and earth-based radar data, including linear troughs, parallel ridges and troughs, and numerous shallow circular structures often containing central mounds. These latter features, 40–1,700 km in diameter, might be impact craters; determining their true origin is important since the presence of impact craters implies an age of several billion years for portions of Venus' surface.

All of these interpretations suggest that Venus had (and perhaps has) tectonism, volcanism, impact cratering, and a complex geologic history. But despite our improved under-

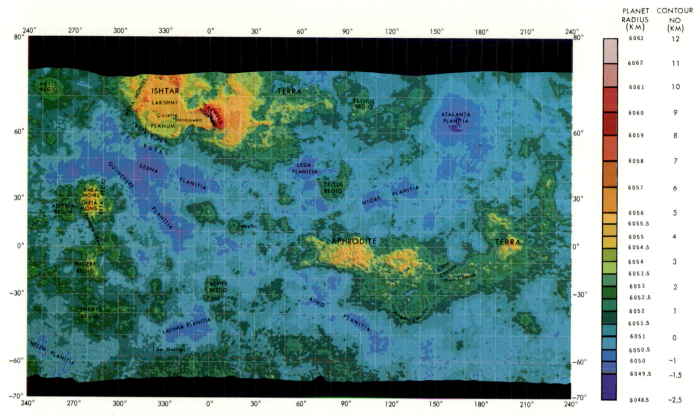

Fig. 10. This computer-generated map uses colored contours to indicate surface elevations on Venus, graduated according to the scale at right (which uses 6,051 km as the mean planetary radius). Features have been named for female figures in mythology and science. Data for the map were gathered by a radar altimeter aboard the Pioneer Venus orbiter; a Mercator projection is used, making features near the poles appear larger than their true relative sizes.

standing of Venus' surface character, the resolution of our observations is still insufficient to determine the geologic processes responsible for the diverse and tantalizing topography, or to compile a geologic map comparable to those of the other terrestrial planets.

THE EARTH

The morphology and geology of the vast majority of the Earth's surface is concealed by liquid water and vegetation. Even though the atmosphere, hydrosphere, and biosphere are not considered terrain units, they are nonetheless significant agents in the modification of the solid crust beneath us. The Earth terrain map appears remarkably different than maps of the other terrestrial planets (Fig. 4). New units appear (folded mountain belts, platform deposits), and some units seen on other planets are either much less widespread (cratered terrain), or much more so (volcanic plains).

The exploration and documentation of the ocean basins revealed their basic geologic and morphologic dissimilarities from the continents and paved the way for the development of the theory of plate tectonics. From this theory has come the realization that the Earth is divided up into a series of rigid lithospheric plates, and that the formation, lateral movement and interaction, and destruction of these plates is responsible for most of the Earth's large-scale structural and topographic features (Fig. 12). Plates collide to produce folded mountain belts, seen in Fig. 13, and deep trenches (where plates dive down into the mantle and are destroyed). Continental rift valleys and vast plateaus of basalts accom-

pany plate breakup. Strings of volcanoes appear where plates are consumed, and new crust forms at the gradually separating mid-ocean ridges.

Despite the apparent uniqueness of many aspects of Earth, several terrain units seem comparable to some seen on the other planets. The distribution of meteorite craters suggests that the ancient rocks of the Earth's surface — the 10 percent comprising the Precambrian shields — are the nearest terrestrial analog to cratered terrain, even though the density of craters is much less and the age much younger than this unit on other planets. The basaltic plains of the ocean floor are the most pervasive terrain unit on the Earth although their mode of origin is different than that of the basaltic plains on other terrestrial planets. These units are among the youngest of Earth's rocks, formed within the last 200 million years. Separate phases of volcanism have built broad basaltic plateaus (Ethiopian and Indian flood basalts, for example) and conical mountains and craters (Mount St. Helens and Hawaii) over the last 65 million years. Polar caps like those on Mars occur on Earth, and they wax and wane with seasons. Unlike the Moon, which has remained largely unchanged for the last half of solar system history, the Earth has a surface constantly in motion, with the positions and relative abundances of terrain units constantly changing.

COMPARATIVE PLANETOLOGY

On the basis of our brief examinations of the terrain types of the terrestrial planets, it is clear that there are some important similarities and differences between these objects. Cratered terrain and volcanic plains are ubiquitous, but their abundances vary among planets. On Earth, Precambrian shields cover a minor fraction of the surface, while volcanic plains dominate over two-thirds of the total crustal area. Yet on the Moon over 70 percent of the surface consists of cratered terrain and only 17 percent is mare lava plains. Mars occupies an intermediate position with abundant cratered terrain and widespread volcanic plains. On the portion of Mercury's surface revealed to date, cratered terrain dominates, but smooth plains occupy approximately 25 percent of the area. And although there is disagreement as to how the Mercurian plains formed, their widespread occurrence (comparable to those on the Moon and Mars) lends support for a volcanic origin.

Cratered terrain is generally acknowledged to be the oldest terrain type on a planet, recording the terminal stages of an object's accretion and the intense bombardment of its newly formed crust. Thus the level of preservation of cratered terrain is a measure of local or global crustal stability over a planet's history. Volcanic units, on the other hand, imply melting in the interior, fragmentation of the crust to provide access to the surface, and outflow of material to build volcanoes or plains. The abundance and type of volcanic units can thus be interpreted in terms of the thermal history of the interior, the state of stress in the crust and lithosphere, and the general level of activity over the several billion years subsequent to crustal formation and the period of heavy bombardment. By comparing the relative abundances of

each unit on a planet we can get a general impression of the extent of evolution of a surface. For example, with over 70 percent cratered terrain and less than 20 percent lava plains, the surface of the Moon appears extremely stable, while the Earth, with the percentages approximately reversed, gives the strong impression of significant evolution beyond the early-formed crust. An actual understanding of the significance of these general impressions requires detailed knowledge of the processes responsible for these terrain units as well as an absolute chronology. Armed with this information, one can then turn to comparing the nature and evolution of the planets.

The ages of most of the terrain units on Earth and the Moon have been determined by field and laboratory investigations, and a framework of radiometrically determined true ages exists for both of these planetary bodies. Since no samples have been returned from Mars and Mercury, our

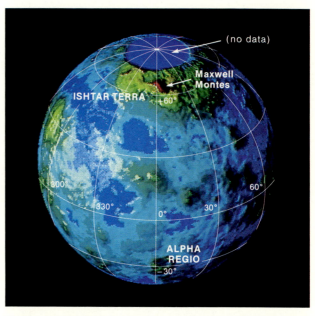

Fig. 11a. **Sophisticated computer graphics were utilized to transform Pioneer Venus radar altimetry information into the sphere seen here, which shows surface features more nearly at their true relative sizes. However, heights have been exaggerated and artificial sunlight added for clarity. Note, for example, the compactness of Maxwell Montes and adjacent structures. This illustration was prepared by P. Ford, E. Eliason, J. Blinn, and W. Brown, under the direction of Gordon Pettengill.**

Fig. 12. **What would the Earth look like if it had no atmosphere, water, or vegetation? Don Davis has rendered the Western Hemisphere in such a situation, showing the recognizable outline of the North American plate, numerous mountain chains, subduction trenches (where one crustal plate dives beneath another), and newly formed crust emerging along the Mid-Atlantic Ridge (near the right edge). Even stripped of its familiar accessories, our planet would still be distinguished by the various tectonic activities that continually modify its surface.**

Fig. 11b. **An artist's impression of Lakshmi Planum and Maxwell Montes on Venus, portrayed with enhanced vertical relief.**

age estimates for their surfaces depend upon the number of craters superimposed on each terrain unit (an indication of how long the unit has existed on the surface and collected craters from incoming projectiles; see Chapter 4). Ages for these two planets thus depend heavily on models of the impact flux in the solar system. Lack of high resolution images of the surface of Venus complicates the problem for that planet, although estimates of the age of part of the planet have been made from the density of circular structures assumed to be impact craters.

Figure 14 compares the estimated distribution of surface ages as a function of surface area for the Earth, Moon, Mars, and Mercury. Evolution of the surfaces of the Moon and Mercury was essentially complete by the end of the first half of solar system history. In contrast, 98 percent of the Earth's present surface was formed in the more recent half and 90 percent within the past 600 million years (the Phanerozoic Era). The surface ages of Mars appear to be distributed in a manner intermediate between those of the Moon and Earth. As noted, the extreme youthfulness of the Earth's surface results from the destruction of old terrain and the creation of new by plate-tectonic activity. An important question in planetary geology is whether surface processes on any of the other planets operate in this style, or whether other processes are responsible for the presence and abundance of surface units. The continued exploration of Venus is important in this respect, offering us an opportunity to compare geologically this Earth-sized planet with our own, in order to see if these similar bodies have experienced similar processes and histories.

SURFACE PROCESSES

Impact craters are not common on the surface of the Earth and those that occur were not commonly recognized as such until recently (Fig. 15). Prior to exploration of the planets, therefore, impact cratering was not thought to be a significant geologic process. At present, we know that the terrestrial planets (perhaps even Venus) are cratered and that the first 600 million years of planetary history (for

Fig. 13. NASA's Seasat satellite recorded this radar image from high above Harrisburg, Pennsylvania (bright area at lower left). Conspicuous are the numerous folded hills that comprise this portion of the Appalachian Mountains. Note that the Susquehanna River, near the bottom, assumed its present course after the deformation of the crust around it. Similar techniques could be used on a spacecraft orbiting Venus to map the planet's surface at a resolution of 1 km.

which the record is missing on Earth) was a period of abundant impact crater formation.

Most geologic processes are thought of in terms of long time scales and modest near-term changes. For example, mountains are built over millions of years and eroded down over additional millions of years. But impact crater formation is dramatically different (Chapter 4). A projectile encounters a planetary surface at velocities measured in kilometers per second. The kinetic energy concentrated at that point can easily equal the total annual heat flow of the Earth! Most of the kinetic energy is expended in fragmentation and shock heating, and ejection of material from the growing crater cavity. Only a tiny portion of the energy is radiated away from the impact point seismically (less than 1 percent), but the total energy can be so large that trains of shock waves are produced with amplitudes several orders of magnitude greater than any known terrestrial earthquake.

Large impact basins exceeding several hundred kilometers in diameter are common on the planets (Fig. 16), and their characteristics reveal much about the nature of the cratering process. The formation of the initial, transient cavity involves the fragmentation and ejection of vast quantities of materials (1-10 million cubic km for the Moon's Orientale basin, some 900 km in diameter). The geometry of the transient cavity is not known, but for large basins it could reach depths of several tens of kilometers. Ejecta is spread over great distances and projectiles thrown from the growing cavity can reimpact to form secondary craters as large as 25 km! The energy involved in the formation of the Orientale basin (seen in Fig. 17) was so great that enough impact-

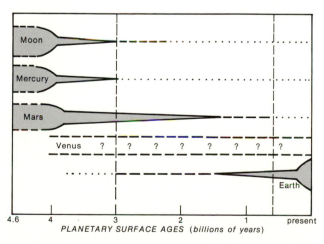

Fig. 14. The ages of the terrestrial planets' surfaces, with gray representing the total surface area of each planet. Most regions on the Moon, Mars, and Mercury are several billions of years old, while two-thirds of the Earth's surface (its ocean basins) formed only within the last 200 million years. Since no global assessment exists yet for Venus, the age of its surface is uncertain.

Fig. 15. **Standing on the outer ring of a Brazilian impact crater known as Serra da Cangalha, geologist John McHone and a guide look south toward the striking inner ring some 5 km away. Pockets of dense vegetation and a small, isolated dwelling sit in the intervening flat sediments, which are 350 m lower than the distant peaks.**

melted material was left within the crater cavity to cover about 200,000 square kilometers — a volume equivalent to the great Columbia River basalt deposits in the northwest United States that took several millions of years to accumulate. The basins then became the locus for mare lava flooding and much of the tectonic deformation of the Moon and planets. Further study of the implications of cratering on early Earth history may provide clues to crustal formation, early continental growth, and the onset of plate tectonics.

Igneous and volcanic processes have also been significant in the evolution of the terrestrial planets as evidenced by the expanse of volcanic plains and constructs seen on their surfaces. Studies of samples from the Earth and Moon and the mapping of similar deposits on other planets suggest that these rocks were formed as a consequence of planetary melting associated with internal radioactive heating. The amounts and rates of lava production varied from planet to planet (and with time on each individual planet), but the influence on the surfaces of all planets was widespread. Formation of the dark lunar maria commenced about 4 billion years ago and ended 1-1½ billion years later. Volcanic activity on Mars and the formation of the great Martian shield volcanoes continued well into the most recent half of the solar system history, and of course volcanic activity on Earth occurs frequently even now (Fig. 18).

Although basaltic plains constitute the most common style of volcanism, the plains on different planets are of different origins. The Earth's ocean basins are created by upwelling and injection of lava at divergent plate boundaries. These new materials are then rifted apart and new material injected in a continuous process known as sea-floor spreading. On the Moon, lavas travel up through a thick crust and emerge along extensive fractures often related to impact basin formation. They pour out into the lowlands, covering the exposed crustal material in a process quite dissimilar to the formation of the Earth's ocean basins. On Mars, many of the volcanic plains are formed from lava flows emerging from the great shield volcanoes and flowing into the surrounding low-lying plains. Thus the formative circumstances of basaltic plains can provide information on the nature of major crustal provinces and environments. Discerning how much lava appeared over time helps to pin down the amount and duration of thermal activity in a planet's interior.

Tectonic activity is also a significant indication of internal dynamics. Deformation of the outer rigid layer of a planet (the lithosphere) can occur in a variety of ways. As mentioned, the Moon's thick global crust shows little in the way of tectonic features, nor does it exhibit any evidence of the major *lateral* movement so typical of terrestrial plate tectonics (Fig. 19). Instead, lunar tectonics are characterized by *vertical* movement downward, due to loading by the thousands of cubic kilometers of lava placed on the Moon's surface.

The extensive lobate scarps on Mercury seem to indicate

Fig. 16. The remarkable Caloris Basin, situated near the terminator in this Mariner 10 mosaic, is a huge impact feature some 1,300 km in diameter. Its broad floor is laced by fractures, often in polygonal patterns, and by sinuous ridges resembling those on the lunar maria. The fractures probably arose when the basin's central area sank slightly from the weight of overyling lava.

Fig. 17. The Moon's Mare Orientale, seen by Lunar Orbiter 4, lies within the Orientale Impact basin. The outermost ring of peaks, called the Cordillera Mountains, is some 900 km across. The impact that formed the basin about 3.8 billion years ago produced large secondary craters up to 2,000 km away.

Fig. 18. In May, 1980, Mount St. Helens in the Pacific Northwest exploded violently, devastating the surrounding region. Shortly thereafter, a NASA aircraft flew over the region and recorded this detailed radar image. About 400 m of the once-dormant volcano's beautifully symmetrical cone is now missing, leaving in its place a gaping crater that stands as a reminder that the Earth, even today, is a dynamic, evolving planet. Photograph courtesy Bruce Hall, Goodyear Aerospace.

global compression and may be the result of cooling and contraction of the planet's radius by 1-2 km. Mars has numerous features comparable to those seen on the Moon, but there is one important exception. The Tharsis region with its concentration of tectonic and volcanic features is a distinctive area unlike virtually any other observed on the terrestrial planets. Whatever the origin of Tharsis — be it deep-seated uplift or long-lasting volcanism — the nature of Martian tectonics is still vertical, rather than the horizontal varieties seen on Earth.

The Earth is presently dominated by numerous laterally moving lithospheric plates and the tectonics associated with their interaction, in contrast to the single lithospheric plates of the smaller Moon, Mars, and Mercury. These planets are dominated by vertical tectonic movement and minor deformation of their single lithospheric plates. However, Venus — an Earthlike planet in many respects — has a surface not seen in sufficient detail to resolve plate tectonic features; major troughs and plateaus unlike those seen on the Moon, Mars, and Mercury may indicate a distinctive tectonic style.

PLANETARY EVOLUTION

Geologic review of the terrestrial planets reveals a number of basic themes and questions and provides a new perspective with which to view the Earth. What are the fundamental processes that formed and modified the surfaces of the terrestrial planets? On the basis of an examination of Earth alone, we never would have listed impact cratering as a fundamental process. The records of other terrestrial planets have revealed its significance, particularly in early planetary history. Exploration of the Earth's ocean basins has forced us to reassess the significance of volcanism as a planetary process. In terms of areal coverage, time duration, and volume, impact cratering and volcanism appear to be the two most significant processes dominating the surface histories of the terrestrial planets. Atmospheric processes have been influential on the surfaces of the Earth, Mars, and perhaps Venus.

Do terrestrial planets share a common early history? The Apollo missions returned samples which provided evidence that the Moon underwent extensive melting of its outer several hundred kilometers, with the source of heat thought to be the intense impact bombardment associated with the terminal stages of planetary accretion. Evidence also exists for the extensive impact bombardment of other planetary surfaces. Were all the terrestrial planets characterized by extensive "magma oceans" early in their histories? Did volatile-rich planets undergo evolutionary paths different from those of volatile-poor planets?

On the basis of our present knowledge of the surface and interiors of the terrestrial planets, we can make some broad generalizations concerning the evolution of planetary bodies, which embody three major influences. The *size of a planet* is an important influence on its subsequent thermal evolution because smaller planets may cool more efficiently due to their higher ratio of surface area to volume. *Planetary chemistry* is also important, for variations in the ratio of iron to silicate materials and in the relative abundances of volatiles and radiogenic heat sources can be significant evolutionary factors. The bulk chemistry of the planets may well vary as a function of distance from the Sun, a vestige of the decreasing temperature and pressure farther from the center of the collapsing solar nebula during planetary formation (see Chapter 20). A third major influence involves the *energy sources* available throughout planetary history and their relationship to one another. Possible sources include energy derived from impacts, tidal interaction between planetary bodies, differentiation of materials in the interior, and radioactivity.

No single factor stands out as a control on the evolution of the terrestrial planets. Continued study of these bodies, particularly Venus, will allow further assessment of the relative importance of each factor and a more complete view of the place of Earth in our solar system.

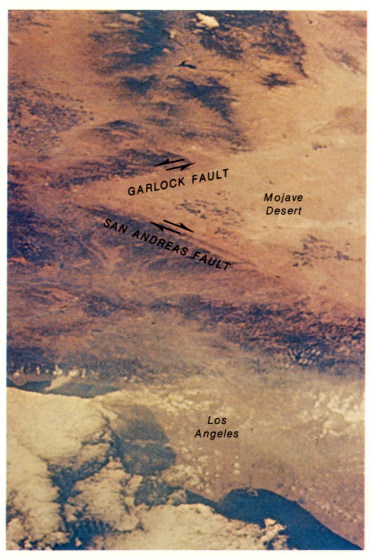

Fig. 19. This Gemini photograph of smoggy southern California clearly shows the intersection of the San Andreas and Garlock faults (arrows indicate the direction of lateral motion). The San Andreas fault system marks the abutment of the Pacific and North American lithospheric plates.

Atmospheres of the Terrestrial Planets

James B. Pollack

The atmospheres of the inner or terrestrial planets of the solar system — Mercury, Venus, Earth, and Mars — show an enormous diversity in their characteristics (Fig. 1). The pressure at the surface of Venus is 90 times that for the Earth, while the pressure for Mars amounts to only 1/150th of the Earth's value. Mercury has almost no atmosphere — it is a million billion times less dense than ours. Most of the gas around Mercury is derived from the rarefied solar wind that flows past the planet. The composition of the Earth's atmosphere has been strongly influenced by the presence of life on its surface, with nitrogen and oxygen being the dominant gaseous species. By contrast, carbon dioxide is the chief gas in the atmospheres of Venus and Mars, while helium and hydrogen are the major components of Mercury's atmosphere (such as it is). Dust storms rage across the surface of Mars, creating an everpresent suspension of dust particles; a dense cloud layer of concentrated sulfuric acid shrouds the entire surface of Venus; and water clouds in all their numerous variations are present over half the area of our world at any given time. It is so cold in the winter polar regions of Mars that seasonal polar caps of dry ice form there; it is so hot at Venus' surface that lead, if present, would melt; temperatures vary from very hot on the day side of Mercury to very cold on its night side; and the temperatures of Earth are just right to permit the occurrence of the most precious form of matter in the universe: life.

Yet, underlying these discordant atmospheric properties,

there are common threads of history and processes. For example, almost all the carbon dioxide once present in the Earth's atmosphere is now locked up in carbonate rocks, such as limestone. The amount of carbon dioxide tied up in these rocks is comparable to that in Venus' atmosphere. "Astronomical variations" of the orbital and axial charac-

Planet	g	P	T	Major Gases	Minor Gases
Mercury	395	10^{-15}	440	He ($\approx$ .98) H ($\approx$ .02)	
Venus	888	90	730	CO_2 (.96) N_2 (.035)	H_2O ($\approx$ 100), SO_2 (150), Ar (70) CO (40), Ne (5), HCl (0.4), HF (.01)
Earth	978	1	288	N_2 (.77) O_2 (.21) H_2O (.01) Ar (.0093)	CO_2 (330), Ne (18), He (5.2), Kr (1.1) Xe (.087), CH_4 (1.5), H_2 (.5), N_2O (.3) CO (.12), NH_3 (.01), NO_2 (.001) SO_2 (.0002), H_2S (.0002), O_3 (.4)
Mars	373	.007	218	CO_2 (.95) N_2 (.027) Ar (.016)	O_2 (1,300), CO (700), H_2O (300) Ne (2.5), Kr (.3), Xe (.08), O_3 (.1)

Table 1. Listed above for the four terrestrial planets are: *g*, the acceleration of gravity at the surface (in cm/sec/sec); *P*, the atmospheric pressure at the surface (in bars); and *T*, the average surface temperature (in degrees Kelvin). Parenthetical values following *major gases* are their fractional abundances by number, depicted graphically in the diagrams below; values following *minor gases* are their fractional abundances for lower atmospheres in parts per million. Except for the rare gases, many minor constituents vary considerably with altitude and sometimes latitude.

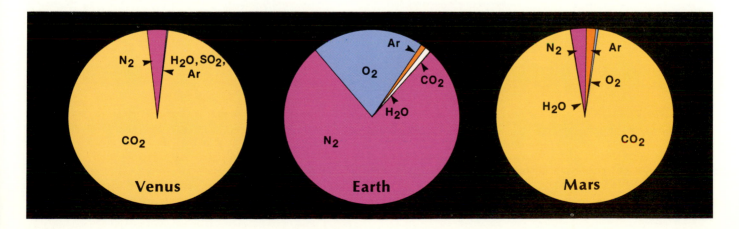

teristics of the Earth and Mars are thought to have caused quasi-periodic climatic changes on both planets, with these variations being responsible in part for the succession of ice ages that have occurred on the Earth over the last few million years.

In this chapter, I first review the current characteristics of atmospheres surrounding the terrestrial planets and the factors responsible for these characteristics. Discussed next are the atmospheres' long-term history and some of the major climatic changes that have occurred. The chapter concludes with a consideration of future efforts that will enhance our understanding of these subjects.

COMPOSITION

Tables 1 and 2 summarize our current knowledge about gaseous and particulate components in the atmospheres of the inner planets. Table 1 displays the surface gravity and temperature, the total atmospheric pressure at the surface, and a measure of the mass of the atmosphere, expressed as *mixing ratios* (the fraction of the molecules they represent). We can estimate the variation in the amount of a given gas among the different atmospheres by comparing the product of the total pressure and its mixing ratio in these atmospheres. Table 2 defines the major types of particles in each atmosphere and gives their vertical and horizontal locations, as well as a measure of the degree to which they interact with sunlight (the optical depth τ). When τ is much less than 1, very little sunlight is scattered or absorbed, while just the opposite is true when τ is large. Later, we will compare the atmospheric properties of Venus, Earth, and Mars.

The amount of water vapor in the bottom portions of the atmospheres of the Earth and Mars is strongly limited by its tendency to condense at moderate and low temperatures. As a result, there is much more water within liquid or solid reservoirs on the surface than in these atmospheres. The oceans and polar ice sheets contain most of the Earth's water, while a water-ice deposit in the north polar region of Mars is the chief water reservoir for that planet. Sunlight warms these reservoirs, causing some water vapor to be released into the atmosphere, where it condenses into clouds in the cooler portions of the atmosphere, and eventually returns to the surface reservoirs. On the average, the amount of water in the lower portions of the atmospheres of the Earth and Mars lies within a factor of two or three of the maximum amount that can be released into an atmosphere at the temperature of the water reservoirs. The temperature at the surface of Venus is so high that there are no reservoirs of water on its surface. All of the planet's water resides in its atmosphere. At high altitudes, where it is much cooler than at the surface, the water vapor in Venus' atmosphere participates in the formation of sulfuric acid cloud droplets.

In all three terrestrial atmospheres, a small fraction of the water-vapor molecules is broken apart by ultraviolet sunlight into hydrogen atoms and hydroxyl radicals, a highly reactive species made of single hydrogen and oxygen atoms. Hydroxyl radicals play an important role in the chemistry of these atmospheres: they help to convert sulfur gases into sulfuric acid vapor and act as a catalyst in the destruction of ozone.

Ultraviolet sunlight converts gases containing oxygen into other oxygen-bearing species. For example, ozone (O_3) is produced in the Earth's atmosphere by a series of chemical reactions that are initiated by sunlight breaking molecular oxygen into two oxygen atoms. These atoms combine with other oxygen molecules to form ozone. Ozone reaches its peak concentration at an altitude of about 30 km above the Earth's surface, in the stratosphere. It helps to shield the ground from biologically dangerous ultraviolet radiation. Small amounts of other gases, such as chlorine atoms and nitric oxide (NO) participate as catalysts in chemical reactions that destroy ozone. There has been concern that our

Fig. 1. At left are spacecraft photographs of the four terrestrial planets: Mercury from Mariner 10, Venus from the Pioneer Venus orbiter, Earth from Apollo 17, and Mars from the Viking 1 orbiter. Venus appears in a false-color version of an ultraviolet image; its cloud details are not apparent in visible light. White clouds hide about half of the Earth at any given time, while clouds of water ice, carbon dioxide ice, and dust enshroud varying portions of Mars. At right, artist Don Davis has portrayed all four terrestrial landscapes as they would appear with the Sun 20° above the horizon.

activities, such as the use of fluorocarbons and fertilizers, may increase the amount of these catalysts in the atmosphere, which could reduce the amount of ozone and hence increase the ultraviolet sunlight reaching the surface. Considerable efforts have been made by the scientific community to predict such changes accurately and thus assess the reality of this potential problem.

For Venus and Mars, carbon dioxide and not molecular oxygen is the chief oxygen-bearing gas. Small amounts of molecular oxygen and trace amounts of atomic oxygen and ozone occur in these atmospheres as a result of a series of chemical reactions that begin with the dissociation of carbon dioxide into carbon monoxide and atomic oxygen by ultraviolet light. In the case of Mars, ozone is most prevalent on the edge of the winter polar region; when the water-vapor abundance is small, few hydroxyl radicals are generated to attack the ozone.

Sulfur gases are present only in trace amounts in the Earth's atmosphere because they dissolve readily in cloud water droplets and because oxygen-rich radicals, water vapor, and other gases convert them into sulfur-containing particles in cloudless regions. Both types of particles eventually fall out of the Earth's atmosphere, taking the sulfur with them. Sulfur-containing particles — sulfuric acid and ammonium sulfate — represent an important component of the non-water particles found in the lowest regions of the Earth's atmosphere, while sulfuric acid predominates at higher altitudes in the stratosphere. By injecting large amounts of sulfur-containing gases, such as sulfur dioxide, into the Earth's stratosphere, volcanic explosions such as those of Mount St. Helens in 1980 can cause a large, but temporary increase in the amount of sulfuric acid there. Human activities, such as the burning of coal, have resulted in an increased input of sulfurous gases into the lower atmosphere, which in turn increases the abundance of sulfur-rich particles there. These particles, together with ones that result from hydrocarbon emissions, are responsible for the smog that exists in many regions of the industrialized world. The high surface temperature of Venus prevents sulfur from being removed from its atmosphere. As a result, large amounts of sulfur dioxide occur in its lower atmosphere and provide source material for the dense layer of sulfuric acid particles that exists at altitudes between about 50 and 80 km. At lower altitudes, the sulfuric acid particles evaporate into water vapor and sulfuric acid vapor; the latter is ultimately recycled into sulfur dioxide. No sulfur-containing gases have been detected in the Martian atmosphere. But they may have been released from Martian volcanoes in the past and converted to sulfur-containing particles that eventually formed the abundant sulfate minerals found on the Martian surface.

TEMPERATURE

What if Venus, Earth, and Mars had no atmospheres? Then their globally averaged surface temperatures would be determined by a balance between the amount of sunlight they absorbed and the amount of heat they radiated to space, with the latter increasing rapidly with higher temperature. The presence of an atmosphere affects both components of this heat balance. On one hand, highly reflective clouds allow less sunlight to be absorbed by the planet, which tends to cool the surface. But the atmosphere absorbs

Planet	Particulate Composition	Altitude (km)	Areal Distribution	Optical Depth (τ)
Mercury	None			
Venus	Concentrated sulfuric acid	50-80	Everywhere	≈ 25
Earth	Concentrated sulfuric acid	12-30	Everywhere	0.003 – 0.3*
	sulfates, silicates, sea salt, organics	0-12	Everywhere, but spatially variable	0.05 – 3
	water	0-12	50% cloud cover	5
Mars	dust	0-50	Everywhere; large temporal changes	0.3 – 6**
	water ice	0-25	winter polar region; morning fog; isolated clouds behind high places	≈ 1
	CO_2 ice	≈ 60	many places	≈ 0.001
		0-25	winter polar region	≈ 1

Table 2. The nature of particulate layers in the atmospheres of the terrestrial planets. Note that in the listings under *optical depth,* **Earth's highest sulfuric acid values (*) occur after volcanic eruptions, and that Mars' maxima due to dust (**) occur during global dust storms.**

some of the heat emitted by the surface and part of this is re-radiated back to the surface, warming it. This situation, termed the *greenhouse effect,* makes the mean surface temperatures of Mars, Earth, and Venus about 5°, 35°, and 500° K warmer, respectively, than they would be in the absence of the thermal infrared opacity of the atmosphere. The surfaces of Earth and Mars experience only a modest increase in temperature because each planet has an atmosphere transparent at some infrared wavelengths, which permits much of the heat emitted by the surface to escape to space.

Carbon dioxide on Mars and carbon dioxide, water vapor, and water clouds on Earth are the chief suppliers of atmospheric infrared opacity. The very large rise in surface temperature for Venus stems from the complete absorption of thermal radiation from its surface, with carbon dioxide, water vapor, sulfur dioxide, and sulfuric acid clouds playing key roles. In part, the differences in the degree of greenhouse warming among the three planets can be attributed to differences in the masses of their atmospheres.

The density and temperature of a planetary atmosphere vary considerably with altitude. Density steadily decreases with increasing height in such a fashion that, at each altitude, a very close balance exists between gravity, which tries to pull the gas down, and gas pressure, which tries to move gas upward from places of high pressure to places of low pressure. In all three terrestrial atmospheres, the density changes by a factor of three with every 10 km of height, although the value of this *scale height* fluctuates with changes of temperature. Density decreases less rapidly in regions of high temperature.

Fig. 2 illustrates how temperature varies with altitude for the three planets. It has been traditional to divide the Earth's atmosphere into four major altitude domains, based on the behavior of the temperature profile, as indicated. For example, in the troposphere, the temperature steadily decreases with increasing altitude, while the opposite is true of the stratosphere. Several processes can transfer heat within

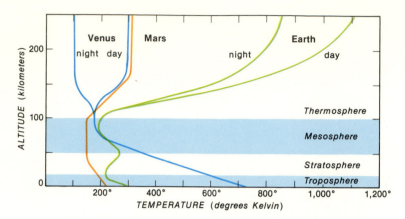

Fig. 2. **A comparison of temperature variations with altitude for Venus, Earth, and Mars. The day-night pairs of curves show the strong diurnal cycle in the upper atmospheres of Earth and Venus. Also indicated are the names given to regions within the Earth's atmosphere.**

the atmosphere, and these, collectively, determine the temperature structure there. Absorption of sunlight and of thermal radiation produced elsewhere tend to warm a given layer of air, while it becomes cooled by its own emission of thermal radiation.

Within the troposphere, heat is transferred from the surface to the atmosphere by the evaporation of water vapor at the surface (which requires heat) and its condensation in the atmosphere (which releases heat). More upward transfer is accomplished by large-scale wave motions and by small-scale turbulence. The latter occurs when the temperature decreases faster with altitude than the *adiabatic lapse rate*. (For Earth, this value varies from about 5-10° K per km, with the maximum corresponding to a dry atmosphere and smaller values to the release of heat by condensing water vapor.) Small-scale turbulence is so efficient in transporting heat that the temperature gradient never exceeds the adiabatic lapse rate by more than a minute amount. In the case of the Earth, the observed rate of change in the troposphere is only about two-thirds of the dry adiabatic value due to the combined effects of water-vapor condensation and heat transport by wave motion. For Mars, the lapse rate in the troposphere is also substantially less than the dry adiabatic value, but here absorption of sunlight by ubiquitous atmospheric dust "stabilizes" the temperature profile. In the case of Venus, the temperature variations with altitude closely match the dry adiabatic value in the lowest 35 km but fall short between 35 km and 48 km, the location of the cloud bottoms, due to an increase in atmospheric transparency at thermal wavelengths here.

The reversal of the temperature gradient between the Earth's troposphere and stratosphere is caused by ozone absorption of ultraviolet sunlight. The temperature profile in the stratosphere reflects a balance between warming due to this effect and cooling by the emission of thermal radiation by carbon dioxide and ozone molecules. At higher altitudes, within the mesosphere, there is less ozone and hence the temperature gradient reverses again due to cooling by carbon dioxide. Because the atmospheres of Venus and Mars have much less ozone than that of the Earth, they lack the equivalent of a stratosphere.

In the highest layer of the Earth's atmosphere, the ther-

mosphere, temperatures increase with increasing altitude. At these altitudes, the sparse numbers of atoms and molecules are less effective at cooling the atmosphere by thermal radiation. Absorption of sunlight at very short wavelengths effectively warms this region, while heat conduction downward by molecules tends to cool it. Much lower temperatures occur in the thermospheres of Venus and Mars because they contain much more carbon dioxide than does the Earth's; the emission of thermal radiation by this gas is very effective in cooling atmospheres. For Earth, atomic oxygen and nitric oxide — not carbon dioxide — produce most of the infrared radiation in the thermosphere. However, they are much less efficient coolants than carbon dioxide.

UPPER ATMOSPHERES

Fluid motions in the lower portions of the atmospheres of the three planets keep the inert gases well mixed. However, at altitudes above about 110 km for the Earth and Mars and somewhat higher for Venus, lower densities make these motions ineffective. Each molecule then assumes a separate variation in abundance with altitude: lighter gases thin out less rapidly with increasing altitude than the heavier gases. The boundary at which fluid motion relinquishes control is termed the *homopause* (or turbopause). On all three planets, the height of the homopause may be determined by the strength of vertically propagating atmospheric "tidal" motions (generated by variations in the sunlight absorbed by the lower atmosphere over the course of a day). The air density is lower at the homopauses of Mars and Venus than at the Earth's, possibly because the tidal motions there are stronger. Absorption of sunlight by dust in the Martian atmosphere excites strong motions there, while absorption of sunlight near Venus' cloud tops may fulfill a similar function.

In the upper regions of an atmosphere, sunlight at very short wavelengths strips electrons from atoms and molecules and creates an ionized layer, the ionosphere. In the case of Mars and Venus, carbon dioxide is the chief victim, while for the Earth it is molecular oxygen and nitrogen. However, due to a series of chemical reactions, ionized oxygen atoms and molecules actually constitute the most abundant ion for all three. Ionized nitrogen oxide is another important constituent of the Earth's ionosphere.

The boundary between interplanetary space and the region dominated by the planet is determined by the nature of the interaction between the planet and the solar wind — a high-velocity, fully ionized gas of very low density that emanates from the Sun. These characteristics are presented in Chapter 3.

METEOROLOGY

Because more sunlight falls on a planet's equator than at its poles over the course of a year, the equator is the warmer place. This latitudinal temperature gradient sets winds into motion that transport heat from warm spots to cold ones. Hence, dynamic atmospheric motions are somewhat self-destructive, by diminishing the temperature differences that drive them. The Martian atmosphere is so thin that heat transport is ineffective in diminishing latitudinal temperature gradients. As a result, temperatures become so cold in the winter polar regions — about 150° K — that carbon dioxide, the major constituent of the Martian atmosphere, condenses on the surface as dry ice. Near the end of the winter in a given hemisphere, this polar ice deposit extends down to a latitude of about 50°. In the northern hemisphere, it totally sublimates away during the subsequent spring and summer. However, a small residual CO_2 ice cap remains throughout some years near the south pole. The atmosphere of the Earth is dense enough that atmospheric motions help to diminish somewhat the equator-to-pole temperature differences that would exist in the absence of these motions. The atmosphere of Venus is so massive that surface temperatures vary by no more than a few degrees between its equator and poles.

The direct response of planetary atmospheres to latitudinal temperature differences is a *Hadley circulation pattern* (named after its discoverer), whereby air rises over the warm regions, moves towards the cooler places at high altitudes, sinks in the cooler regions, and returns back to the warmer places at low altitudes (Fig. 3). However, on a rotating object, such as a planet, there exists an inertial force called a *Coriolis force* that acts to deflect air perpendicularly to its direction of motion. Consequently, Hadley circulation has an east-west component to its motion as well as a north-south component.

On the Earth, the trade winds at equatorial latitudes correspond to the low-level return flow of the Hadley cell there. Instability prevents a single Hadley cell from spanning the entire distance from the equator to the pole on the Earth; even a small perturbation to the flow pattern results in a drastic change. Rather, each hemisphere has three cells, with the middle one — the Ferrel cell — circulating in a thermodynamically indirect sense, that is, air rises at the cold end and sinks at the warm end. The equatorial Hadley cell is driven primarily by heat released during the condensation of atmospheric water vapor rising from the ocean. Much of

this condensation takes place in a narrow latitudinal belt called the Intertropical Convergence Zone, occurring within clusters of cumulus clouds that draw on moisture brought into the ICZ by trade winds. Due to the increasing strength of the Coriolis force at higher latitudes, the winds blow mostly in an east-west direction at mid- and high latitudes, rather than north-south.

Fluid-dynamic instabilities also cause a further subdivision of the large-scale motions into smaller-scale "eddy" motions, such as the progression of cyclones (lows) and anticyclones (highs) that constitute an important component of the weather at mid-latitudes on the Earth (Fig. 3). The instability that causes these particular disturbances is called a *baroclinic instability,* which occurs when the latitudinal temperature differences become too large. Baroclinic waves — the eddy motions resulting from this instability — play a major role in transporting heat poleward at mid-latitudes and in transporting heat vertically upward, thus decreasing the vertical temperature gradient or lapse rate within the troposphere. They also transport momentum toward a narrow latitudinal region in the upper troposphere and thus help to maintain the high velocity east-west moving "jet stream." Typically about six pairs of highs and lows girdle the Earth in this latitudinal belt.

The Earth's topography (for example, large mountain ranges) and the temperature differences between the oceans and neighboring continents set up another type of wave pattern, standing waves (Fig. 3). In contrast to baroclinic waves, these waves do not propagate horizontally, although they can transport heat and momentum vertically. They tend to have longer wavelengths — typically it takes only one to three wavelengths to girdle a latitude circle. Standing waves in the troposphere transfer momentum into the stratosphere

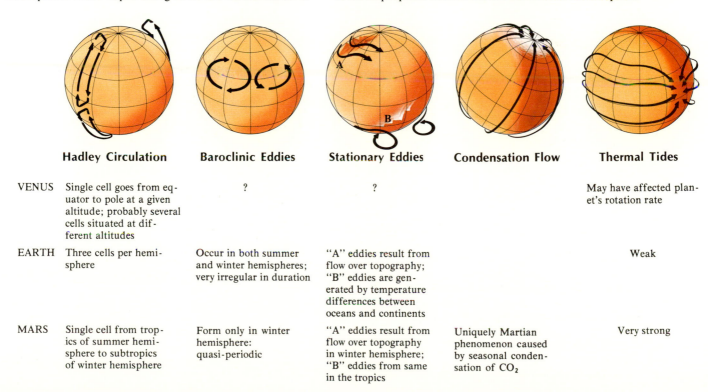

	Hadley Circulation	Baroclinic Eddies	Stationary Eddies	Condensation Flow	Thermal Tides
VENUS	Single cell goes from equator to pole at a given altitude; probably several cells situated at different altitudes	?	?		May have affected planet's rotation rate
EARTH	Three cells per hemisphere	Occur in both summer and winter hemispheres; very irregular in duration	"A" eddies result from flow over topography; "B" eddies are generated by temperature differences between oceans and continents		Weak
MARS	Single cell from tropics of summer hemisphere to subtropics of winter hemisphere	Form only in winter hemisphere: quasi-periodic	"A" eddies result from flow over topography in winter hemisphere; "B" eddies from same in the tropics	Uniquely Martian phenomenon caused by seasonal condensation of CO_2	Very strong

Fig. 3. Major circulation patterns within the atmospheres of terrestrial planets appear schematically above a comparison of their characteristics for Venus, Earth, and Mars. Arrows on the globes indicate the direction of flow, and each type of circulation is described in the text.

and so influence the circulation patterns there. In summary, then, the winds in the Earth's troposphere can be roughly divided into three major components: a longitudinally averaged or mean circulation mainly in the tropics, and baroclinic and standing eddies at higher latitudes.

Because Mars rotates at a rate very similar to that of the Earth — the length of a Martian day is 24 hours, 37 minutes — its circulation patterns are expected to crudely resemble those of the Earth. Thus, for example, a Hadley cell is thought to be present at equatorial latitudes, and baroclinic waves have apparently been observed by the Viking lander meteorology experiment at mid-latitudes during certain seasons. Also, standing waves set up by *very* large elevation differences (up to 25 km) are predicted.

However, some fundamental differences exist between the circulation patterns in the lower atmospheres of these two planets. Because there are no oceans on Mars, the entire surface of the planet responds very rapidly to changes in incident sunlight. During the summer season in a given hemisphere, the hottest place on Mars is not at the equator, but displaced to a tropical or subtropical latitude in the summer hemisphere, where the Sun is directly overhead at noon. As a result, a single Hadley cell spans both sides of the equator at this season, rather than there being a separate Hadley cell for each hemisphere, as occurs for the Earth. Warm air rises in the summer hemisphere of Mars and sinks in the winter hemisphere. For the Earth, the Hadley cell is confined to equatorial latitudes.

Another consequence of the essentially instantaneous response of the entire Martian surface to incident sunlight is that strong temperature differences are present only at mid-latitudes in the winter hemisphere. Thus, baroclinic waves are expected and observed only there. Baroclinic waves on Mars appear to be somewhat more regular than on the Earth in the sense of having a more well-defined period. In this regard, weather may be more predictable for Mars. As on the Earth, baroclinic waves transport momentum towards (and thus help maintain) a strong eastward moving jet that exists in the winter hemisphere close to the boundary of the seasonal CO_2 ice cap.

Because CO_2 is a major constituent of the Martian atmosphere, its condensation at high latitudes in the winter hemisphere results in the pressure being somewhat lower there than elsewhere. This latitudinal pressure gradient sets up a strong, planetwide circulation that transports heat and momentum as well as mass toward the region of the forming polar cap (Fig. 3). Because of this "condensation flow," the mean circulation is the dominant wind component at all latitudes. In particular, baroclinic waves play a less central role than for the Earth in transporting heat poleward at mid-latitudes. The formation and dissipation of the seasonal CO_2 ice caps cause the atmospheric pressure to vary by about 20 percent from one season to the next.

As a result of having a thin atmosphere composed largely of a good infrared radiator, CO_2, temperatures in the bottom portion of the Martian atmosphere decrease by significant amounts — tens of degrees Kelvin — from the daytime when the sun is shining, to the nighttime when it is not. These diurnal temperature variations drive "thermal tidal" winds that propagate in the direction of motion of the Sun in the Martian sky (see Fig. 3). Much smaller temperature differences and hence much weaker tidal winds

are generated in the more massive atmosphere of the Earth.

There are very interesting relationships between dust and winds on Mars. When the wind speed close to the surface exceeds a critical threshold value — about 50 to 100 m per second — sand grains about 100 microns across undergo a skipping type of motion called saltation. Upon impact with the surface, they place smaller dust grains — a few microns in size — into suspension in the atmosphere, where they remain for months before settling back out. Dust particles may be removed preferentially in the winter polar regions by acting as growth sites for CO_2 snowflakes, which because of their large size fall more rapidly to the ground. About 100 local dust storms occur over the course of a Martian year in places where the winds are particularly strong, such as the edge of the seasonal CO_2 cap, where thermal gradients drive winds; or subtropical highlands, where the tidal winds are enhanced.

Once or twice a Martian year (equal to about two of ours), a local dust storm grows to global proportions, thereby totally enveloping the planet in a dense shroud of dust. This dramatic growth in the size of a dust storm may be due to a positive feedback relationship between dust and the winds, first suggested by Conway Leovy and the author. Increasing the amount of dust in the atmosphere tends to strengthen the tidal and Hadley winds, which in turn raise dust over a larger area. However, once the dust loading becomes too great, winds near the surface tend to weaken because most of the sunlight is now being absorbed at high altitudes.

The threshold wind speed needed to set particles into motion decreases with increasing atmospheric pressure. Hence, lower wind speeds are required to raise dust into the Earth's atmosphere than into the Martian atmosphere and even smaller speeds are needed on Venus. However, vegetation protects the soil on much of the continental terrain of the Earth, so dust storms occur mostly in desert regions, where the vegetation cover is sparse. The "dust bowl" that occurred in the central United States during the 1930's was preceded by a loss of the vegetation cover. While the wind speeds are quite small close to the surface of Venus — about 1 m per second — they may be large enough to set sand into motion there.

Despite the very slow rotation rate of Venus (one day there is about 117 of ours), the winds in its lower and middle atmosphere blow primarily in an east-west direction. While the Coriolis force is undoubtedly too small to be responsible for this reorientation of the planet's winds, another inertial force due to the winds themselves — centrifugal force — may play a key role in this regard. The east-west winds on Venus blow in the direction of the planet's rotation at all altitudes, increasing from a very modest 1 m per second near the ground to a phenomenal 100 m per second close to the cloud tops. In contrast to the Earth and Mars, where such high velocities are encountered in only a narrow zone of latitudes, these rapid winds occur over essentially the entire globe of Venus. The transport of momentum by a combination of the Hadley circulation and eddies from the dense lower atmosphere to the much more rarefied cloud top region is believed to be responsible for generating the high-velocity winds. Much more modest wind speeds of 5-10 m per second characterize the north-south component near the cloud tops, with the winds at this level blowing from the equator to the poles.

 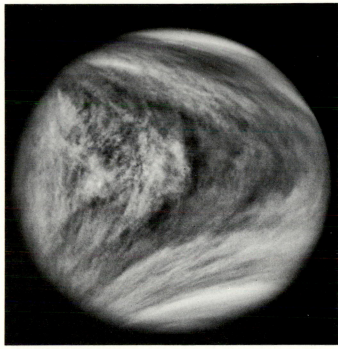

Fig. 4. The clouds of Venus, as photographed by Mariner 10 in 1974 (left) and by the Pioneer Venus orbiter in 1979 (right). Both images were obtained in ultraviolet light to bright out details in the upper cloud layer, which appears very bland at visible wavelengths. Although Venus rotates once every 117 days (as seen from the Sun) in a retrograde direction (right to left in these views), its atmosphere near the cloud tops whirls around in only four days. In doing so, it also moves from equator to pole in a Hadley circulation pattern, giving rise to the chevron-shaped markings seen here.

Important differences exist between the wind pattern near the cloud tops determined in 1974 from photographs obtained by Mariner 10 and the pattern found in 1979 by the Pioneer Venus orbiter (Fig. 4). Certain mid-latitude regions exhibited enhanced east-west winds in 1974, whereas a much smoother trend was found in 1979. Thus, the circulation patterns on Venus may undergo natural oscillations on a time scale of years.

While thermal tidal winds are quite weak in the massive lower atmosphere of Venus, they may, nevertheless, have had an important influence on the current rotation rate of the planet. As noted above, Venus rotates more than a hundred times more slowly than does the Earth. Also, it rotates in the opposite direction so that sunrise occurs in the west and not the east. Because Venus is closer to the Sun, gravitational tidal forces exerted by the Sun on the solid body of the planet may have markedly reduced Venus' rate of rotation over its lifetime of 4.6 billion years. For Venus' current sense of rotation, atmospheric thermal tides would have acted to increase the rotation rate, with their strength steadily increasing as the gravitational tides slowed Venus down. According to Anthony Dobrovolskis and Andrew Ingersoll, the current rate of rotation may be determined chiefly by a balance between these two opposing forces. If so, the large mass of Venus' atmosphere may have permitted the atmospheric tides to oppose the body tides effectively.

ATMOSPHERIC EVOLUTION

We next consider the sources of planetary atmospheres, the factors that determine their chemical makeup, and possible variations of atmospheric composition with time. The terrestrial planets formed about 4.6 billion years ago within a giant disk of gas and dust, the solar nebula. Due to gravi-

tational attraction and low-velocity collisions, dust grains aggregated into larger and larger solid bodies (planetesimals) that eventually became the four terrestrial planets (see Chapter 20). Conceivably, the dust grains and planetesimals that formed the inner planets contained a very small amount (about 0.01 percent) of volatile compounds. When parts of the planet heated up, gases were released from volatile-laden minerals and ultimately were vented into its atmosphere. For example, water may have been present initially in hydrated minerals such as serpentine. Molecular nitrogen and carbon dioxide may have originated from nitrogen and carbon atoms that were part of the chemical makeup of substances, such as the nonbiological organic compounds prevalent in certain classes of meteorites. Gases containing nitrogen and carbon were formed from this raw material when it became heated and chemical reactions took place within it and between it and other gases, such as water. We will refer to this theory of the origin of planetary atmospheres as the "accretion" hypothesis (Fig. 5a).

Alternatively, it is conceivable that the terrestrial planets formed out of material that was essentially free of volatile-containing compounds and hence their atmospheres were obtained from some external source. During or after planetary formation, the planets may have obtained an atmosphere by gravitationally capturing and retaining gases from the primordial solar nebula (the "solar nebula" hypothesis); or by capturing some of the solar wind that has flowed past them over their lifetimes (the "solar wind" hypothesis); or by colliding with volatile-rich comets and asteroids (the "comet-asteroid" hypothesis). These are depicted in Figs. 5b-d. The hydrogen and at least some of the helium in Mercury's current atmosphere are thought to have been derived from the leakage into its magnetosphere of a small fraction of the solar wind flowing past it. Some of the helium may also come from the decay of certain long-lived radioactive isotopes in its interior. A fraction of the helium produced in this manner may make its way through a network of fissures to the surface. The heavily cratered surface of the Moon attests to the substantial number of small stray bodies encountered by objects in the inner solar system over their lifetimes. Thus, the "comet-asteroid" hypothesis would seem reasonable on *a priori* grounds.

It is possible to distinguish among these four competing theories for the origin of planetary atmospheres by comparing the amount of primordial rare gases remaining in the atmospheres of Venus, Earth, and Mars. The term "primordial" means that we exclude rare gases produced from the decay of radioactive isotopes in the planet's interior. Rare gases of intermediate atomic weight, namely neon, argon, and krypton, are particularly useful because they are too heavy to escape readily to space from the top of these atmospheres and are very difficult to incorporate into surface rocks and sediments. The fractional amounts of certain primordial rare gases (in particular the ratio of neon-20 to argon-36) are very similar for all three planetary atmospheres, but distinctly different from that characterizing the Sun's atmosphere. Since the relative abundances of rare gases in the primordial solar nebula and the solar wind are expected to be quite similar to those of the solar corona, the observed abundance pattern for planetary atmospheres is inconsistent with that expected for both the solar-nebula and solar-wind hypotheses. However, a potentially viable hybrid of the solar-wind and accretion hypotheses will be discussed shortly.

The absolute abundance of primordial argon decreases by about 70 times — almost two orders of magnitude — from Venus to the Earth and by another several orders of magnitude from the Earth to Mars. (Absolute abundance is the total mass of a given gas relative to the mass of the whole planet.) Since Venus and Earth have about an equal chance of encountering comets and asteroids, the comet-asteroid hypothesis implies that the absolute abundance of their primordial rare gases should be about the same, in obvious contradiction of the observed amounts. Therefore, by elimination, only the accretion hypothesis remains intact.

Studies of rare gases in meteorites show that two principal mechanisms incorporated gases into these objects' parent bodies. Some "gas-rich" meteorites obtained large quantities of rare gases from solar wind ions embedded within the surfaces of their parent bodies. Other meteorites like carbonaceous chondrites are rich in rare gases that were preferentially trapped in certain minerals (carbonaceous material, for example) as they formed and grew in the solar nebula. The ratio of primordial neon to argon in these trapped gases closely resembles that found in the Earth's atmosphere, yet in the "gas-rich" meteorites this ratio is much larger and more closely matches the solar value.

The large excess of primordial neon and argon on Venus relative to the Earth can be explained by the accretion hypothesis in one of two ways. Perhaps the planetesimals that ultimately formed Venus were preferentially irradiated by an early, intense solar wind that became almost totally absorbed before reaching planetesimals in our general vicinity. Alternatively, since Venus formed closer to the center of the nebula than did the Earth, the pressure in the solar

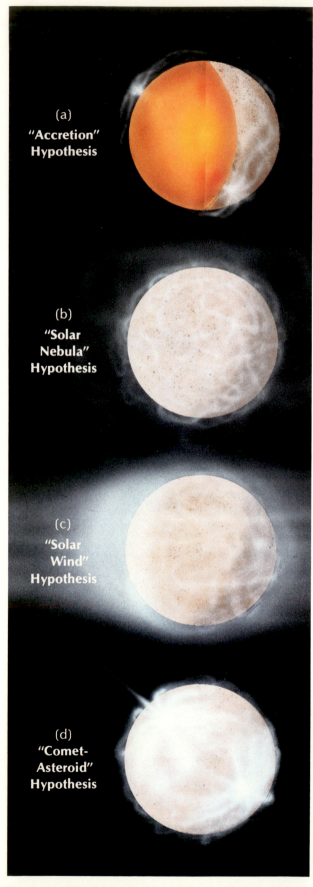

Fig. 5. As discussed in the text, these four models represent possible origins for the terrestrial planets' atmospheres.

nebula should have been higher near Venus' location and hence a larger amount of rare gases should have been adsorbed by the grains that formed it. Similarly, the grains that formed the Earth should have been richer in rare gases than the ones that formed Mars. In order to explain the similarity in the relative abundances of primordial neon and argon among the three planets, one must suppose that rare gases stuck to planet-forming grains at times when the temperature of the nebula was similar in all three locations. This need not have occurred simultaneously; the times could have corresponded to epochs when chemical compounds were formed in the solar nebula that were particularly effective in trapping rare gases. The comparable neon-to-argon ratios on the Earth and Venus favor the latter explanation: higher pressures near the forming Venus forced it to adsorb more gas. But the krypton-to-argon ratio on Venus may be a better match to the Sun's than the Earth's. So the idea of an intense period of solar wind irradiation may be the correct one. In any event, the accretion hypothesis (in its most general sense) accommodates the rare-gas patterns observed now on Venus, Earth, and Mars.

We can also compare the amount of other volatile species that have been vented into the atmospheres of the three outer terrestrial planets over their lifetimes. In so doing, we try to count volatiles that have condensed on the surface, such as the water in the Earth's oceans, or that have been locked into rocks, such as the carbon dioxide contained in carbonate rocks. In both of these examples, much more material exists in the Earth's volatile reservoirs than is contained as gas in its atmosphere at the present time, although at one time or another all this material was in the atmosphere. In contrast to the situation for the rare gases, roughly comparable amounts of nitrogen and carbon dioxide have been vented into the atmospheres of the Earth and Venus. This difference between the rare gases and the other volatiles is due to the former being incorporated onto the surfaces of planet-forming grains and planetesimals, while the latter were incorporated as chemical compounds. Apparently, the two planets formed with comparable amounts of nitrogen- and carbon-containing compounds.

Mars may have formed with fractional abundances of nitrogen, carbon, and water comparable to the Earth's, but smaller percentages of these gases have made their way into its atmosphere over the planet's lifetime. We suspect that the interior of Mars, a less massive body, has not been heated as much as ours has. Still, enough water to form an ocean or ice blanket about 10-100 m deep has been released, along with perhaps several tens to several hundred times the amount of carbon dioxide now in its atmosphere.

Age-dating of rocks brought back from the Moon by the Apollo astronauts indicates that the outer layers of the Moon experienced elevated temperatures during its first several hundred million years. It seems likely that the more massive inner planets had even warmer early histories. Thus, most of their volatiles were probably outgassed during the first billion or so years of their 4.6-billion-year lifetimes.

The chemical makeup of the gases vented into the early atmospheres was a function of the temperature of the rocks from which they were released and the oxidation state of the rocks, particularly that of their iron compounds. Under conditions plausible for the early planets, the released gases would consist chiefly of water vapor, carbon dioxide,

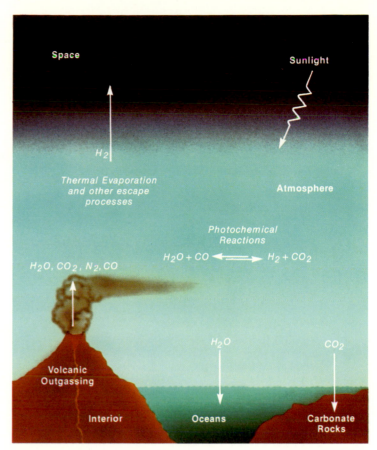

Fig. 6. **Sources and sinks of atmospheric gases. New or "juvenile" gases are introduced by volcanoes. Nitrogen and rare gases, among others, tend to accumulate in the atmosphere over time, while certain species become depleted through various loss processes: hydrogen escapes to space, water condenses out, carbon dioxide takes part in chemical weathering, and so on. Thus the steady-state abundances of atmospheric gases may be markedly different from the mixtures initially spewed out by volcanoes.**

carbon monoxide, molecular hydrogen, and molecular nitrogen. Hydrogen and oxygen were released by the partial dissociation of water at the hot-rock temperatures, but the oxygen was very efficiently reincorporated back into the rocks. Thus, no significant amount of free oxygen is expected to be present among the gases initially vented into planetary atmospheres, while molecular hydrogen and carbon monoxide represent the major reduced compounds.

Once gases began to accumulate in planetary atmospheres they were subjected to a variety of loss processes and chemical transformations that strongly affected the steady-state composition of the atmospheres (Fig. 6). Some gases were effectively lost to the surface, provided its temperature was not too high. On the Earth, Mars, and perhaps early Venus, almost all the released water vapor condensed to form liquid oceans, ice deposits, or both. Also, through the weathering of surface rocks, some carbon dioxide was converted into carbonate rocks. Hydrogen is so light that it can rapidly escape from the Earth's present atmosphere to space, with the escape rate being proportional to its abundance in the atmosphere. If hydrogen was able to escape effectively at early times from all three planets, then its abundance in these atmospheres may have been determined by a dynamical balance between its rate of release from the planetary interior and its rate of escape to space at the top of the

	Volatile Inventory	*The First 10^9 Years*	*The Next 3.5×10^9 Years*	*Present*
MERCURY	$CO_2 \leqslant$ few mb		*all volatiles lost to space*	minute amount of H_2 and He from solar wind and interior
VENUS	$CO_2 \approx 90$ b $H_2O/CO_2 \approx 1$ $^{36}Ar \approx 2$ mb	$CO_2 \approx 1$ b $H_2O/CO_2 \ll 1$	*runaway greenhouse* *almost all water lost*	$CO_2 \approx 90$ b $H_2O/CO_2 \ll 1$
EARTH	$CO_2 \approx 60$ b $H_2O/CO_2 \approx 10$ $^{36}Ar \approx .03$ mb	$CO_2 \approx 1$ b $H_2O/CO_2 \ll 1$ $H_2 \approx 1$ mb	*more $CO_2 \longrightarrow$ rocks rise of O_2*	$CO_2 \approx .3$ mb $O_2 \approx 200$ mb $N_2 \approx 800$ mb
MARS	$CO_2 \approx 2$ b $^{36}Ar \approx .0001$ mb	$CO_2 \approx 1$ b $H_2O/CO_2 \ll 1$ $N_2 \approx 30$ mb	*loss of CO_2 and N_2 to space and surface*	$CO_2 \approx 7$ mb $N_2 \approx .2$ mb

Fig. 7. The amounts of gases in the terrestrial planets' atmospheres have varied considerably over the history of the solar system. Each evolutionary path includes critical times in the development of each atmosphere.

atmosphere. For typical outgassing rates based on the current totals of water and carbon dioxide, hydrogen is expected to have constituted about 0.1 percent of the early atmospheres. Thus, these atmospheres were composed chiefly of nitrogen and carbon dioxide, with minor but significant amounts of water vapor, carbon monoxide, and hydrogen. It is within such an atmosphere that the first chemical steps that led to the formation of life on the Earth may have taken place, with lightning and solar ultraviolet light providing the energy source to convert these simple gases into more complex organic molecules.

The atmospheres of Venus, Earth, and Mars have evolved to a more oxidized state with time (Fig. 7), because the volcanic outgassing rate has gradually slowed down and the lost hydrogen has not been fully replenished. Moreover, some water vapor has been converted by sunlight and chemical reactions into hydrogen, which eventually escaped to space, and into oxygen, which oxidized the remaining reduced gas species in the atmosphere. As a result, at present the atmospheres of Venus and Mars contain only trace amounts of reduced gases, such as hydrogen and carbon monoxide. These small quantities are generated as a result of the action of sunlight on carbon dioxide, the major gas in these atmospheres. These transformations also produce small amounts of free oxygen. The large quantities of oxygen in the Earth's atmosphere today are almost entirely due to the accumulation of oxygen produced by photosynthetic organisms. Significant quantities of oxygen first appeared here about two billion years ago (long after the origin of life) and probably increased steadily over at least the next billion or billion and a half years. The delayed appearance of oxygen in the Earth's atmosphere relative to the beginnings of life was probably due to a combination of a delay in the occurrence of oxygen-producing photosynthetic organisms and the uptake of oxygen generated at early times by reduced compounds in the oceans and on the land. Accompanying the rise of oxygen in the Earth's atmosphere was a steady increase in the amount of ozone.

Planetary mass and distance from the Sun have influenced the size of the atmospheres of Venus, Earth, and Mars. In part, Mars' smaller size has, in itself, limited the mass of volatile compounds outgassed to the surface, as previously discussed. Also, its outer layers cooled more rapidly so that the level of volcanic and tectonic activity on Mars in recent times has been much reduced over earlier levels. This reduction implies that once atmospheric gases are locked into surface rocks there, it is very unlikely that they will be recycled into the interior and heated up, and the confined gases returned to the atmosphere. Such recycling does occur on the Earth on a time scale of hundreds of millions of years, and it keeps the Earth's atmosphere resupplied with carbon dioxide and other gases (Fig. 8). Mars' lower gravity has permitted gases such as nitrogen to escape into space at significant rates. While nitrogen is too heavy to reach escape velocities at the temperatures in the Martian thermosphere, Michael McElroy has shown that it can achieve the required speed as a result of chemical reactions at the top of the thermosphere between ions and neutral species. Finally, Mars' greater distance from the Sun than that of the Earth gives it a colder surface. As a result, larger fractions of certain gases are segregated out of the atmosphere into near-surface reservoirs. For example, vast quantities of carbon dioxide may now be withheld from the atmosphere by adsorption on the surfaces of tiny grains buried up to hundreds of meters deep in the Martian regolith.

Venus' massive atmosphere may simply result from its proximity to the Sun. Were we hypothetically to move the Earth to Venus' distance, it too would acquire a massive atmosphere, due to what is called the "runaway" greenhouse effect. The amount of water vapor in a planet's atmosphere increases very rapidly as the surface temperature increases, which enhances the efficiency of greenhouse warming; that, in turn, further boosts the surface temperature. If a planet is close enough to the Sun, all of the greenhouse "windows" (through which heat emitted by the surface can escape to space) are closed, trapping enough heat to drive all water into the atmosphere. Thus, the oceans disappear and the surface temperature rises to a very high value. Under these conditions, carbon dioxide remains in the atmosphere, as no carbonate rocks are formed. The difference in the amount of sunlight falling on Venus and Earth — about a factor of two — is sufficient to cause such a scenario for Venus.

Venus, with its torrid surface, could not possibly possess

oceans of water. Yet observations have shown that its atmosphere contains only about 0.001 percent of the water in Earth's oceans. If one assumes that both planets began with roughly equivalent fractions of water, the obvious question is: Where did it all go on Venus? Key loss processes might include the dissociation of water vapor by ultraviolet sunlight; the reaction of water vapor with the early large amounts of carbon monoxide to produce carbon dioxide and hydrogen; and the reaction of water vapor with molten rocks near the surface. In all three cases, the hydrogen produced by the chemical breakdown of water escapes to space. Apparently, these processes were effective for Venus — but not for the Earth — because much more water vapor resided in Venus' atmosphere and because Venus' surface is so hot. Evidence that Venus once had much more water than it does today is given by a hundredfold enhancement of atmospheric "heavy" water (which contains deuterium) over the amount in our atmosphere. Because deuterium has twice the mass of ordinary hydrogen, it escaped to space less readily and became concentrated in Venus' atmosphere as the initial water was lost.

Mercury has an almost negligible atmosphere at present because whatever gases escaped from its interior were efficiently removed to space. One important loss mechanism for heavy gases such as carbon dioxide and nitrogen involves their ionization by ultraviolet sunlight and subsequent convection through the magnetosphere to the solar wind, which carries them away. However, the solar wind cannot pick up more mass than a fraction of its own mass. Even over a span of 4.6 billion years, the amounts of nitrogen and carbon dioxide that can be removed by this and other processes are considerably less than that contained in the initial volatile inventories of the Earth and Venus. Thus, Mercury has no appreciable atmosphere today possibly because it outgassed a smaller fraction of its volatile content than did these other planets. Furthermore, it was endowed with less volatile material to begin with. However, if the release of gases from its interior occurred preferentially at very early times, as seems likely, it may once have had a much more substantial atmosphere (Fig. 7).

CLIMATIC CHANGE

Important changes in climate have occurred on the Earth on time scales ranging from years to aeons. Additional evidence suggests that other planets, particularly Mars, have experienced climatic changes. We concentrate here on factors that have caused such changes on longer time scales.

Our Sun, a rather ordinary star, produces energy by slowly converting hydrogen to helium in its deep interior through a series of nuclear reactions. Partial conversion of the Sun's hydrogen to helium over the 4.6-billion-year age of the solar system has gradually increased the solar luminosity with time by several tens of percent. This long-term change in solar output, by itself, would have resulted in dramatic changes in the climate of the terrestrial planets. During the early history of the solar system, lower levels of luminosity may have permitted Venus to have a moderate surface temperature, oceans of water, and perhaps even incipient life forms. However, at some point, the increase in solar output caused a runaway greenhouse to occur and high surface temperatures resulted.

The lower solar luminosity in the past could have caused the Earth to experience a deep ice age during its early history and Mars to experience an even colder climate than it does today. In fact, if no other factors intervened, the Earth's oceans should have been entirely frozen more than two billion years ago. Yet there is sound geologic evidence for the existence of liquid oceans on Earth for the last 3.8 billion years and of life for almost that long. Similarly, the presence of ancient, dry river valleys on Mars suggests that the cli-

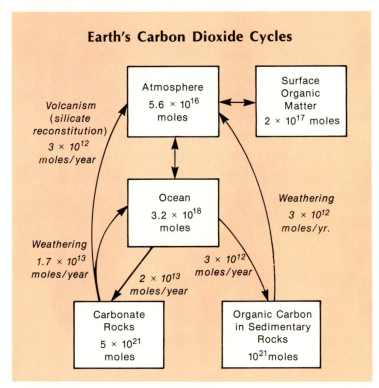

Fig. 8. **This diagram traces the cycle of carbon dioxide for the Earth, based on calculations by James Walker. Although this gas finds its way into the atmosphere, oceans, biosphere, and crustal rocks over hundreds of millions of years, at any given time almost all of it occurs in the last of these reservoirs.**

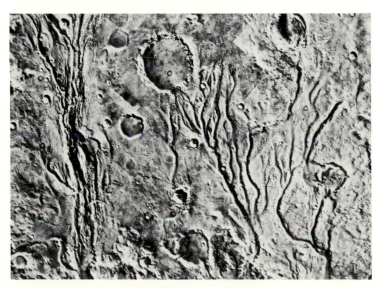

Fig. 9. **A Viking photograph of channels on Mars. Bearing strong resemblance to Earth's river beds, these features may have been carved by running water, even though no liquid water now exists anywhere on the Martian surface.**

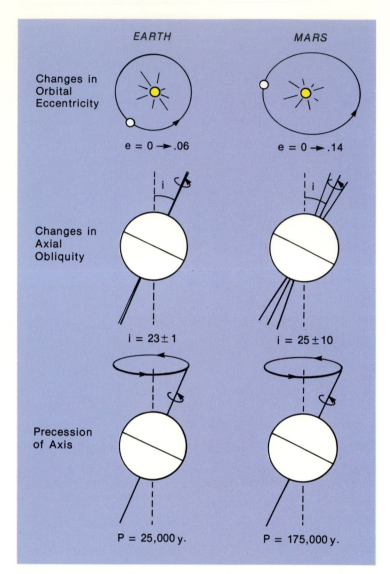

EARTH   MARS

Changes in
Orbital
Eccentricity

e = 0 → .06   e = 0 → .14

Changes in
Axial
Obliquity

i = 23 ± 1   i = 25 ± 10

Precession
of Axis

P = 25,000 y.   P = 175,000 y.

Fig. 10. This diagram illustrates quasi-periodic changes in the orbital and axial characteristics of Mars and Earth. Note how much greater the variations are (in most cases) for the latter. Such fluctuations are far smaller and less important for Venus.

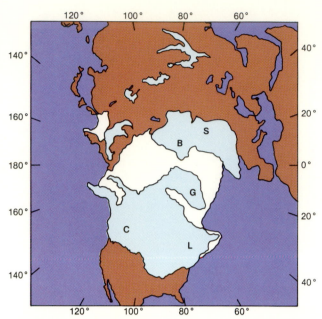

Fig. 11. The maximum coverage of the Earth's Northern Hemisphere by ice during the last million years. Continental ice sheets, indicated by colored areas, are: *B*, Barents Sea; *S*, Scandinavian; *G*, Greenland; *L*, Laurentide; and *C*, Cordilleran. Sea ice is indicated in white. It is apparent all of Canada and much of the northern United States were covered at some time. The last glacial maximum, which occurred only about 18,000 years ago, covered about 90 percent of the area mantled by ice here.

mate was more clement in its distant past (Fig. 9). Thus there is a paradoxical relationship between our knowledge of past climates and the expected effects of decreased solar output.

Since a solid foundation exists for the predicted temporal variation in the solar output, the resolution of this "solar luminosity" paradox lies in the consideration of other factors that could have counteracted the cooling effect of decreased solar output in the past. In particular, a stronger greenhouse effect at earlier epochs could have permitted the Earth and Mars to have moderate surface temperatures then. Since the early atmospheres of the planets may have been somewhat reducing, Carl Sagan and George Mullen have suggested that ammonia may produce such an effect. They point out that as little as 10 parts per million of ammonia would be necessary. However, practically no ammonia should have been outgassed by volcanoes into the early atmospheres. Moreover, ammonia is easily broken down by ultraviolet radiation into nitrogen and hydrogen,

and it is very difficult to recombine these two back into the parent molecule. Therefore, ammonia is probably not the agent involved in the early greenhouse enhancement.

A second and perhaps more plausible possibility assumes that elevated concentrations of carbon dioxide existed at that time. To warm Mars' surface above the freezing point of water (thus allowing the resulting liquid to flow across the landscape), I estimate that some 200 times more carbon dioxide needed to be present in the atmosphere. For the Earth, Robert Cess, Tobias Owen, and V. Ramanathan calculate that 10-100 times more CO_2 was necessary. Such increases are not unreasonable. As discussed above, the amount of carbon dioxide that was outgassed from Mars' interior over the planet's history has been estimated to be up to several hundred times the amount currently in its atmosphere. Furthermore, most of this gas came out during the early history of Mars. Probably, carbon dioxide in the early atmosphere of Mars was much more abundant than the amount observed today, with this excess gradually declining with time by incorporation into surface and subsurface reservoirs (such as carbonate rocks and the regolith). In fact, CO_2 accumulation in these reservoirs, especially in carbonate rocks, may have been greatly accelerated once the temperature rose above the freezing point of water. In other words, the factor that permitted moderate temperatures on Mars — elevated amounts of carbon dioxide gas — may have set the stage for its own demise. Unlike the situation for the Earth, almost all of the carbon dioxide that is lost to surface reservoirs on Mars remains there. It is not eventually recycled back into the atmosphere by reservoir burial and heating. As a result, the atmospheric pressure on Mars has probably declined steadily over most of its history.

Fig. 12. A detailed photograph of layered or laminated terrain near Mars' south pole, taken by the Mariner 9 spacecraft. This landform lies exposed on steep slopes, appearing as a series of stripes (due to illumination) that suggests terraced topography.

Over periods of hundreds of millions of years, carbon dioxide levels in the Earth's atmosphere may have been determined by a balance between its enrichment by volcanic outgassing and its depletion by the weathering of continental silicate rocks into carbonate rocks (Fig. 8). The continual shifting of large blocks of the Earth's surface (*plate tectonics*) eventually transports these carbonate rocks into our planet's hot interior, where they decompose and release carbon dioxide back into the atmosphere. Thus, CO_2 is not permanently lost by weathering. An enhanced level of carbon dioxide in the Earth's early atmosphere could have resulted from increased volcanic activity then, and from a negative feedback relationship between CO_2 concentration and temperature. As pointed out by James Walker and James Kasting, the weathering rate of rocks, and hence the rate of loss of carbon dioxide from the atmosphere, may decrease with decreasing temperature.

Due to gravitational forces exerted by other planets, the Moon, and the Sun, the orbital eccentricity of Mars and Earth (a measure of the deviation of their orbits from perfect circles) and the tilt or obliquity of their axes of rotation (with respect to a line perpendicular to their orbital planes) undergo quasi-periodic variations. These variations occur on time scales of 10,000 to 100,000 years for the Earth and 100,000 to 1,000,000 years for Mars, with the variation of eccentricity and obliquity being considerably larger for Mars. In addition, both have rotational axes that precess, or revolve, continuously — much like a spinning top (Fig. 10).

These changes in orbital and axial characteristics result in variations in the seasonal and latitudinal distribution of sunlight, which can lead to climatic changes. For example, the polar regions receive more sunlight at large obliquities than at low ones. Also, large eccentricities increase the sunlight falling on one hemisphere during summer. Studies of sedimentary deposits in the Earth's oceans show that the succession of ice ages and ice-free epochs over the last million years has occurred with discrete frequencies consistent with these astronomical variations, which, therefore, have probably played a key role in causing these recent dramatic changes in climate. During ice ages, giant ice sheets form at subpolar latitudes on continents of the Northern Hemisphere and expand to mid-latitudes (Fig. 11). Because only Antarctica is situated at high southern latitudes and has always been covered with snow over the last tens of millions of years, there has been little change in land area covered by ice in the Southern Hemisphere. When the obliquity is small, high latitudes receive less sunlight over a year. Moreover, if summer in the Northern Hemisphere occurs when the Earth is farthest from the Sun, less sunlight is available during the northern summer to dissipate snow that fell during the previous winter. These two conditions, occurring simultaneously, tend to initiate ice ages.

Prior to several million years ago, no major ice advances occurred in the Northern Hemisphere for almost three hundred million years, even though the astronomical cycles maintained similar amplitudes over that entire period. Thus, they are not the only stimuli to the icing-over of our planet. Another factor might be *continental drift*, the slow shifting of Earth's continents relative to the poles and to each other in response to internal forces. The positioning of land masses near the poles, such as Antarctica and Greenland, permits the formation of permanent ice covers upon them. This may be one condition necessary for the onset of an ice age, for the high reflectivity of ice decreases the amount of incident sunlight that can be absorbed by the ground and thus makes for cooler temperatures. Also, some continents may have to be situated at subpolar and middle latitudes to permit the formation of large ice sheets.

The presence of layered deposits in the polar regions of Mars (Fig. 12) implies that periodic climatic changes have also occurred there. Since the astronomical variations are much larger for Mars than for the Earth, they are a likely cause of these changes. The polar layered deposits or "laminated terrain" may reflect quasi-periodic fluctuations in the deposition of dust, water ice, and carbon dioxide ice in the polar regions of Mars, with the first two materials constituting the permanent deposits of the laminated terrain. As an example of how astronomical variations can influence the rate of deposition of these materials, consider the effect of changing obliquity on dust deposition. During times of high obliquity, latitudes near the poles experience higher temperatures. As a result, less carbon dioxide is adsorbed onto the surfaces of particles in the subpolar regolith, and the atmospheric pressure increases. So lighter winds are needed to set particles into motion, and a greater dust loading in the atmosphere results. Hence, at times of high obliquity, the dust deposition rate in the polar regions may increase. Similarly, at times of low obliquity the dust deposition rate may decrease and, in fact, even vanish, since supersonic winds (a very unlikely possibility) may then be required to set sand into motion.

The layered deposits occur only in the polar regions of Mars because water ice, a key ingredient, is stable only at high latitudes. Water ice helps to cement the dust particles together and thus stabilizes the deposits against erosion. Astronomical variations also modulate the amount of water ice deposited in the polar regions. The layered appearance of the laminated terrain may reflect different proportions of dust and ice (or their deposition rates) over the period of the astronomical changes.

Because its axis of rotation is almost perpendicular to its orbital plane, and because its atmosphere has a large heat capacity, Venus does not experience any appreciable seasonal change in climate. Hence, orbital variations have little effect on Venus' climate.

So far we have discussed climatic changes for which both the cause and the effect operate over extended periods of time and vary in a continuous fashion. There are, however, certain singular events that happen very abruptly and that may result in a short-lived, but catastrophic change. These events are collisions with the largest of the stray bodies — Apollo objects — that cross the orbital plane of the Earth (see Chapters 4 and 10). There is some evidence that such collisions may have set in motion events that abruptly terminated the presence of many life forms on the Earth, such as the extinction of the dinosaurs and — more profoundly — 75 percent of all species that existed on the Earth 65 million years ago.

The Apollo objects are small bodies that range in size from a fraction of a kilometer to about 10 km. It is inevitable that some of them will collide with the Earth. A collision with an Apollo object results in the formation of a crater whose diameter and depth are about 20 and 5 times larger than the diameter of the object, respectively. The object is almost totally vaporized and about 100 times its mass is ejected from the crater. Some of the material, particularly very fine grains, are thrown into the stratosphere, where they may remain for weeks to years.

Luis and Walter Alvarez and their colleagues have obtained evidence that the extinction of the dinosaurs and many other species 65 million years ago coincided precisely with the time the Earth was struck by a giant Apollo object. The evidence was obtained by analyzing the chemical composition of sedimentary layers that were formed around the time of this extinction event. They studied the abundance of the metal iridium, since the Earth's crust has about ten thousand times less of it than do most meteorites. Thus, an enhancement of iridium in a section of the sediment provides a marker for the presence of extraterrestrial material. Exactly at the time of the extinction event, they found a very sizeable increase in the iridium content of sedimentary layers from several diverse locations and suggested that this enrichment was due to the impact of a very large Apollo object. If so, the extinction could have been caused by the large amount of very fine debris that was thrown into the stratosphere. For a period of months, much less sunlight would have reached the surface of the Earth. As a result, photosynthesis may have been suddenly turned off over the entire world. With no new food produced for this period, many marine species (for which the carbon cycle has a relatively short duration) that depend directly or indirectly on the plants would have perished along with many plants. The suspended dust may also have blocked sunlight so effectively as to sharply lower temperatures on the continents and for a time to prevent animals from seeing. These two factors may have contributed to the extinctions of land animals, including the dinosaurs.

A QUICK RECAP

Twenty years ago, our knowledge about the atmospheres of the other terrestrial planets was rudimentary. For example, while carbon dioxide had been detected in the atmospheres of Venus and Mars from ground-based telescopic observations, it was not known that this gas was the major constituent of these atmospheres. And it was not clear whether Mercury had an appreciable atmosphere. As a result of spacecraft missions to these planets and sophisticated ground-based studies, we now have a much more detailed, if still incomplete, knowledge of these other atmospheres, and attempts can be made to examine the ways in which common processes have affected their character.

In this chapter, we have illustrated the power of comparative planetology for answering some very fundamental questions. For example, comparisons of the abundances of rare gases in the atmospheres of Venus, Earth, and Mars have indicated that these atmospheres resulted chiefly from the outgassing of volatiles contained in the material that formed these planets. Due to a poorer endowment of volatiles and a less efficient outgassing of them, Mercury has been unable to make up for the loss of gases to space; its atmosphere is negligible at present. The dating of lunar rocks has shown that the outer layers of the Moon were quite hot early in its history. By analogy, the atmospheres of the terrestrial planets came largely into existence at an early time. These early atmospheres were probably mildly reducing, thus providing the raw ingredients for the formation of life on the Earth. As time passed, all three atmospheres evolved toward a more oxidized state, as hydrogen escaped to space. However, the presence of life on the Earth, in particular oxygen-generating plants, led to the rise of large amounts of free oxygen in our atmosphere, beginning some two billion years ago; as a by-product of this rise of oxygen, ozone accumulated in the stratosphere and shielded the surface from biologically damaging solar ultraviolet radiation.

Due to solar, astronomical, and atmospheric variations, the climates of Venus, Earth, and Mars have undergone large changes over time. A reduced solar output during the early history of the solar system may have permitted a much milder climate on Venus. However, as the Sun's luminosity gradually increased, a runaway greenhouse occurred, leading to very high surface temperatures and the emplacement of almost all of Venus' outgassed volatiles into its atmosphere. Despite the early, low solar luminosity, the Earth and, at times, Mars had benign climates, probably due to an enhanced greenhouse effect. Substantially elevated amounts of carbon dioxide may have played a key role in this regard. Quasi-periodic changes in orbital and axial characteristics caused both the Earth and Mars to experience fluctuating climates in recent times, with the Earth undergoing a succession of ice ages and ice-free periods and Mars having large changes in its seasonal cycles of dust, water, and carbon dioxide, which led to the generation of polar sedimentary layers. Collisions with large Apollo objects may be responsible for sudden extinction events that characterize the record of life on the Earth.

We have now reached the point where the activities of the most advanced life form on our planet — human beings — represent potential hazards to our own climate. It is perhaps fortunate that this coincides with our first overtures to explore the solar system. The histories of these other bodies offer forceful reminders of how drastically planetary atmospheres and climate can change. These other atmospheres also constitute huge natural laboratories that can be used to test our understanding of the basic processes at work in our atmosphere and our ability to predict the changes that could affect this oasis of the solar system.

The Moon 7

Bevan M. French

In 1969 over half a billion people witnessed the "impossible" coming true as Earthlings first walked on the surface of the Moon. For the next three years, people of many nations watched as one of the great explorations of human history unfolded on their television screens. Between 1969 and 1972, supported by thousands of scientists and engineers back on Earth, 12 astronauts explored the lunar surface. Protected against the airlessness and the killing heat of the lunar environment, they stayed for days, and some of them traveled many kilometers across its surface in Lunar Rovers. They made scientific observations and set up instruments to probe the interior of the Moon. They collected thousands of samples of lunar rocks and soil, thus beginning our first attempt to decipher the origin and geological history of another world from actual samples of its crust (Fig. 1).

The initial excitement of success and discovery has passed. But here on Earth, lunar scientists are still unraveling the immense treasure of new knowledge returned by the Apollo program. We now have rocks collected from nine different places on the Moon (Fig. 2). The six Apollo landings returned a collection weighing 382 kg and consisting of more than 2,000 separate samples. Three automated Soviet spacecraft in the Luna series also landed on the Moon and returned with small but important samples totaling about 310 grams (about 11 ounces) from the eastern edge of the Moon. Luna 24, the most recent of these (August, 1976), drilled into the lunar surface and brought back a core of material 160 cm long.

Instruments placed on the Moon by the Apollo astronauts operated for as long as eight years. They detected moonquakes and meteorite impacts, measured the Moon's motions, and recorded the heat flowing out from the interior. Cameras on Apollo spacecraft obtained so many accurate photographs that we now have better maps of parts of the Moon than we do for some areas on Earth. Special detectors, mounted near the cameras, measured weak X-rays and radioactivity emanating from the lunar surface. From these measurements, we have been able to determine the chemical composition of about one-quarter of the Moon's surface, an area the size of the United States and Mexico combined. By comparing data acquired in orbit with analyses of the returned samples, we can draw conclusions about the composition and nature of large areas of the Moon.

Thus, in less than a decade, science from the Apollo Program has changed our natural satellite from an unknown and unreachable object into a familiar world. The following paragraphs describe the principal scientific results from this remarkably productive period of lunar exploration.

THE NATURE OF THE MOON

The Lunar Surface. Long before the space program, since the time of Galileo, observers could see that the Moon's surface was complex. Using earth-based telescopes, they distinguished the level maria and the rugged highlands. They recognized circular craters, stark mountain ranges, and deep winding canyons or rilles.

All of these lunar landscapes are covered by a layer of fine

Fig. 1. Apollo 17 astronaut Harrison ("Jack") Schmitt uses a special rake to collect centimeter-sized rock chips from the Moon's Taurus-Littrow Valley. Schmitt, before his election to the U. S. Senate, earned distinction as the only geologist to roam the lunar surface.

powder and broken-up rubble about 1 to 20 m deep. This layer (the *regolith*) is usually called the lunar soil, although it contains no water or organic material, and it is totally different from soil formed on Earth by the action of wind, water, and life.

The lunar soil could only have formed on the surface of an airless body. It built up over billions of years of continuous bombardment by large and small meteorites, most of which are so small they would have burned up if they had entered the Earth's atmosphere.

These meteorites form craters when they hit the Moon. Tiny particles of cosmic dust produce microscopic craters perhaps one micron across (Fig. 3), while the much rarer impact of a large body may blast out a crater many kilometers, or miles, in diameter (Fig. 4). Each of these impacts shatters solid rock, scatters materials around the crater, and stirs the soil (a process called *gardening*). As a result, the regolith is a well-mixed sample of a large surface region. Individual samples of lunar soil have yielded rock fragments that originated hundreds of kilometers from the collection site.

However, the lunar soil is more than ground-up and reworked rock. As the boundary layer between the Moon and outer space, it absorbs matter and energy coming from the Sun and the rest of the universe. Tiny bits of cosmic dust and high-energy atomic particles that would be stopped high in the Earth's protective atmosphere rain continually onto the lunar surface, where they become trapped and preserved in the soil.

The Moon's Composition. Before the first Moon rocks were collected, we could perform laboratory tests on only two types of solar system bodies: our own planet Earth and the meteorites that occasionally fall to Earth from interplanetary space. Now we realize that the Moon is chemically different from both the Earth and meteorites, although it is more like the former.

Rocks that make up the Moon are similar enough to their terrestrial counterparts that we can use the same terms to describe both groups. No sedimentary rocks, like limestone or shale (which are deposited in water), were found on the Moon. Instead, the lunar specimens are all igneous, formed by the cooling of molten lava. They consist of the same minerals that comprise terrestrial lavas — silicates like pyroxene, olivine, feldspar, and tridymite, and the iron-titanium oxide ilmenite. Three new minerals, never found on Earth, occur in small amounts within lunar rocks: tranquilityite (for the Apollo 11 site), armalcolite (for the three Apollo 11 astronauts), and pyroxferroite.

The *maria* (dark regions that suggest features like "The Man in the Moon") are low, level areas covered with layers of basaltic lava (Fig. 5), a rock similar to the lavas that erupt from terrestrial volcanoes in Hawaii, Iceland, and elsewhere. The light-colored parts of the Moon (the *highlands*) are higher, more rugged regions that are older than the maria. These latter areas are made up of several different kinds of rocks that cooled slowly deep within the Moon. Again using terrestrial terms, we call these rocks gabbro, norite, and anorthosite. They tend to contain more of the mineral feldspar (a calcium-aluminum silicate) and less ilmenite than do maria lavas.

Moon rocks and Earth rocks possess some basic differences, however. It is easy to tell them apart by chemical analysis or by examining them under a microscope (Figs. 6 and 7). The most obvious difference is that lunar rocks have no water at all (while almost all terrestrial samples contain at least a percent or two of water). They are therefore very well preserved, because their minerals never reacted with water to form clays or rust. The crystals in a 3½-billion-year-old Moon rock look fresher than those from a water-bearing lava just erupted from a terrestrial volcano.

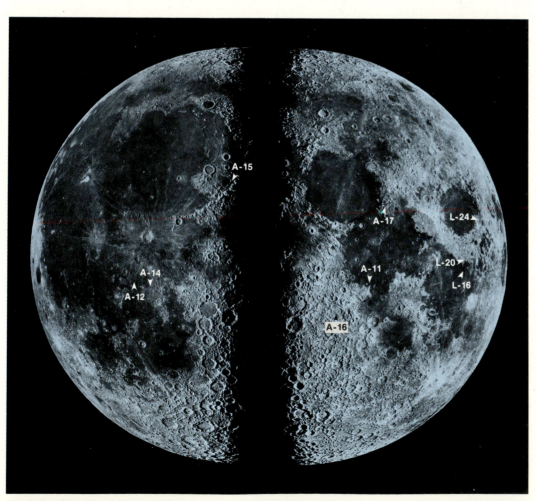

Fig. 2. Landing sites of the nine American (Apollo) and Soviet (Luna) sample-return missions are identified on a pair of Lick Observatory photographs taken in 1937.

Another important difference is that samples returned from the Moon formed where there was almost no free oxygen. As a result, some of the iron in these rocks was not oxidized when the lunar lavas formed; it still occurs as small metallic crystals (Fig. 8).

Because Moon rocks have never been exposed to water or oxygen, any contact with the Earth's atmosphere could "rust" them badly and destroy much of their scientific value. For this reason, the returned Apollo samples are carefully stored in a special Planetary Materials Laboratory at NASA's Johnson Space Center in Houston, Texas, seen in Fig. 9. The samples are kept in an atmosphere of dry nitrogen, from which all oxygen and water have been removed. The air in the laboratories is carefully filtered to remove all contaminating terrestrial dust. No more lunar material than necessary is sent out to scientists for analysis, because such material will then be exposed to normal terrestrial air.

Lunar rocks consist of the same chemical elements that

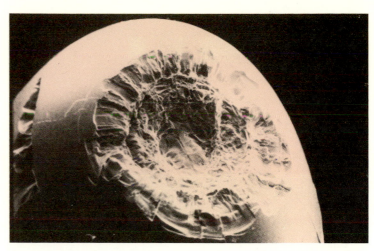

Fig. 3. Less than 1 mm wide, a spherule of glass from a lunar sample exhibits a tiny crater on its tip. Note the pattern of radial and concentric fractures, a miniature of that found in craters many kilometers across.

Fig. 4. Mare Orientale, as viewed by Lunar Orbiter 4. This 750-km feature, which from Earth appears along the Moon's limb, was formed some 4 billion years ago when an object perhaps 25 km across struck the lunar surface.

Fig. 5. This chunk of lunar lava, called a vesicular basalt, was collected by the Apollo 15 astronauts. As molten rock reached the surface, gas that had been dissolved in the melt under pressure formed the bubbles seen here, which froze in place when the rock cooled.

Fig. 6. Sifting through Apollo 17's orange soil, lunar scientists discovered these spherules and flecks of colored glass, some 0.3 mm across.

make up Earth rocks, although the proportions are slightly (but significantly) different. They contain more of the common elements calcium, aluminum, and titanium than do most rocks beneath us. Rarer elements like hafnium and zirconium, which melt at high temperatures, are also more plentiful in lunar rocks. However, other elements like sodium and potassium, which have low melting points, are scarce in lunar material. Because Moon rocks exhibit these differences among high- and low-temperature elements, we believe that the material that formed the Moon was once much hotter than material that formed the Earth. This heating, which may have occurred even before the material came together to form the Moon, boiled off the low-temperature elements and left the Moon enriched in the higher-temperature ones.

The chemical composition of lunar material also varies from place to place, through a sorting out of compounds called *differentiation,* that occurred within the early, partially molten Moon. Thus the rocks of the light-colored highlands are rich in calcium and aluminum, while the lavas of the dark-colored maria contain less of those elements and more titanium, iron, and magnesium.

The Moon's Interior. Even before the Apollo 11 landing, we had learned that the interior of our satellite was not uniform. Five unmanned Lunar Orbiter spacecraft, which photographed the Moon during the 1960s, detected concentrations of mass (*mascons*) beneath the more circular lunar maria (for example, Imbrium, Serenitatis, Crisium, and Nectaris). This extra mass may be associated with layers of volcanic lava that fill these basins. We now know that such lavas appeared three to four billion years ago. The fact that mascons are detectable today implies that the lunar interior remained both rigid and cool enough to support the extra mass of accumulated lava over long periods of time.

For nearly eight years, seismometers placed on the lunar surface by Apollo astronauts recorded the tiny vibrations caused by meteorite impacts on the surface and by small "moonquakes" deep beneath it. Just as on the Earth, these vibrations provide clues to the nature of the material through which they pass, so that scientists analyzing the records can determine what the Moon's interior is like.

We quickly learned that beneath its surface the Moon is quiet — at least in comparison with the Earth. The seismometers detected only about 3,000 tremors each year, while a similar instrument on the Earth would sense hundreds of thousands. All of these moonquakes were very weak by terrestrial standards, each releasing only about as much energy as a firecracker does. In total, our satellite generates less than one ten-billionth of the seismic energy released by Earth. Moonquakes occur at a depth of about 600-800 km, much deeper than almost all the quakes on our own planet. Certain kinds of moonquakes occur at about the same time every month, suggesting that they are triggered by repeated tidal strains as the Moon moves in its orbit around us.

A picture of the interior of the Moon (Fig. 10) has been slowly put together from thousands of recorded moonquakes, meteorite impacts, and the deliberate impacts of discarded Apollo rocket stages onto the Moon. The lunar interior is not uniform but divided into a series of concentric

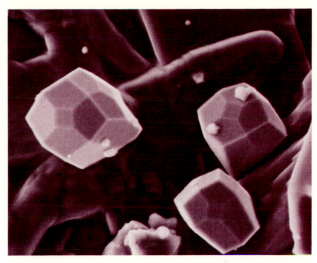

Fig. 8. Only a few thousandths of a millimeter wide, these gemlike crystals of metallic iron line the walls of a small cavity in a lunar sample. After condensing some 3 billion years ago from a hot vapor that contained no water or free oxygen, the fragile crystals remained unchanged on the Moon's airless surface.

layers just as the Earth is, although the layering is different. The outermost part of the Moon is a crust about 60 km thick, probably composed of calcium- and aluminum-rich rocks (gabbros and anorthosites) like those found in the highlands. Beneath this crust is a thick layer of denser rock (the mantle) which extends down to more than 800 km, the level at which moonquakes occur. The nature of this mantle is not known; scientists think it may be rich in olivine, $(Mg,Fe)_2SiO_4$. The outer part of the Moon is much colder, thicker, and more rigid than the outer part of the Earth. For this reason, the Moon shows no evidence of tectonic activity in its crust, in contrast to the horizontal motions of continental and oceanic plates that are the moving force in the geological activity of our own planet.

The deep interior of the Moon is still unknown. Some evidence suggests that the Moon may still be hot and even partly molten inside. It may also contain a small iron core at its center, just as the Earth does. But the lunar core, if it exists, is so small (less than about 1200 km in diameter) that it has not been detected by any of the Apollo investigations.

The Moon does not now have any detectable magnetic field as does the Earth, so the most baffling and unexpected result of the Apollo program was the discovery of preserved (remanent) magnetism in many of the old lunar rocks. The cause of this magnetism is still debated. One explanation is that the Moon once possessed a magnetic field, generated by a small iron core in its center, and that the field somehow disappeared after the old lunar rocks formed. Other suggestions include the generation of strong magnetic fields by large meteorite impacts or by the previous existence of a more active Sun.

One reason we have been able to learn so much about the Moon's interior is that the instruments placed on the lunar surface operated much longer than expected. Some of the instruments, originally designed for a one-year lifetime, actually operated from 1969 and 1971 until 1977, when the last surviving instruments were shut off. This extended operation provided information that we could not have ob-

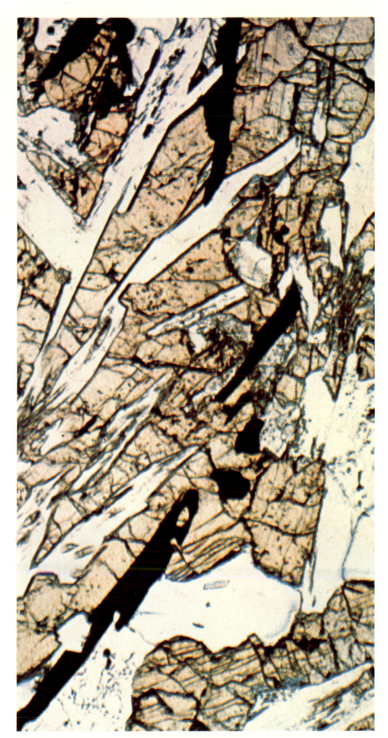

Fig. 7. By shining polarized light through paper-thin slices of a lunar rock, geologists can learn much about its crystal structure and mineral composition. Here a basalt from Apollo 12 exhibits crystals up to 1 mm across; their size and shape suggest rapid cooling typical of lavas emerging onto the surface. Three minerals can be identified: feldspar (clear), pyroxene (light brown), and titanium-rich ilmenite (black).

tained from shorter records. For example, the long lifetime of the heat-flow experiments set up by the Apollo 15 and 17 missions has made it possible to determine more accurately the amount of heat escaping from the Moon. This heat flow is a basic indicator of the temperature and composition of the lunar interior. The new value, 2 microwatts per square centimeter, is about two-thirds of the value calculated from earlier data and about one-third the amount of heat now coming out of the inside of the Earth. As a result, we can now produce better models of the lunar interior.

As explained previously, the seismometers used to monitor moonquakes also recorded vibrations from the impacts of small meteorites on the lunar surface. The Moon's quiet interior allowed these instruments to detect the arrival of a meteorite with a mass of only 10 kg (an object about the size of a grapefruit) — anywhere on the Moon! Our diehard detectors registered between 70 and 150 impacts each year, caused by meteorites of from 100 g to 1,000 kg. Smaller debris, down to the size of dust grains, strikes the Moon more frequently. Folding all of this information together, we have derived an accurate relationship between the current impact rate and the size of meteorites impinging on the Moon.

Moreover, some small meteorites seem to travel in groups. Several such swarms, composed of meteorites weighing a few kilograms each, struck the Moon in 1975. The detection of such events has provided us with new information on the distribution of meteorites and cosmic dust in the solar system.

The long lifetime of the Apollo instruments also made some cooperative projects possible. Our instruments were still making magnetic measurements at several Apollo landing sites in 1970 and 1973 when, elsewhere on the Moon, the Soviet Union landed similar instruments attached to automated lunar roving vehicles (Lunokhods). By making simultaneous measurements and exchanging data, American and Soviet lunar scientists obtained a better picture of the magnetic properties of the Moon.

Is there life on the Moon? Despite careful searching, nei-

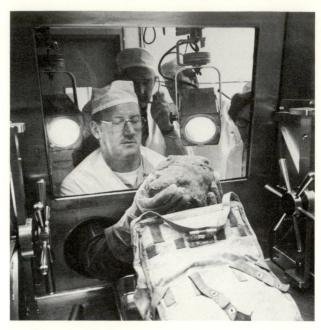

Fig. 9. A technician at NASA's Lunar Receiving Laboratory in Houston inspects "Big Mulie," collected by the crew of Apollo 16 during their first walk on the lunar surface.

ther living organisms nor fossil life have been found in any lunar samples. The lunar rocks are so barren of life that the quarantine period for returned astronauts was dropped after the third Apollo landing.

No water has been found on the Moon, either free or combined chemically in its rocks. Water is a substance that is necessary for life as we know it, making it very unlikely that life could ever have originated on the Moon. Furthermore, lunar rocks contain only tiny amounts (a few parts per million) of the carbon and carbon compounds essential to our life forms. Moreover, most of this carbon is not native to the Moon but is brought to the lunar surface in meteorites and as atoms blasted out of the Sun.

TIME AND THE MOON

Prior to the return of lunar samples, we had no means to determine the Moon's age and thus develop a clearer understanding of its evolution. The decay of natural radioactive elements like uranium, thorium, and potassium-40 gives us a series of clocks for measuring the age of a rock — or an entire planet — but this delicate analysis must be done in a laboratory here on the Earth. When performed on meteorites, such measurements indicate that the formation of the solar system took place about 4.6 billion years ago. Similar evidence in both lunar and terrestrial rocks suggests that the Earth and Moon may also have formed at that time. Yet the oldest known rocks on our planet are only 3.8 billion years old, perhaps because older rocks have been destroyed by the Earth's continuing volcanism, mountain-building, and erosion.

The Moon rocks fill in some of this time gap between the Earth's oldest preserved rocks and the formation of the solar system. Lavas from the dark maria are the Moon's youngest rocks, but they are as old as the oldest rocks found on Earth, with ages of 3.1 to 3.8 billion years. Rocks from lunar highlands are even older. Most highland samples have

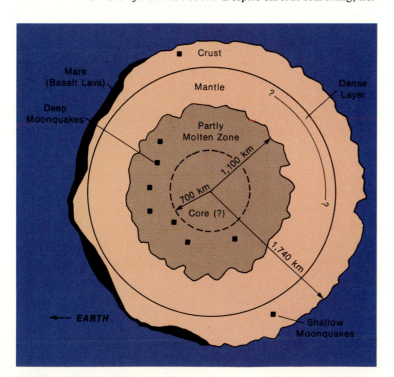

Fig. 10. A schematic diagram of the lunar interior. Apollo seismometers have identified both shallow and deep-seated moonquakes during many years of "listening." Note the concentration of mare material on the Moon's Earth-facing hemisphere.

ages of 4.0 to 4.3 billion years. Some Moon rocks preserve traces of even older lunar events. Studies of these specimens indicate that widespread melting and chemical separation took place in the lunar interior about 4.4 billion years ago, or not long after the Moon formed.

One of the techniques used to establish this early phase of lunar history is a new age-dating method (involving the elements neodymium and samarium) that did not exist when the first Apollo samples were returned in 1969. The combination of new instruments and careful protection of the lunar samples from contamination thus make it possible to achieve a more precise understanding of the early history of the Moon.

Even more exciting is the discovery that a few lunar rocks seem to record the body's actual formation. Some tiny green rock fragments collected by the Apollo 17 astronauts have yielded an apparent age of 4.6 billion years. Early in 1976, another Apollo 17 crystalline rock was found to have the same ancient age, and a few more of these primordial rocks have been identified since then (Fig. 11). Such pieces may be some of the first material that solidified from the once-molten Moon.

Aside from these revealing glimpses, the first few hundred million years of the Moon's lifetime were so violent that few traces of this era remain. Almost immediately after the Moon formed, its outer layers were completely melted to a depth of several hundred kilometers, either by the rapid infall of accreting planetesimals or by the rapid decay of a short-lived radionuclide such as aluminum-26. While this molten layer gradually cooled and solidified into different kinds of rocks, the Moon was bombarded by huge asteroids as well as smaller objects. Some of the asteroids must have been the size of small states, like Rhode Island or Delaware, and their impacts created huge basins hundreds of kilometers across, of which two called Imbrium and Orientale are the youngest and best preserved.

This catastrophic bombardment tapered off about 4 billion years ago, leaving the lunar highlands covered with huge overlapping craters and a deep layer of shattered and broken rock (Fig. 12). As the bombardment subsided, heat produced by the decay of long-lived radioactive elements like uranium and thorium began to melt the inside of the Moon about 200 km below its surface. Then, for the next half billion years (from about 3.8 to 3.1 billion years ago), great floods of lava rose from within and poured out over the surface, filling in the large impact basins to create the dark areas of the Moon seen from Earth today (Fig. 13).

As far as we know, our satellite has been quiet since those last lavas erupted more than three aeons ago. Since then, its surface has been altered only by occasional meteoritic impacts, by the atomic particles that stream out from the Sun and stars, and (more recently) by the tread of astronauts. Erosion on the Moon proceeds more slowly than anywhere on Earth; it requires tens of millions of years there to wear away a layer only one millimeter thick. For this reason, features formed on the Moon almost three billion years ago are well preserved today — if the astronauts had landed one billion years earlier, their view would have been much the same as it was during their actual visits. In fact, the lunar surface changes so slowly that footprints left by the Apollo astronauts will remain sharp for millions of years.

The Moon thus emerges as a kind of museum world, in

Fig. 11. By comparing traces of certain radioactive elements and their decay by-products, lunar scientists have estimated the age of this exquisite Apollo 17 sample to be 4.6 billion years. It may thus have been one of the first rocks to solidify when the Moon formed. Called a troctolite, it consists mainly of pale gray feldspar and yellow-brown olivine crystals.

Fig. 12. Typical of material covering the lunar highlands, this breccia could be aptly called the "rubble of ages," for it contains the crushed fragments of many different rocks. The sample shown here, collected by Apollo 16 astronauts, is about 4 billion years old; it records the Moon's early bombardment by innumerable chunks of interplanetary debris.

marked contrast to the ever-changing Earth. Our planet still behaves like a youngster: its interior has remained hot, powering the volcanism and mountain-building that have gone on continuously as far back as we can decipher the rocks. According to new geologic theories, even the present ocean basins are less than about 200 million years old, having formed by the slow separation of huge, moving plates that make up the Earth's crust.

Before we explored the Moon, there were three main suggestions to explain its existence: it formed near the Earth as a separate body; it separated from the Earth; and it formed somewhere else and has been captured by the Earth.

Lunar scientists still cannot decide among these three theories, but we now have a great deal of information against which all of the theories can be tested. We have learned that the Moon formed as a part of our solar system and that it has existed as an individual body for 4.6 billion years. If it did separate from the Earth, it must have done so right at the beginning of the solar system and not at any later time. Some scientists consider separation from the Earth unlikely because many basic differences in chemical composition exist between the two bodies, such as the apparent absence of water on the Moon. But the various theories are still evenly matched in their strengths and weaknesses. We will need more data and perhaps some new theories before the origin of the Moon is settled.

THE MOON AND THE SUN

When the astronauts collected lunar samples, they were also collecting solar samples. The airless surface of the Moon is continually exposed to the solar wind, a stream of atoms boiled into space from the Sun's atmosphere. Since the Moon formed, the lunar soil has trapped billions of tons of these atoms from the Sun. The soil also contains traces of cosmic rays produced outside our own solar system. These high-energy atoms, probably produced inside distant stars, leave permanent tracks when they strike particles in the lunar soil (Fig. 14).

By analyzing the soil samples returned from the Moon, investigators have been able to measure directly the chemical composition of matter ejected by the Sun and thus learn more about how the Sun operates. A major surprise was the discovery that the material in the solar wind is not the same as that in the Sun itself. The ratio of hydrogen to helium atoms in the solar wind reaching the Moon is about 20 to 1. But the ratio of these atoms in the Sun, as measured with earth-based instruments, is only 10 to 1. Some unexplained process in the Sun's outer atmosphere apparently operates to eject the lighter hydrogen atoms in preference to the heavier helium atoms.

Samples of lunar rocks and soil preserve material that came out of the Sun millions to billions of years ago. It is reassuring that the ancient Sun seems to have been much like the Sun of today (discussed in Chapter 2). The ratio of hydrogen to helium and the amount of iron in our solar "samples" show no change for at least the past few hundred thousand years. Moreover, present-day phenomena like the solar wind and solar flare eruptions apparently were occurring more than a billion years ago.

Yet lunar scientists find one puzzling difference between

Fig. 13a. The Moon, 3.9 billion years ago. Shortly after its formation, the Moon solidified and its surface began to record collisions with fragments of material left over from the accretion of the solar system. The largest of these impact features was the multiringed Imbrium Basin, shown in the upper left quadrant. The large, fresh-looking crater on the basin's floor exists today in ruined form as Sinus Iridum. At that time, most of the lunar surface was saturated with craters; portions of that ancient landscape remain today as the lunar highlands. This evolutionary sequence was prepared by Don Davis and Don Wilhelm, and was supplied by the U. S. Geological Survey.

Fig. 13b. The Moon, 3.1 billion years ago. Soon after the barrage of meteorites ended, deeply shattered rock beneath the impact sites allowed magma from the lunar interior to reach the surface and flood across the basin floors. Although most of the maria were formed in this manner during the 800 million years preceding this view, the youngest flows seen on the Moon today are probably only about 2½ billion years old. The maria appear dark because they contain more of such elements as iron and magnesium than do other surface rocks.

today's Sun and the one that shone on the Moon when it was young: the solar wind's nitrogen content has changed drastically during the past three aeons, both in total amount and in isotopic ratio. (Isotopes are atoms of the same chemical element which have different atomic weights.) Two kinds of solar-derived nitrogen, with atomic weights of 14 and 15, were analyzed from lunar rocks whose surface-exposure ages could be measured independently. The ratio of nitrogen-15 to nitrogen-14 has apparently increased by about 30 percent during the last three billion years. No mechanism has yet been found to explain such a significant change nor do we understand what other changes may have occurred during that period.

Clues to the ancient history of the Sun may soon emerge from ongoing work with lunar samples. During the Apollo 15, 16, and 17 missions, three long cores of lunar soil were obtained by driving hollow tubes into the regolith, penetrating as much as 3 m deep. The layers of soil in these cores contain a well-preserved history of the Sun-Moon relationship that may extend back for 1½ billion years. No single terrestrial sample contains such a long record, and no one knows how much can be learned when all three cores are carefully opened and studied.

REMAINING MYSTERIES

Despite the great scientific return from the Apollo program, many questions about the Moon remain unanswered.

Chemical Composition of the Whole Moon. For all we have learned, our conclusions about the global chemistry of the Moon are still mostly theories. Only nine lunar sites have

Fig. 14. A solar fingerprint. Only a few hundredths of a millimeter long, these tracks in a grain of lunar soil were created by fast-moving charged particles ejected from the Sun and other stars. The photograph is courtesy R. L. Fleischer, from *Science 167,* 1970, 568-71.

been sampled, six with Apollo and three with Soviet Luna spacecraft. The chemical analyses made from orbit cover only about a quarter of the lunar surface. We still know little about its far side and nothing whatever about the poles.

The Moon's Asymmetry. Orbiting Apollo spacecraft used a laser device to measure accurately the heights of features over much of the lunar surface. From these careful measurements, we discovered that the Moon is not a perfect sphere. It is slightly egg-shaped, and the small end of the egg points toward the Earth.

There are other differences between the front and back hemispheres of the Moon. The front (Earth-facing) side is covered with large dark maria, but the far side is almost entirely composed of light-colored, rugged, and heavily cratered terrain identical to the highland regions on the front side, with only a few patches of mare material. The Moon's upper crust is also uneven. On the side facing us, where most maria occur, the lunar crust is about 60 km thick, but on the back side it swells to a thickness of more than 100 km (Fig. 10).

We still lack good explanations for these observations. Perhaps some connection exists between Earth's gravity and the Moon's synchronous rotation in its orbit. Perhaps lava erupted primarily on the front side because the crust was thinner there or because more heat-generating radionuclides collected in that hemisphere. These differences could tell us a great deal about the Moon's early years, if we could understand them.

Questions for Future Lunar Missions. Is the Moon's interior molten? We know that there were great volcanic

Fig. 13c. The Moon today. Many bright, rayed craters like Copernicus and Tycho have appeared since the formation of the maria. Others, such as Eratosthenes, predated the final flows, and portions of their ray systems now lie buried under lava. Moreover, the rays become gradually less visible as new impacts churn the surface. Note that large areas of the maria have lightened considerably, principally due to overlapping mantles of ejecta from fresh craters. Barring some unforeseen cataclysm, the Moon will retain its present appearance for millions of years.

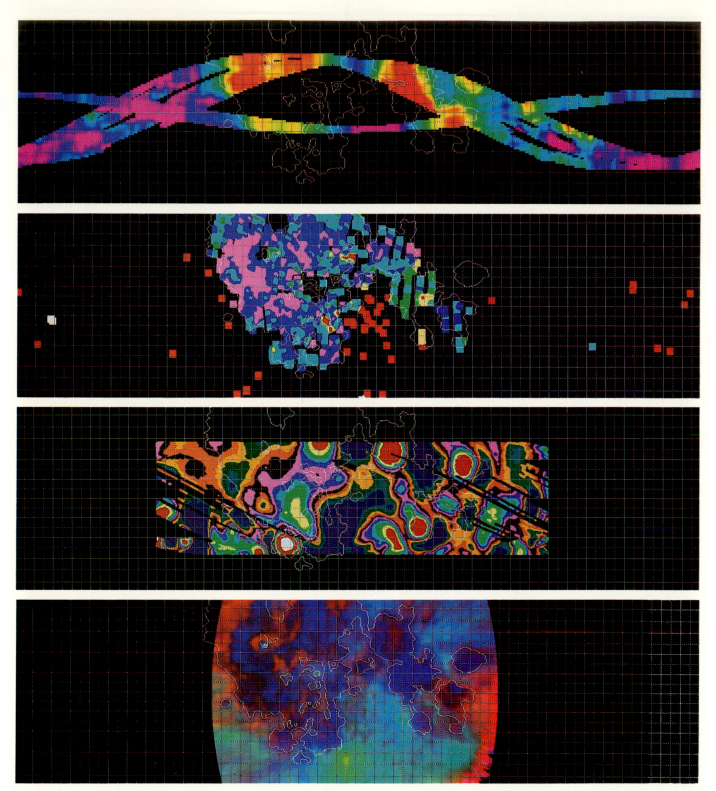

Fig. 16a. While orbiting the Moon, spectrometers aboard Apollos 15 and 16 measured the gamma-ray flux from the lunar surface. The color scale grades from pink through blue to orange-red, corresponding to radiation counts of from about 19 to 21 "hits" per second. Higher values, mostly in the maria, indicate concentrations of iron and titanium. **Fig. 16b.** Relative ages of the lunar maria and selected highland regions, based on measurements of crater freshness (such as rim slopes and erosion features). By matching these data with Apollo samples, the following absolute time scale has been determined: pink, younger than 3.0 billion years; dark blue, 3.0-3.3; turquoise, 3.3-3.6; green, 3.6-3.75; yellow, 3.75-3.85; red, older than 3.85.   **Fig. 16c.** Computer analysis of tiny velocity changes experienced by Apollo 15 in its lunar orbit has yielded this map of gravity anomalies on the Moon. Contours in dark purple to brown indicate below-average values, while pink through red and white show those above average. Note that the strongest fields occur in circular maria, revealing the presence of mass concentrations or *mascons*. **Fig. 16d.** The color of the lunar surface has been exaggerated tremendously in this computer rendition of a ground-based photograph. Some correspondence is evident between certain hues and specific features like the maria. These four maps, the work of some 30 lunar scientists, were supplied by the U. S. Geological Survey, courtesy Laurence Soderblom. Mercator projection is used, and the smallest grid rectangles are five degrees on a side.

eruptions on the Moon billions of years ago, but how long did they continue? To understand lunar history completely, we need to find out if the inside of the Moon is still hot and partly molten. More information about the heat flow coming out of the Moon may help provide an answer.

Does the Moon have an iron core? This question is critical to solving the puzzle of ancient lunar magnetism. At the moment, we have so little data that we can neither rule out the existence of a small iron core nor prove that one is present. If we can determine more accurately the nature of the Moon's interior and make more measurements of the magnetism on the lunar surface, we may find the answer to this intriguing question.

How old are the youngest lunar rocks? Samples collected from the Moon solidified as little as 3.1 billion years ago. We cannot determine our satellite's complete thermal history until we know whether these eruptions were the last or whether volcanic activity continued for a much longer time. Studies of photographs and crater densities suggest that lavas only 2 billion years old may exist in the Marius Hills and in western Oceanus Procellarum. Until samples are returned from this region and their ages measured, this part of the Moon's history will remain concealed.

Is the Moon now really "dead"? Unexplained occurrences of reddish glows, clouds, and mists on the lunar surface have been reported for over 300 years. These "lunar transient events," as they are called, remain unexplained and continue to be observed. It is important to determine their origins; if they truly exist, they may indicate regions where gases and other materials are still escaping from inside the Moon.

THE ONCE AND FUTURE MOON

Not so long ago, the Moon was an inaccessible light in the sky. Now it has become a landscape, a place to visit, a new world to explore. It has become a planet in its own right, with a character and history of its own. Similar to the Earth in some ways, the Moon is strikingly different from it in others. Instead of being a living, active planet, the Moon is a museum world in which are preserved the secrets of the origin and the development of planets. The Moon is also a space probe that has sampled the Sun and the rest of the universe for billions of years, allowing us to examine their histories as well.

We have only just begun to learn about our natural satellite. Some of the most important and exciting lunar samples remain to be studied in detail, and data from the Apollo instruments are still helping to refine our theories and models. Several laser reflectors placed on the Moon by astronauts allow us to measure variations in the Earth-Moon distance (about 400,000 km) with an accuracy of a few centimeters. This is accomplished by timing the round-trip journey of a laser beam fired from the Earth and bounced off these reflectors (Fig. 15). From such measurements, repeated over years, we learn more about the motions and inner structure of the Moon.

These data may also help us to monitor the slow movements of the Earth's crustal plates, thus using the Moon as a benchmark to study our own planet. We have plenty of time; the laser reflectors, which need no power, will work for centuries, perhaps millenia, before being slowly covered by a layer of lunar dust.

The exploration of the Moon has given us new abilities and new confidence for the future. We know that we can return there any time we wish, to collect, preserve, and analyze more of its rocks and to answer the many questions that remain. We can even send machines back to the Moon in place of astronauts. The Lunar Polar Orbiter mission, once considered by NASA and now being actively studied by the European Space Agency, could circle the Moon from pole to pole, affording complete global data on its chemical composition, radioactivity, gravity, and magnetism. Such a spacecraft could determine heat flow from orbit and even search for deposits of water, perhaps still present as ice in the permanently shadowed regions of the lunar poles.

Fig. 15. The Moon, seen here during a 1979 eclipse, is little more than a one-second trip away for this laser pulse fired from Lure Observatory, on the island of Maui, Hawaii. Its target is a panel of high-efficiency reflectors left on the lunar surface by Apollo astronauts. Photograph by Paul Ely, Institute for Astronomy, University of Hawaii.

Fig. 16. The full Moon rises over a cloud-veiled Earth in a photograph by the crew of Skylab 3.

Other robot spacecraft, like the Soviet Luna landers, could return samples from areas never before visited: the far side, the poles, or the sites of the puzzling transient events.

In the future, the Moon could become more than a place to visit (Fig. 16). Our successful ventures into space have generated increasingly serious discussions of ideas that, only a few years ago, were found only in science fiction —

permanent bases on the Moon, lunar telescopes to probe deeper into the universe, even mining and processing of the regolith to support space-based industry. The concept of using the Moon's unique environment — for science, exploration, and industry — is now much more realistic than the notion of simply *getting* there seemed little more than a generation ago.

8 Mars

Harold Masursky

Of all the planets, Mars has had the greatest hold on modern imaginations. Early this century, Percival Lowell's dogged observation of this planet from Flagstaff, Arizona, revealed light and dark markings resembling those described by Giovanni Schiaparelli a few decades earlier. Schiaparelli called what he had seen *canali,* an Italian word for channels, but Lowell was convinced that the features were the artifacts of an advanced Martian civilization. He believed the planet's inhabitants constructed the channels, which he called canals, to irrigate their desert plains using water from Mars' polar caps. Lowell's canals became the object of widespread attention and controversy, and Edgar Rice Burrough's stories of John Carter and his adventures on Mars did much to capture the popular imagination. Until spacecraft first visited the red planet, the idea of a Martian civilization remained alive in science-fiction stories, broadcasts, and films.

The first close reconnaissance of Mars, by the Mariner 4 spacecraft in 1965, signaled the start of a barrage of investigative space expeditions that has been exceeded in scope and intensity only by the exploration of the Moon and Earth. Mariner 4's 22 low-resolution images revealed a Mars very much like the Moon, with a bleak and cratered landscape. In 1969, Mariners 6 and 7 flew past the planet and relayed hundreds of photographs that seemed to confirm the dead, cratered, Moonlike character of the Martian surface. Mariner 9 assumed an orbit around Mars in 1971 for the purpose of photographically mapping the entire planet. After waiting out a months-long global dust storm that preceded its arrival, the spacecraft sent back images of craters, to be sure, but also of great volcanoes, enormous canyons, sand dunes, channels bearing a striking resemblance to Earth's riverbeds, and intricately layered deposits at the poles.

As Mariner 9 proceeded to record surface detail at a resolution of 0.1-1 km, our image of a cratered, static Mars was replaced by one of a dynamic and fascinatingly complex world that called out for further exploration. To satisfy our collective curiosity, four Viking spacecraft arrived at Mars in 1976 (Fig. 1) to scrutinize the surface in even greater detail, both from orbit and on the ground, and to seek an answer concerning the centuries-old question of whether life existed there (see Chapter 9).

All the Viking spacecraft worked well, long past their expected three months. The Viking 1 orbiter continued operations for four years, and the second orbiter for two years; together these spacecraft collected more than 55,000 images before depleting the pressurized gas used to maintain their orientation in orbit. The Viking 1 lander continues

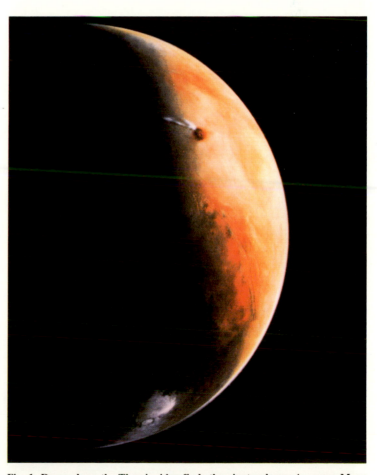

Fig. 1. Dawn along the Tharsis ridge finds the giant volcano Ascreaus Mons tufted with plumelike water-ice clouds on its western flank. Part of the vast Valles Marineris canyon system is seen at center, and near the bottom lies Argyre, a large impact basin here dusted with frost. This Viking 2 photograph was taken in August, 1976.

Fig. 2. Two weeks after landing on Mars in July, 1976, the Viking 1 lander returned this spectacular 100-degree panorama from its touchdown site in western Chryse Planitia. Early-morning sunlight backlights most of the scene, delineating drifts of fine-grained sediments and shadowing a pair of large rocks at left nicknamed ''Big Joe'' by mission scientists. Just to the right of the lander's white meteorology boom is a drift that has been scoured to the point where internal stratification shows. The rocky area at right may contain exposed patches of bedrock. Big Joe, about 2 m across and 8 m from the camera, appears below in another image taken some eight months later. The surface color results from oxidized iron, and dust suspended in the atmosphere lends a peach hue to the sky.

to operate and may well monitor the weather and surface changes on Mars until the early 1990's. Lander 1 set down in a region known as Chryse Planitia (22°.27 N, 47°.97 W) on July 20, 1976, and lander 2 arrived at Utopia Planitia (47°.57 N, 225°.74 W) on September 3rd. From these locations, the spacecraft recorded more than 1,400 photographs and large quantities of other data concerning the character, composition, and organic content of the Martian surface, and the composition and climatic patterns of the lower atmosphere.

The first lander sits in a moderately cratered, low-lying plain near the mouth of a large outflow channel. Craters up to 600 m across are visible near the horizon; these are probably the source of rocks littering the surrounding scene. Rocks near the landing site occur in a variety of sizes and colors. Some bedrock may also be exposed, and fine-grained material occasionally collects into large drifts (Fig. 2). Lander 2, situated 1,500 km farther north, is in a

Fig. 3. Thousands of rocks litter the Viking 2 landing site in Mars' Utopia Planitia. Their porous texture may be the result of gas bubbles formed as the rocks cooled or of erosion by windblown dust. The largest boulder, near the center of the mosaic, measures about 0.6 by 0.3 meter. A narrow trough, running from upper left to lower right, is part of a polygonal network that resembles ''patterned ground'' observed in Earth's polar regions. The horizon appears tilted because the spacecraft landed with one footpad on a rock and sits inclined 8° from level ground.

region of fractured plains about 200 km south of the large crater Mie; it probably sits atop a lobe of the crater's ejecta blanket. This site is very flat and sparsely cratered (Fig. 3), subdivided by polygonal fractures and troughs that criss-cross the area. While no drifts are visible and rocks cover a higher proportion of the surface than at the other site, rocks surrounding the Viking 2 lander look remarkably similar to those near lander 1, with numerous pits giving them a spongy appearance. The pits may be vesicles (holes once filled by bubbles of gas in cooling rocks) or the erosional consequence of scouring and plucking by wind-blown dust. Yet subtle differences distinguish the two sites, and they may have experienced dramatically contrasting geologic histories. During the course of a full Martian year (a little less than two Earth years), the landers recorded only slight variations in the brightness and color of the surrounding surface, which can be explained by the deposition and removal of dust and, for lander 2, frost (Fig. 4).

Five experiments sampled the soil within reach of a retractable arm, a little more than 3 m long, aboard each spacecraft. The arms themselves were used to test the properties of rocks and soils, which ranged in character from loose piles of drift material near lander 1 to rocks at both sites that were too hard to be chipped or scratched. Three biological investigations and a gas-chromatograph/mass-spectrometer searched for life forms on the planet; the results are described in the next chapter.

Most of the data concerning the composition of the surface came from X-ray fluorescence spectrometers, which showed that small particles were remarkably similar at both sites. Silicon and iron account for about two-thirds of the samples' content. Moreover, surprisingly high concentrations of sulfur were found — 100 times more than in the Earth's crust — while potassium occurs at only one-fifth of the proportion in terrestrial rocks. Magnets on the sampling arms picked up small collections of particles probably composed of the iron-rich mineral maghemite. In general, the surface material at both sites can be described best as iron-rich clays — apparently the rusty color of Mars is caused by just that: rust.

Both landers monitored the weather (atmospheric pressure, temperature, wind velocity, and direction) for more than a full Martian year. Temperatures ranged from a low of 150° K at the location of the more northerly lander 2 to a high above 250° K at the first lander. On any given day, the diurnal variation at each site spanned 35-50° K, although dust storms tended to moderate temperatures. In the Martian northern summer, atmospheric pressure dropped toward its minimum value because frost condensed on the southern winter polar cap and in doing so withdrew gases from the atmosphere. At such times, the weather near the spacecraft became rather monotonous: diurnal variations were very consistent, and winds averaged a mere 1-2 m per second. But toward autumn, day-to-day variations increased (especially at the second lander), and a regular sequence of cyclones and anticyclones passed over the more northerly site at a rate of about once per week. Some of these moving systems may be associated with fronts such as the so-called north polar hood (a seasonal haze of suspended CO_2 ice crystals).

A major contribution of the Viking orbiters was mapping the entire planet at a resolution of 150-300 m and selected

Fig. 4. Patches of frost surround the Viking 2 lander in this view taken late in the spacecraft's first Martian winter. The frost probably arose from water vapor transported to the site during dust storms; dust-water aggregates precipitated when carbon dioxide froze out of the atmosphere and adhered to them. Here the frost layer is perhaps no more than a few hundredths of a millimeter thick.

areas at 8 m — bettering the Mariner 9 coverage by a factor of 12. They found that the residual north polar cap consists of water ice and the south polar cap probably frozen carbon dioxide. Moreover, the orbiters made a detailed determination of the composition of Mars' thin atmosphere.

MARS' WATER ENIGMA

Although the outer-planet satellites are composed largely or entirely of water ice, Mars is the only known planet other than the Earth where water has flowed on the surface at times in the past. Because it is essential for life and is involved in many geologic processes, water provides a focus for examining the history of the surface of Mars. Here we will examine its numerous interactions with the rocks and atmosphere. On Mars, water and wind have modified the basic rock-forming processes driven by heat from the planet's interior and have altered the cratering record left by impacting cosmic debris.

Where did the water come from? Scientists believe that the early atmospheres of the inner planets were lost during a brief, intensely energetic phase of the Sun's development. Water and carbon dioxide are the most abundant gases

Fig. 5. The huge shield volcano Olympus Mons, together with its surrounding aureole of fractured and eroded rock, is seen in this Viking mosaic prepared by the U. S. Geological Survey. Clearly visible are the volcano's complex summit caldera and the scarp that skirts the mountain's base.

emitted by terrestrial eruptions, and the same is probably true elsewhere. So Mars' current water supply must have come from the volcanoes that dot the planet's surface. Mars not only has volcanic mountains but also lava plains that resemble the mare basins on the Moon.

The largest volcano in the solar system, Olympus Mons (Fig. 5), is about 600 km in diameter and 26 km tall. In contrast, Hawaii's Mauna Loa, the largest comparable structure on Earth, stands only 8 km above the floor of the Pacific Ocean basin. The region containing Olympus Mons appears relatively bright to telescopic observers, who have called it Nix Olympica (Snows of Olympus) because of the clouds that frequently form in its vicinity. The peak's great *summit caldera,* a crater formed by repeated collapse after eruptions, and its radially spreading lava flows confirm a volcanic origin. Olympus Mons is but one of four huge volcanoes situated in the uplifted Tharsis region situated amid the lowlands of Mars' northern hemisphere (see Chapter 5). All of these have the broad, gently sloping shape character-

istic of terrestrial shield volcanoes built up from low-viscosity basaltic lavas. We can tell that Olympus Mons grew over a considerable period of time, since various portions of its flanks show different ages (determined by the density of craters there). Older, heavily wind-eroded volcanic rocks form looping patterns surrounding the mountain that are collectively termed the aureole.

One of the most puzzling features concerning Olympus Mons is the great cliff or scarp, more than 4 km high, that rings its base (Fig. 6). Faulting may have played a role in the formation of both this cliff and the aureole; these features are the objects of continued study, as are the apparently very recent lava flows seen cascading over the scarp's edge.

Further evidence of volcanism on the surface can be found in Viking images. In parts of the ancient, heavily cratered crust that dominates Mars' southern hemisphere, many of the craters and intercrater plains are partially flooded with younger lava flows derived from the quartet of Tharsis volcanoes (Fig. 7). In addition to the young volcanic shields, many places in the ancient uplands have older volcanic centers (Fig. 8) with radiating flows and lava channels that bear many superimposed impact craters. These volcanic craters like Tyrrhena Patera have very low relief (about 1 km), differing strikingly from the great shields in the Tharsis region which stand 10 to 25 km high.

Mars' surface is replete with the evidence of volcanism. The features described here undoubtedly supplied large amounts of water and other volatiles to the young Martian atmosphere, although today we find no trace of liquid water there. In fact, the atmospheric pressure on Mars is so low that any liquid water placed on the surface would evaporate with explosive suddenness.

Where is the water stored? A portion of Mars' water resides in the north polar ice cap (Fig. 9) which is composed of water ice, according to infrared measurements. The cap has a striking spiral pattern that is probably an effect of radial wind erosion and the sublimation of water and carbon dioxide from the surface. Dirty ice lies on top of beautifully layered rocks that probably are deposits of silt, clay, and tiny particles of silicate rock intermixed with ice. Computations may soon determine the volume of ice stored in the polar regions.

Fig. 6. Now-solid volcanic flows cascade down the slope of Olympus Mons and over the steep cliff at its base in this image from the Viking 1 orbiter.

Fig. 7. These ancient impact craters in Mars' southern highlands have been partially flooded by lava flows emanating from the Tharsis volcano Arsia Mons, 1,500 km away. The largest crater, Pickering, is 120 km across.

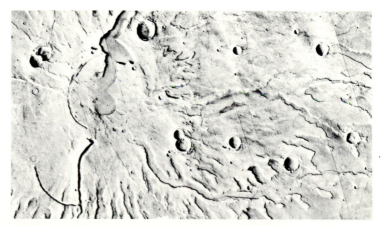

Fig. 8. Tyrrhena Patera, in Mars' southern highlands, is one of several ancient, eroded volcanic centers found on the planet. Radial flows and lava channels, with superimposed impact craters, surround the deteriorating central crater.

Fig. 9. From its high-inclination orbit, the Viking 2 orbiter photographed the involved spiral of Mars' north polar water-ice cap seen at left. Visible in the image reproduced above is a cliff in the Martian north polar cap; strikingly layered deposits of dust and ice have been eroded, overlain by fresher deposits of ice, then eroded again. The resulting contact between the two sequences of deposition, which is also apparent, is called an unconformity.

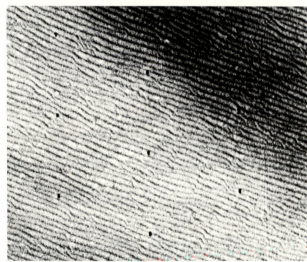

Fig. 10. An enormous dune field surrounds the ice of Mars' north polar cap. The dunes here align roughly north-south, and the vague circular forms are probably buried craters.

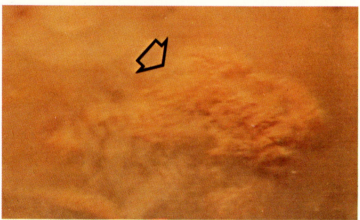

Fig. 11. Dust storms, such as this one scurrying across the floor of the Argyre basin, carry micron-size particles high into Mars' atmosphere and occasionally cloak the entire planet.

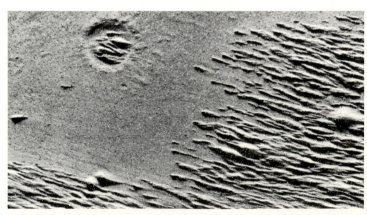

Fig. 12. Many large dunes have streamlined forms that resemble inverted boat hulls. These wind-eroded forms, called yardangs, are common on Mars and have also been observed on the Earth.

Fig. 13. These dark streaks probably formed when wind flowing around and over craters produced turbulence that removed a thin layer of bright dust to expose the darker surface below.

Adjacent to the north polar cap are vast dune fields (Fig. 10) which display many different dune types. The fine-grained material carried aloft in dust storms (Fig. 11) settles out onto the layered deposits in the polar region, possibly incorporated within snowflakes. This material is then eroded and spread outward from the pole by wind, and clumps of sand-sized material become shaped into dunes that form the north polar *erg* or sand sea. This same phenomenon is seen in sand-grain-sized aggregates of silt and clay found in the dunes adjacent to dry lakes in the southwestern United States.

Eolian or wind-formed features are so numerous in other parts of Mars (Fig. 12) that winds clearly have been the dominant force modifying the surface in the planet's recent past. Dunes are frequently found in channels, canyons, and craters. In many areas of Mars, bright and dark streaks, formed by eolian redistribution of dust, are common (Fig. 13). In some cases these represent deposits of bright or dark material; in other cases bright dust has been carried away to expose the darker underlying rocks. These same mechanisms, working on a larger scale, account for many of the light and dark markings apparent from Earth, including Lowell's canals (Fig. 14).

Wind erosion also carves grooves on the sides of mesas in the great equatorial canyons. Layered deposits show clearly on the sides of the mesas where they are exposed by wind erosion and mass wasting (downhill movement caused by gravity and probably seismic shaking). The canyons are enormous; Ius Chasma (Fig. 15) is about 7 km deep, almost three times the depth of the Grand Canyon. Smaller faults, scarps, and troughs abound as well. In many cases craters are aligned along the troughs (Fig. 16). These craters have been interpreted as volcanic vents — common on the Earth and Moon — or as drainage centers along the fault zones, down which surface material flows.

The Martian *regolith*, or soil, and rocks near the surface provide the most likely reservoir for the remainder of Mars' water, which may reside in the form of ice. Peculiar lobes of ejecta that surround many impact craters (Fig. 17) resemble wet-rock avalanches and mud flows on Earth. These ejecta blankets seem to confirm, when aided by infrared measurements, that water is stored in the regolith and near-surface rocks as water or ice.

Water has probably flowed on the surface; this is strongly indicated by the many channel and valley forms observed by Mariner 9 and Viking. Three types of channels occur — large, medium, and small. The large channels originate in the southern highlands area fringing Chryse Planitia and flow into the northern lowlands. They reach lengths of more than 1,000 km, and some are more than 100 km wide in places. These channels fan out widely, in many cases as multiple channels (Fig. 18), and disappear into the lowlands

Fig. 14. Detailed Viking pictures provided the basis for this painting of Mars and its satellite Deimos by Don Davis. The Martian terminator crosses two major impact basins, Isidis (*upper*) and Hellas. Classically recognized dark markings like Syrtis Major (on Isidis' western edge), Sinus Sabaeus, and Sinus Meridiani (together resembling the stem and bowl of a pipe) contain bright and dark patches as well as wind streaks.

Fig. 15. Ius Chasma, part of the Valles Marineris canyon system, is about 7 km deep. This contour map has been prepared using stereo photography and uses a contour interval of 200 m.

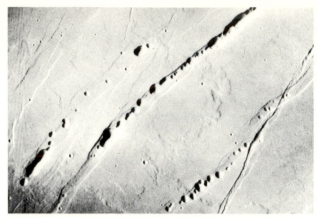

Fig. 16. The pits that lie along these linear troughs north of Tharsis may be volcanic vents, although they are probably not the source of the extensive lava flows evident in this image. An alternate theory holds that they are really drainage holes for material running along the troughs.

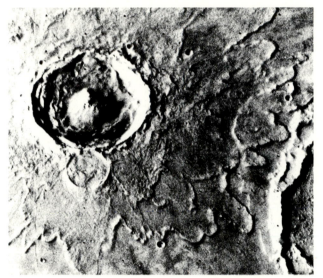

Fig. 17. Yuty, an impact crater some 20 km across, has a lobate, layered ejecta blanket that likely formed when subsurface ice melted during the impact and rendered the surface material into a fluidlike mass that splattered outward.

with no evidence for standing water in lakes or seas. Standing water would soon create wave-cut cliffs and sand bars such as those distinctive in pictures of ice-age lakes in Utah and Nevada. But these are not seen on Mars. The water must have vanished by evaporation and by soaking into the rocks and soils as does water in modern terrestrial deserts.

Channels commonly emerge from fractured jumbles of rock called *chaotic terrain* (Fig. 19), which may have formed when volcanic heat melted subsurface layers of ice, causing overlying rock to collapse. The resulting catastrophic release of water could well have carved the channels. Additional water may have come from the artesian flow of subsurface aquifers. The past depth of water can be measured in those places where channels have cut into the ejecta blankets of craters (Fig. 20).

Intermediate channels are commonly sinuous and have many tributaries with blunt ends, sometimes called box canyons (Fig. 21). The blunts ends of the channels are probably evidence of "spring sapping" — water flows out of springs at the base of the cliff, which then retreats headward to extend the channel. Intermediate-size channels are so widespread that it is difficult to account for them by local heating. Perhaps these streams formed during a period of climatic change that warmed the entire planet.

Dendritic networks of small channels are even more common (Fig. 22). Early analyses suggested that only rainfall could have caused this widespread phenomenon. However, the blunt ends of these channels in highly detailed pictures indicate that sapping and underflow may be required to form both small channels and their intermediate-size relatives. Perhaps the precipitation of water as snow (rather than as rain) was responsible: the snow would melt, soak into the ground, and form the springs to cut the channels. Small channels' ubiquitous nature might also indicate some previous climatic change — a subject of very active debate at this time.

Many channels have been modified heavily by wind action. Dunes mask the floors and wind erosion has reversed the relief of small inner channels making the channels stand as ridges (Fig. 18). Similar reversed relief is seen in the Little Colorado River in Arizona where fine-grained channel deposits are coherent enough to stand as ridges while surrounding terrain becomes eroded by wind and water. Beautiful detail shows in the Viking pictures where small young faults cross the channels (Fig. 23), and many small impact craters are visible on the channel floors.

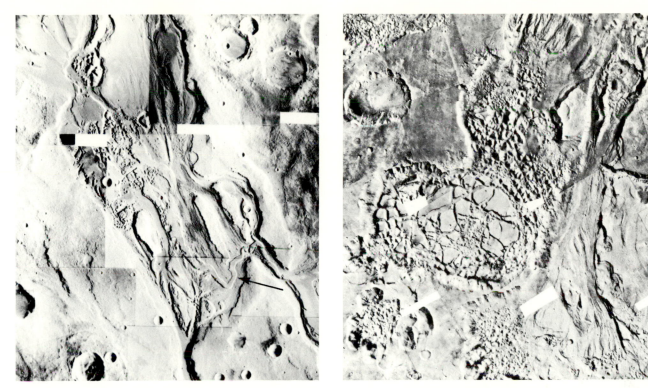

Fig. 18 (*left*). Running water probably carved Mangala Vallis, which runs from Mars' southern highlands northward into lowland areas. The feature's mouth contains a complex network of channels and streamlined islands. Wind erosion has left the floor of an inner channel standing as a ridge (arrow) by stripping away the surrounding, less-resistant material. Fig. 19 (*right*). Tiu Vallis emerges from an area of chaotic terrain and extends northward into Chryse Planitia, about 1,000 km away.

Fig. 20. Viking 1 image shows part of Maja Vallis in Chryse Planitia. Water was dammed behind the ridge at top, flowed over the low points, and cut into the ejecta surrounding the impact crater Domore.

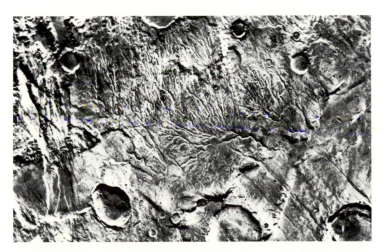

Fig. 22. Branching, dendritic channel networks, like these in Mars' southern highlands, are common. The largest neighboring craters are some 35 km in diameter.

Fig. 21. A typical intermediate-size channel, located about 4,000 km northeast of the Viking 1 lander, displays characteristic sinuous form and branching tributaries with blunt ends.

Fig. 23. The ridge crossing this channel floor is a small, young fault. Impact craters on the floor allow the channel's age to be estimated.

Ages derived from crater densities can be determined for many channels and volcanic units. As mentioned, water cannot flow on the Martian surface without evaporating explosively due to the low atmospheric pressure. Previous theories held that only in Mars' early history, when the atmosphere was presumably denser, could stream channels have formed. Unfortunately, images of the surface obtained before Viking were not good enough to show craters on the floors of the channels; age estimates based on crater densities could only be made for the areas around the channels, making a determination of their formation chronology uncertain at best.

However, late in the Viking mission, exceptionally high-resolution images were obtained that enabled new crater measurements of the actual channel floors. These new re-

Fig. 24. A Viking image showing a stream channel cut into layered rocks. Locations like this would be fruitful sites for exploration by an unmanned roving vehicle, which could collect samples and return them to Earth for further examination.

sults clearly show that there are lava flows both older than the oldest channels and younger than the youngest ones — the channels apparently possess a wide range of ages that lie somewhere between very ancient and very recent times. There are also many volcanic units of intermediate age that have been cut by channels. Many episodes of both volcanism and stream flow have apparently taken place, although they are not necessarily connected by a cause-and-effect relationship.

FUTURE STUDIES

After so much investigation of Mars by proxy, we have reached the point where the actual return of surface samples is necessary to make the jump to the next level of understanding. Compositional and age data for the surface materials, obtainable only by analysis of samples in terrestrial laboratories, would provide a factual foundation for our speculation.

A rover would be very useful to extend the area from which samples could be acquired. One potentially good collecting site is a stream channel (Fig. 24). Here a rover could collect not only from a sequence of layered rocks (such as lava flows in place) that could be radiometrically dated, but also the soils that lie between the rock layers. Perhaps these soils, with their clays and other alteration products, can be tied to changing climatic conditions — either cold and dry, or hot and wet — and thus to the times of channel-cutting episodes. It may be possible to find soils in which organic materials may have been preserved due to a less hostile environment. In terrestrial rocks, red shales with little organic material alternate with organic-rich shales and coal beds. Similar alterations may have occurred in the Martian climate, producing soils adjusted to varying environments. Either manned or unmanned rovers could traverse about 10 km and collect these rocks and their associated soils. If we could send a rover with a range of 200 km to the great canyons of Mars, it could perhaps collect from a section of crust several kilometers thick. Much of the geologic history of Mars would be recorded in these sequences of layered rocks. A similar record is beautifully preserved in the Grand Canyon on the Earth, where contrasting environments are easily deciphered.

By dating the glacial and interglacial times on Mars from returned samples, we could compare them with similar events on the Earth. Perhaps the warm and cold times on the two planets have coincided. If so, the climatic changes may be due to variations in the output of the Sun. This determination would solve one of the great mysteries of Earth science: how glaciations start and stop. If they are different, then the two planets' histories would be the result of effects that are still not understood. Future exploration of Mars may resolve this and other mysteries. But there is an axiom in planetary exploration that certainly holds true for Mars: the more one learns, the more questions emerge to be answered.

Life on Mars?

Gerald A. Soffen

The question of life on Mars remains enigmatic. We still do not know the answer.

And so this subject continues to stimulate many scientists. To some, it is an extension of the historically fascinating but outdated ideas posed by Giovanni Schiaparelli, Alfred Wallace, and Percival Lowell. To others, it is a link towards understanding the origin of life on the Earth, and to still more, it sheds light on the question of our aloneness in the cosmos. It is a question so profound that Viking scientist Norman Horowitz reflected, "The discovery of life on Mars would be hailed as one of the most significant discoveries of the 20th century."

To many biologists, rather than asking "Is there life on Mars?", the more important question is, "If there is life on Mars, is it of different origin than terrestrial life?" Most of them believe that all terrestrial life is the result of a single sequence of events. Earth cooled, and chemical evolution — helped by an abundance of solar energy and simple mate-

rials — led to the formation of an organic broth. When combined with a complex inorganic chemical milieu, in a watery solution, these components led to some very involved macro-molecules and then (here our story is very incomplete) there appeared some kind of self-replicating organism.

At this point, we better understand the story. This "biological entity" began to evolve through natural selection. Certain advantageous mistakes in its ability to reproduce turned our infant organism into a better competitor in the world around it. To make more of itself, it had only to use the surrounding sources of energy. But alas, the second law of thermodynamics began to catch up, and the survival of the new creature became dependent upon some very difficult "discoveries": a food source, an efficient internal chemical machinery, rapid responses to changing milieu, and environmental crises!

Terrestrial life made it. At least one form made it all the way, eventually spawning the 500,000 species observed today. But why only one beginning? Was there a successful second genesis on Earth? We do not believe so. The evidence is strong that all contemporary life came from that one occurrence. All terrestrial life has the same biochemistry, uses the same genetic code, and the same unique set of organic *stereoisomers*. (Asymmetric organic molecules come in pairs, something akin to the mirror images seen in Fig. 1; life uses only one of these, the so-called "left-handed" set, and rejects the other.) The best explanation for this is that once life was initiated, the process of evolution was so effective and rapid that no other beginning had a chance to succeed. One form rapidly became dominant, and its success prevented the others'. There is a good deal of conjecture here, but given our state of knowledge, this is where we stand.

And so our solar system had this one event, "life on the Earth," and we counter with, "Did it happen elsewhere, as a separate event from our own beginning?" The cosmologist who thinks about the extent of life in the universe is faced with a singular known event, and the discovery of a single bacterium of indisputably different origin would rock the intellectual world.

Actually, we have found organic "stuff" in interstellar space. Carbonaceous chondrites, meterorites that fall to the

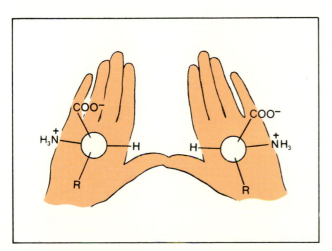

Fig. 1. This schematic illustration by C. Pellegrino and J. Stoff demonstrates the meaning of "left-" and "right-handed" amino acids. The large circles represent carbon atoms and *R* stands for carbon-hydrogen chains of atoms. In 1979, biologists S. C. Bondy and M. E. Harrington pointed out that certain left-handed organic compounds on Earth are selectively absorbed by bentonite clay over their right-handed (and biologically uncommon) counterparts. Such preferential binding may have influenced Earth's early stereochemistry.

Earth, contain several percent organic material that is loaded with amino acids, and in the laboratory the precursor chemicals of living things are synthesized with relative ease. And so, with high hopes, we anticipated finding the organic precursors of life — or perhaps some primitive organisms — on Mars.

We are limited in what we understand to be "life." There is no universal agreement, but most biologists would recognize terrestrial life by two of its characteristics: the ability to reproduce, and the ability to evolve through natural selection. While we have not exhausted all possible chemistries, we are struck by the fact that terrestrial life contains the most abundant elements of the cosmos — hydrogen, oxygen, carbon, and nitrogen.

Among the various environments in the solar system, Mars remains the most likely host for terrestrial-type life. In the Viking program, eight years of goal-setting, hardware specification, computer programming, data collection, and analysis by scores of scientists and engineers resulted in an almost perfect set of experiments, based on our limited assumptions. *The Viking missions did not find a terrestrial type of life at either of the two landing sites.* This evidence may suggest that Mars is lifeless, but science demands a more rigorous proof. Thus, we still do not know if life exists on Mars.

Many of the Viking instruments could have detected life. The orbiter camera could have seen cities or the lights of civilization (Fig. 2). The infrared mapper could have found an unusual heat source. The water-vapor sensor could have detected watering holes or moisture from some great metabolic source. The entry mass spectrometer could have identified gases that were wildly outside the limits of chemical equilibrium (as oxygen is on Earth). Seismometers could have detected a nearby elephant. But Viking's search utilized three main tools on each lander: a pair of cameras, the pyrolytic gas chromatograph and mass spectrometer (GC/MS), and the biology experiments. These are pointed out in Fig. 3.

Lander Cameras. Two cameras were used to take pictures from the base of the lander to the horizon, in a complete azimuthal circle, at all times of the day and a few at night for two complete Martian years. Conscientiously examined for any signs of life, the pictures contain countless interesting forms, several subtle changes in the terrain, and some very suggestive colors — but nothing to make us believe that we have detected life at either of the landing sites. One unusual marking, known as the "B" on the rock in Fig. 4, was a delightful interruption during the mission's difficult schedule, but no scientist seriously considered it anything more than an odd coincidence of geometric form and lighting.

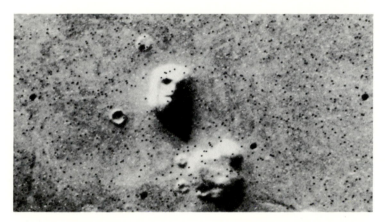

Fig. 2. A sign of civilization? The Viking 1 orbiter took this picture of mesa-like landforms in the Elysium region of Mars on July 25, 1976. Shadowing on one of the mesas (above center) has given it the appearance of a human head. The feature, about 1.5 km across, was nicknamed "The Face of Mars."

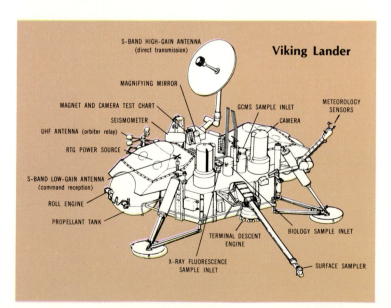

Fig. 3a (left). This schematic diagram points out various equipment on the lander. Fig. 3b (right). A model of the Viking lander tries out its 3-m-long sampling arm, seen extended to the left. The scoop on the end delivers handful-sized samples of soil to three experiment hoppers.

The Pyrolytic GC/MS. Both the landers obtained several samples of surface and subsurface material that were heated in steps to 773° K; effluent gases were then analyzed with a combined gas chromatograph and mass spectrometer to detect indigenous organic compounds. The instrument measured carbon dioxide and water absorbed in the soil and could detect concentrations as low as one part per million (and down to tenths of this for some compounds). But no organic molecules were detected by either of the spacecraft from either landing site — a remarkable result in light of our expectation that organic matter derived from falling meteorites would linger on the surface.

The Biology Experiments. The Viking landers contained three biology experiments (Fig. 5), each based on a different set of premises. This was done to broaden the search as much as possible. Because the lander did not have wheels, only samples collected by an extendable scoop from an area of about 12 square meters could be analyzed. These areas are shown in Fig. 6.

The Pyrolytic Release experiment was a physiological test for any Martian life able to assimilate and reduce carbon dioxide or carbon monoxide. This was tried in several modes: in both dark and light (as in the photosynthesis of terrestrial life), in dry surroundings, and in humid ones. The strength of this experiment is that it makes very few assumptions about Martian life; in fact, it assumed only that CO_2 or CO was needed as a carbon source. The experiments were done using these two gases labeled with radioactive carbon-14, which is very easy to detect. Special precautions were taken to prevent obtaining a falsely positive result. The experiment employed heat sterilization of the samples as a control to distinguish biological uptake from chemical reactions which simulate biology. This was the one experiment that appeared to give a weakly positive result on the first attempt. Unfortunately, to the consternation of the scientists, the effect could not be repeated on Mars. Subsequent laboratory experiments have demonstrated possible scenarios that could explain that first unusual result. It is probably related to some unusual iron compound in the sam-

ples, but we still do not fully understand the high reading.

Another experiment was called Labeled Release. This experiment also used radioactive ^{14}C as a sensitive tracer, but here it was in the food supply. An organic nutrient broth with several microbial "goodies" was inoculated with the surface samples. Presumably, Martian microbes would eat the food containing the organics, and breathe out $^{14}CO_2$ or ^{14}CO which could be measured. Here again, we got a strange result. Part of the organic mixture was converted to CO_2 but then the conversion stopped (this is unlike terrestrial microbes that continue to feed until the supply runs out, or until the toxic products of their own metabolism kill them). The sample was given a second and third injection of the organic nutrient, but to no avail! The resulting reaction, shown in Fig. 7, does not resemble the way terrestrial life handles organic material. The LR experiment also used heat sterilization to distinguish chemistry from biology, and to our chagrin, heat seemed to destroy whatever was converting the organic nutrient. Results of this experiment by itself could be interpreted either by a biological or chemical explanation, although the Martian biology, if it exists, would have to metabolize organic material in a manner very different from terrestrial life. The answer lies in the results of the third biology experiment.

In the Gas Exchange experiment, a very rich organic

Fig. 4. One of the rocks near the Viking 1 lander appears to have *B* etched onto its face. This is an illusion caused by shadowing of surface grains on the rock, or possibly by a dark stain produced by weathering.

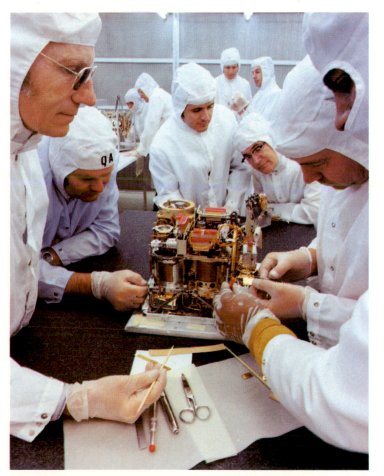

Fig. 5. This TRW photograph shows a "surgical team" of technicians swarming around the Viking biology instrument. The compact package occupies less than one cubic foot of space yet contains thousands of individual components.

Fig. 6a. At one point, Viking 2's sampling arm pushed aside a rock to obtain soil not exposed to large doses of ultraviolet radiation. It was thought that the protected soil would provide a safe habitat for Martian organisms.

nutrient broth (which we called "chicken soup") was combined with Martian soil and the chamber's headspace monitored for gaseous metabolic changes. In order not to shock the Martian microbes, who live in a very dry world, "they" were first humidified prior to the inoculation with only water vapor. Here, the results were most remarkable. When the sample was humidified, it released an extraordinary burst of oxygen. This was not anticipated, and no terrestrial soils have ever done this. Next, the actual nutrient broth was added; very little happened, except that a small amount of CO_2 was evolved. No more explosive oxygen, and no continuing exchange of gases due to biological metabolism. These data suggest that some kind of oxidizing substance exists on the Martian surface. A great deal of speculation (and some laboratory work) supports the idea that the oxidizing agent could be some kind of iron peroxide or superoxide. An important point is that such an interpretation explains why the organic compounds used in the labeled release experiment were changed, and the results resembled a strange kind of biological reaction. One (or maybe two) of the organic compounds in the nutrient is rather sensitive to these kinds of oxidizing agents. During Viking's tests, the oxidizing agent was used up so that subsequent addition showed no effect.

Combining these results, the biology team made the statement that "the Viking results do not permit any final conclusion about the presence of life on Mars." However, a recent panel of the National Academy of Sciences published a report that takes a less circumspect view. It concluded that Viking results have lowered the possibility of life on Mars, and that further exploration of the question must await samples of Mars returned to Earth laboratories.

The experiments to detect life on Mars have spawned a brilliant line of research in chemical evolution to understand the origin of life on the Earth. It now appears that the organic world of carbon chemistry was very tightly coupled to the inorganic world of aqueous and gaseous chemistry. Reactions involving salts, ions, metals, and especially the early formation of clays were part of the events that led to the first large polymers of inorganic molecules, then to information-bearing molecules, and finally to the self-replication that is considered the hallmark of living organisms.

I have devoted 20 years to thinking about the question of life on Mars. In my role as the Viking project scientist dur-

Fig. 6b. This partial view of the landscape surrounding the Viking 1 lander includes a series of trenches excavated by the surface sampler. The tall white mast is topped with a suite of miniature weather instruments.

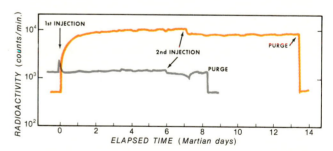

Fig. 7. This graph shows results from Viking 1's Labeled Release experiment, for both active (orange) and sterilized (gray) samples of Martian soil. The curves record the accumulation of radioactive carbon monoxide and carbon dioxide after the injection of nutrient labeled with carbon-14.

ing the primary mission, I began with a very optimistic view of the chances for life there. I now believe that it is very unlikely, but one doubt lingers: we have not visited the polar regions. I have always believed that in the search for life you must go where the water is. The permanent polar caps of Mars are frozen water and would act as a splendid "cold finger" of the planet to trap organic molecules. The oxidizing agents should be absent . . . and who knows what those lucky future explorers of Mars will find there?

Asteroids

Clark R. Chapman

Asteroids are a multitude of small planets orbiting the Sun at distances ranging from inside the Earth's orbit to beyond Saturn's. They predominate in a large torus (the "main belt") which has a volume exceeding the spherical volume of space inside the orbit of Mars. While asteroids are individually small, they are ubiquitous throughout the inner solar system. And the discovery of Chiron, orbiting beyond Saturn, hints at the possible existence of additional asteroid belts in the outer solar system.

Until the past decade, these bodies were dismissed as uninteresting planetary "dregs" or as "the vermin of the skies." Belatedly, we now realize they hold important clues concerning the nature of the planetary system and its earliest history. And public curiosity about asteroids, sparked by several "disaster" movies, has paralleled a developing appreciation by scientists of the relevance of asteroids to the evolution of life on Earth and to the dawning age of human exploration and utilization of the planets.

Interdisciplinary research is shaping a picture of the origin and cosmogony of asteroids. The physical and chemical natures of asteroids are being revealed by ground-based astronomical techniques. Petrological, chemical, and isotopic measurements of meteorites — the probable fragments of some asteroids — are being refined and synthesized. And theories of planetary formation are becoming more focused so that the role of asteroids is now more clearly understood.

Apparently, the asteroids, or their precursors (which may have resembled Ceres, the largest asteroid), were planetesimals just like those growing elsewhere in the solar nebula during planetary accretion. Before they could form an asteroidal planet, however, they were gravitationally perturbed into tilted, elongated orbits. Instead of slowly accumulating into a planet from circular orbits, asteroids began to impact each other at speeds of kilometers per second, often resulting in catastrophic fragmentation and disruption rather than coalescence. The process of collisional destruction continues at a diminishing rate today.

What kinds of gravitational perturbations affected planetesimals in the asteroidal zone more so than those in other planetary zones? We cannot be sure, but two plausible scenarios have been advanced. Several scientists have hypothesized that a very large planetesimal in Jupiter's zone was

gravitationally "scattered" by a close approach to that giant, partly formed planet into an eccentric orbit that penetrated the asteroidal region (Fig. 1). It would have made close passes to most of the asteroids before once again encountering Jupiter and being ejected from the solar system. If the body were as large as Mars or even the Earth, as seems consistent with the planetesimal hypothesis of planetary formation, its close passes to asteroids would have stirred up their velocities. Most early asteroids presumably were destroyed, either by collisions with a swarm of such Jupiter-scattered planetesimals accompanying the planet-sized body, or by collisions with each other.

An alternate idea holds that distant gravitational forces from Jupiter itself were responsible for stirring up and depleting the asteroids during the period when the Sun was

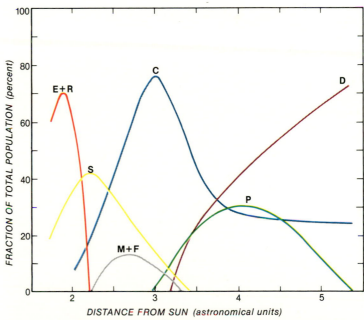

Fig. 1. A census of asteroids of different compositions shows that they are systematically distributed with respect to distance from the Sun. Edward Tedesco and Jonathan Gradie used the results of a survey made in eight colors to derive the approximate fraction of asteroids above a certain size (determined by observational limitations and varying with distance) within each class. The letter designations, which refer to different spectral types, are discussed on page 103.

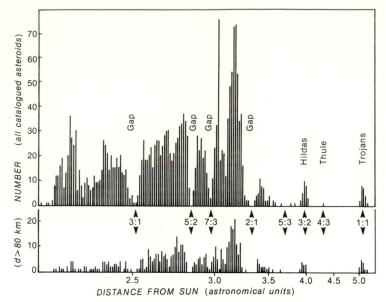

Fig. 2. **Histograms of the number of asteroids at different distances from the Sun. The top one shows all catalogued (numbered) asteroids, but it is biased in favor of small, bright bodies inside 2.5 astronomical units (AU), and against very dark asteroids beyond 3.2 AU. The lower, nearly complete sample of bodies larger than 80 km diameter gives a truer picture of the distribution. Arrows show commensurabilities with Jupiter's orbital period; for example, a Kirkwood gap exists at 2.5 AU where an asteroid orbits the Sun three times for every revolution Jupiter makes. Different orbital ratios are associated with other Kirkwood gaps, but closer to Jupiter the reverse is true: there are isolated *groups* of asteroids at ratio values of 3:2, 4:3, and 1:1.**

Fig. 3. **The long trail of asteroid Ra-Shalom is obvious in its discovery plate, obtained by Eleanor Helin on September 10, 1978, during an extensive search for Apollo or Earth-crossing asteroids. Hale Observatories photograph.**

divesting the primordial solar system of its nebular gases. We see today numerous spaces (the *Kirkwood gaps*) and other lacunae in distributions of asteroidal orbital elements that are due to commensurabilities and resonances with Jupiter (Fig. 2). Resonant effects occur at fixed locations today, but might have swept through the asteroidal region while the solar system was losing mass early in its history. Such resonant interactions with Jupiter might have pumped up asteroidal velocities. Either way, it seems likely that massive Jupiter was responsible for the absence of an asteroidal planet.

The reservoir of asteroids persists today, despite small leakages which result in stray bodies such as the *Apollo* and *Amor* asteroids (which occupy temporary Earth- and Mars-crossing orbits, respectively), and the meteorite fragments that fall onto the planets and their moons. Presumably, there were many small bodies left over throughout the solar system after the planets formed. But those in orbits that made close approaches to planets risked planetary impact (the final dribble of accretion) or ejection from the solar system. Only in orbits far from any planet or from strong resonant locations could remnant bodies have remained until now. The asteroid belt is such a place. Within the orbit of Mercury is another, if material ever condensed and accreted there in the first place. A few other bodies — such as the Trojans, orbiting ahead of and behind Jupiter in its orbit — are protected in special resonant orbits. Still other remnants (for example, Chiron) may exist in the vast volumes of space between the outer planets.

One final population of remnant planetesimals is the comets, which were ejected from their place of origin (pre-

sumably in the outer solar system) by close planetary encounters (see Chapter 17). Those that did not quite escape the solar system have been preserved for aeons in the deep-freeze of outer space. Chance perturbations by passing stars and subsequent encounters with the outer planets bring a few comets into the inner solar system, where their ices sublimate and boil, producing the flashy comas and tails that are their hallmarks. After a cosmically short time, a comet's volatiles are depleted, and it dies. If it contains a solid core, that body is then termed an "asteroid"; it is called an Apollo-Amor type asteroid if its orbit crosses or approaches the orbit of the Earth (Fig. 3).

We can only speculate about whether asteroids and comets might originally have been the same type of planetesimal. Most asteroids remain in nearly their original orbits, where any surface ices would long since have sublimated away. Prior to entering the inner solar system, "new" comets are better preserved than asteroids, but all trace of their origin has been lost due to their orbital wanderings. Most scientists believe that comets and their remnant cores were formed mainly in the outer solar system while most asteroids were formed in the asteroid belt, with a few asteroids implanted into the belt from other places in the inner solar system. It is therefore rather ironic that these two groups of remnant planetesimals, with very different histories, are roughly equally represented among the temporary Earth-crossing asteroids and meteoroids from which we obtain our meteoritic samples. We still aren't sure which meteorites come from comets and which from asteroids.

Comets and asteroids still impact the Earth and other planets. Earlier populations of remnant planetesimals from distant places presumably were the last to coat the surfaces of the nearly completely formed planets. They may have contributed preferentially to planetary crusts and hydrospheres. Certainly their impacts formed craters, which remain the dominant topographic features on all but the most geologically active terrestrial bodies (see Chapter 4). There have been conjectures about whether volatile-rich comets and asteroids might have helped life get started on Earth. Certainly life has been affected by these bodies, even if Fred Hoyle is wrong about their being responsible for epidemics. Mass extinctions (possibly including dinosaurs') 65 million years ago apparently resulted from the impact of an Earth-crossing asteroid: trace elements in certain Cretaceous-Tertiary clays bear an extraterrestrial signature.

It is uncertain what role asteroids may play in our future. There is a remote but real probability of a small asteroidal-impact disaster in the foreseeable future. An unexpected impact could be mistaken for a nuclear attack, with frightening consequences. More likely, however, asteroids will serve as humanity's stepping stones to the planets. Searches made for small, Earth-approaching asteroids have found that some are easily accessible from high-altitude orbits. From them, future astronauts may mine materials necessary for living and working in space. Materials expected to be abundant within some asteroids include water, organic compounds, and metals; it will be far cheaper to mine them from nearby asteroids than to hoist them up from Earth.

It is widely supposed that since asteroids are so small, they have been spared the kinds of processes that have virtually destroyed records of primordial history on the Earth and Moon — the extreme heat, pressure, chemical alteration, and crustal motions that accompany the life cycle of a large planet evolving in response to its vast reservoirs of primordial and radioactive heat. Are the asteroids primitive planetesimals, arrested in their growth and preserved intact for four and a half aeons? Or have their clues about planetary origins been disturbed as well? In short, what are asteroids like and what can we infer about their evolution?

We turn first to the astronomical evidence, gleaned from study of the time-variable behavior and spectral characteristics of reflected and emitted radiation coming from their tiny, distant, starlike images. Asteroid sizes are now well established (see Fig. 5 on the next page). The largest, Ceres, is about 1,000 km across and constitutes about 30 percent of the mass of all asteroids combined. Vesta and Pallas are each between 500 and 600 km in diameter. Still smaller ones are increasingly numerous, grading down to countless kilometer-size asteroids and smaller boulders too small to detect, unless they happen to pass very close to the Earth.

Asteroid shapes and configurations are a matter of active controversy. It had long been supposed that asteroids

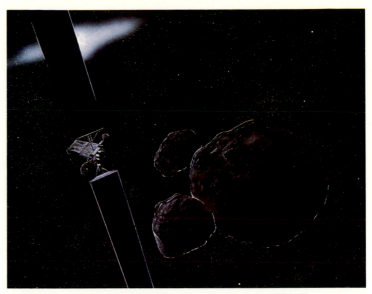

Fig. 4. A future mission to the asteroid belt could utilize a spacecraft powered by solar-electric propulsion. Low-thrust engines operating for months at a time could maneuver payloads into long-duration rendezvous for extensive investigation of promising objects. Multiple asteroids or those thought to be rich in metals would rate high as candidates for such a mission. Painting by Charles Wheeler.

ranged from roughly spherical bodies to elongated and irregular objects bespeaking a fragmental origin. This was inferred from variations in asteroid brightnesses as they spin with rotational periods of roughly nine hours (extremes range from less than three hours to several days). Recent evidence suggests that some asteroids — especially those previously believed to be the most elongated bodies — may in fact be double or multiple bodies (Fig. 6). In some instances, asteroidal satellites were "detected" by the occultation of a star as an asteroid passed between Earth and the star; unfortunately, such observations are either poorly confirmed or unconfirmed. Several asteroidal light curves are better explained in terms of eclipsing binaries or "contact" binaries rather than by elongated single bodies.

Fig. 6. The large Trojan asteroid Hektor is a very elongated body and may, in fact, be a tidally deformed contact binary. Two views of a possible configuration for Hektor are shown in this painting by William K. Hartmann.

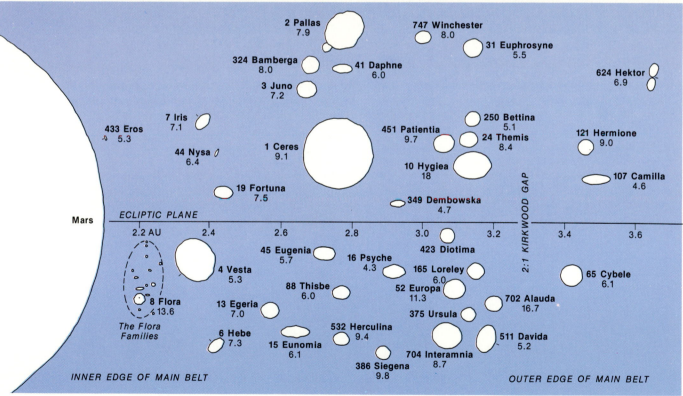

Fig. 5. This representation of the physical properties of interesting asteroids includes all 33 asteroids larger than 200 km in diameter. They are portrayed in their correct relative sizes and shapes (the limb of Mars is shown for comparison); colors, albedos, and polar obliquities (if known) are indicated. The bodies are positioned at their correct relative distances from the Sun. Asteroids located near the top or bottom of the diagram occupy relatively eccentric or inclined orbits (or both), while those shown near the ecliptic plane move in relatively circular, noninclined orbits. Rotation periods, in hours, are given in the lower panel. Among the special smaller asteroids included are all members of the Flora families larger than 15 km in diameter, but this illustration would be hopelessly cluttered if it were to show all asteroids of comparable sizes: an estimated 1,150 asteroids, in the main belt alone, have diameters larger than 30 km (only five Flora family asteroids attain that size). Note the contact-binary Trojan asteroid Hektor and the possible satellite of Pallas. Painting by Andrew Chaikin.

Undeniable evidence for binary asteroids may exist in recent speckle images of Pallas and Victoria. The technique of speckle interferometry has been successful in "beating" Earth's limited seeing conditions, in recording images of very close binary stars, and in resolving the disks of certain red giants and the asteroid Vesta. Speckle images of Pallas and Victoria show large satellites orbiting those asteroids; it is doubtful that the speckle results are erroneous. It remains to be determined what fraction of asteroids are binary or multiple systems.

Fifty years ago, Nicholas Bobrovnikoff made an important discovery about asteroids: they differ in color. The full import of Bobrovnikoff's discovery was not appreciated until the past decade, when asteroid spectrophotometry blossomed. Spurred by new detector technology, spectra of sunlight reflected from asteroids have been measured from the ultraviolet out to the mid-infrared, where heat radiation from the warm surface of an asteroid begins to overwhelm the reflected component. Not only do the visible colors of

asteroids differ, but many asteroid spectra exhibit absorption bands that are due to different minerals and hydrous compounds; some examples appear in Fig. 7. The diversity of colors and spectra imply different surface compositions among the asteroids. Efforts have been made to determine asteroidal mineralogies and to relate them to the different types of meteorites that are being studied by cosmochemists (see Chapter 19). If meteorites could be linked to particular asteroids, then we could tie the meteoritical inferences about primordial environments and events to specific bodies in particular parts of the solar system.

There are two chief types of meteorites: those whose nonvolatile chemical elements occur in roughly cosmic abundances and hence are inferred to be relatively unaltered condensates from the primordial solar nebula and those highly enriched or depleted in certain elements in ways that imply they were created after primordial material was greatly modified by processes of "planetary" evolution. (One such

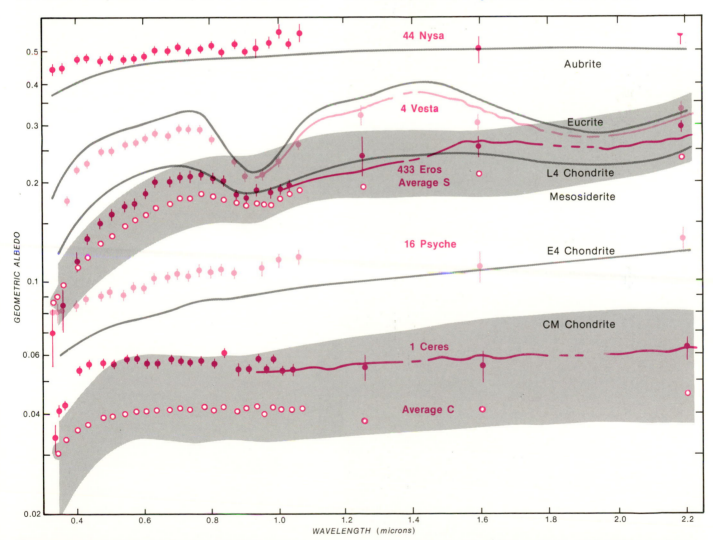

Fig. 7. Reflectance spectra of asteroids and meteorites are compared for wavelengths ranging from visible through infrared. Asteroid data, shown in shades of red, consist of points with error bars (from filter spectrophotometry), lines (from Fourier spectroscopy), and open circles (average values for the S- and C-type asteroids, which together comprise about 90 percent of all such objects). Laboratory measurements of meteorite powders are shown in black; two classes occupy the ranges of values indicated by gray bands. Investigators deduce surface mineralogy primarily from the shapes of these curves rather than from the objects' precise albedos: S-type asteroids appear to have spectra intermediate between the mesosiderites and L4-type ordinary chondrites; 433 Eros is more like an ordinary chondrite than the typical S-type asteroid. Evidently, the diverse mineral assemblages found in our meteorite collections are also represented in the asteroid belt.

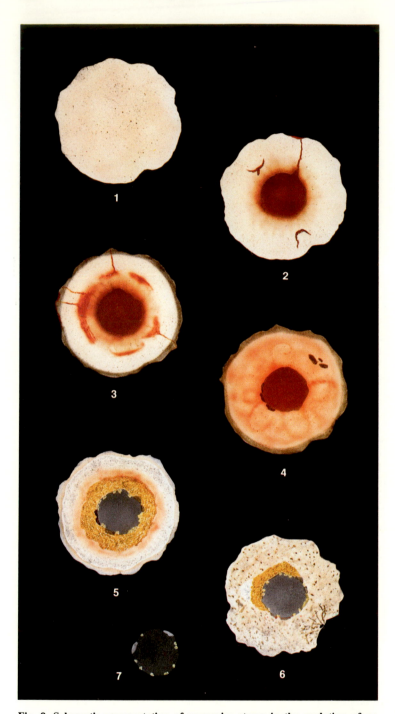

process was the melting and physical segregation of material, known as differentiation, which occurred early in the Earth's history; metals sank to form a core and lighter materials floated and cooled to form the crust. See Fig. 8.) Most primordial types of meteorites are called chondrites while most of the geochemically altered meteorites are found in the iron, stony-iron, and achondrite classes.

An important conclusion about asteroids is that *both* the primordial and altered mineralogies are represented in the main belt population. At least one asteroid, Vesta, has a surface mineralogy similar to the so-called basaltic achondrites — meteorites similar in physical properties and chemistry to the basaltic lava flows common on the surfaces of the Earth and Moon. Many other asteroids are extremely black and show infrared absorption bands indicative of the water-rich, carbonaceous mineralogy of the most primitive meteorites — the carbonaceous chondrites. Therefore, some asteroids were apparently melted and geochemically differentiated, just like the larger terrestrial planets, while the chemistry of other asteroids has been preserved more-or-less intact. It remains a profound mystery how some asteroids could have been so altered while others of similar size in nearby orbits could have escaped modification.

The simple fact that some small asteroids were melted after they formed establishes important constraints on the thermal conditions and heating processes operating in the early solar system. It had been supposed that the high surface-area-to-volume ratio of small bodies would have radiated away any internally or externally generated heat and kept them cool. Evidently some bodies encountered much more intensive heating than can be ascribed to any of the traditional sources of heat. Hence, more exotic sources of heat must be considered, including an intense pulse of heat due to the decay of short-lived radionuclides. Aluminum-26, for example, may have been synthesized in a supernova explosion that helped trigger the onset of solar system formation. But in order for heat from its decay (into a stable form of magnesium) to melt an asteroid, aluminum-26 had to be incorporated into an accreting body within only a few million years of its synthesis. In a way, then, the telescopic detection of several absorption bands in Vesta's spectrum may be thought of as helping to establish the time scale for the formation of the solar system.

About three-fourths of the asteroids are extremely dark (with typical geometric albedos of 3.5 percent) and have

Fig. 8. Schematic representation of successive stages in the evolution of an asteroid that is heated early in its history. The original body of primitive composition (*1*) is heated to the point that constituent iron separates and sinks to its center, forming a core (*2*). Partially melted rock from the mantle floats to the surface through cracks in the crust, erupting onto the surface as basaltic lava flows (*3*). Fractured by impacts, sections of the basaltic crust may founder through the mushy, partially molten, convecting mantle to float at the outer edge of the core (*4*). As heat radiates away, the body cools, the iron solidifies, heavy crystals of olivine form and accumulate in the deep interior, and magmas solidify into an ever-thickening crust (*5*). Repeated collisions fragment the mantle and crustal rocks into a "megaregolith" (*6*) and ultimately eject the rocks exposing the iron core (and any imbedded rocks) to space. Most asteroids were not heated beyond the first two stages; Vesta cooled after reaching stage *3*, while Dembowska may be a body in stage *6*. Some smaller M- or S-type asteroids (*7*) may be the parent bodies for iron and stony-iron meteorites. Models for such parent bodies have been discussed for more than two decades; this is the scenario offered by the author and Richard Greenberg.

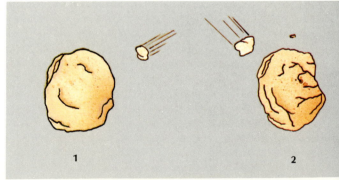

Fig. 10. Stages in the fragmentation history of a moderately large asteroid. Originally composed of strong rock (*1*), the asteroid is cratered (*1, 2*), and then catastrophically fragmented by a more

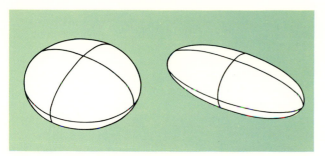

Fig. 9. Asteroids may be structurally weakened by collisional fragmentation. Stuart Weidenschilling and his colleagues think such bodies may form conglomerations which are quasi-stable when pieces reassemble following particularly large impacts. If they are spinning rapidly, they might adopt the flattened, oblate shape of a Maclaurin spheroid (left). If collisions add more angular momentum to the system, the equilibrium shape is a Jacobi ellipsoid (right). Still more angular momentum would make the conglomerate body unstable, perhaps resulting in a contact binary (like Hektor; see Fig. 6), a more separated binary, or a multiple configuration.

mineralogies analogous to carbonaceous chondrites, so far as we can tell. There are significant spectral variations among these so-called *C-type asteroids,* reflecting mineralogical differences of uncertain nature. One low-albedo type of asteroid is unusually reddish; such *D-type* spectra are common for bodies beyond the outer edge of the main belt, including many Trojans and some of Jupiter's small outer satellites.

Roughly one-sixth of the asteroids have moderate albedos (typically 16 percent) and reddish colors. The spectra of such *S-type asteroids* imply assemblages of iron- and magnesium-bearing silicates (pyroxene and olivine) mixed with pure metallic nickel-iron. Unfortunately, researchers cannot yet determine the proportion of metal to silicates to better than a factor of three. Thus the S-types could be analogous to two radically different types of metal-bearing stony meteorites: the ordinary chondrites, relatively unaltered primitive meteorites believed to have formed closer to the Sun than the carbonaceous chondrites or the stony-iron meteorites, which are enriched in metal and other compounds due to extensive melting and geochemical fractionation within a parent body.

A third type of asteroid spectrum is also ambiguous. So-called *M-type asteroids* have moderate albedos. Their spectra exhibit the signature of metallic nickel-iron, with no hint of silicate absorption bands. It is expected that these may be metallic asteroids, perhaps the remnant cores of differentiated precursor bodies stripped of their rocky mantles and crusts by asteroidal collisions. But one type of stony meteorite also has the spectral signature of nickel-iron: enstatite chondrites are primitive meteorites formed in a highly reducing (oxygen-poor) environment, perhaps close to the Sun or deep within a protoplanet. They consist of grains of nickel-iron imbedded in a clear matrix of the magnesium-rich silicate enstatite. Since enstatite is colorless and lacks absorption bands, an enstatite chondritic asteroid would have a spectrum just like that of a nickel-iron asteroid. There is hope that this ambiguity will soon be resolved. New radar soundings already show that the M-type asteroid 16 Psyche is an unusually metal-rich body. The existence or absence of pure metallic bodies in the asteroid belt will probably not remain a mystery for long.

The eight types of asteroids show a remarkable distribution with distance from the Sun (Fig. 1). Two dark types of asteroids have been newly defined, the P-types beyond the main belt and the F-types. Some small, bright asteroids near the inner edge of the belt are called E-type and R-type. While examples of most types span a large range of solar distances, there is a clear progression from E's and R's to S's in the inner belt, to M's, F's, and C's in the outer belt, with P's occurring most often near 4 AU and D's dominating among the Trojans. It is not known whether this gradation with distance from the Sun reflects primordial properties of the condensing solar nebula or arises instead from later evolutionary processes. In addition to the common types, several percent of asteroids are oddballs of one sort or another, like the olivine-rich body Dembrowska or basalt-covered Vesta.

The preceding summary represents nearly all we know about what asteroids are like and what they are made of. Of course, we can analyze theoretically the processes that have been affecting asteroids during the life of the solar system and speculate on the implied geologic morphologies of asteroids. However, until a spacecraft visits an asteroid, no one can know what surprises await us. (Since such a multi-asteroid rendezvous mission (Fig. 4), or even a simpler visit to a single Earth-approacher, is not planned until the 1990's at the earliest, let me proceed with speculations.)

It is difficult to imagine that planets' radioactively driven

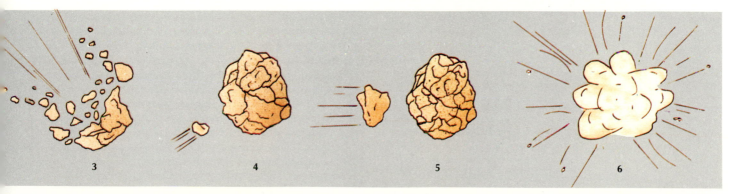

energetic impact (3). Most of the ejecta fail to reach escape velocity, and the body is reassembled (4). Later impacts further fragment the body, converting it into a gravitationally bound pile of boulders. Finally (5, 6), a sufficiently gigantic collision occurs to completely disrupt and destroy the asteroid; its remnants then become scattered through space.

internal "heat engines" (which keep some of them actively evolving aeons after they formed) can be relevant to the comparatively tiny asteroids. In all probability, asteroids are cold, dead, atmosphereless bodies whose destinies are shaped solely by exogenic forces. Their orbits may be occasionally perturbed by gravitational interactions with each other or distant planets. Their surfaces are certainly bombarded by micrometeorites, solar wind particles, and cosmic rays. And their surface temperatures rise and fall on rapid day-night cycles as they spin in the feeble light of the distant Sun, never getting much warmer than 200° K. By far the most important effects on asteroids, however, are due to those rare events when they encounter another asteroid hurtling through space on an intersecting orbit.

Despite the tiny sizes of asteroids and the immensity of the torus of space through which they revolve about the Sun, there are enough asteroids traveling sufficiently fast that major collisions are inevitable during the life of the solar system for all but a lucky few asteroids. The typical collision velocity is about 5 km per second. One can calculate the total energy involved in the largest impacts a typical asteroid might be expected to experience. The energy is much more than sufficient to fracture and fragment a body having the material strength of rocks. Only bodies with the strength of iron might be expected to survive more-or-less unscathed.

Unless an asteroid is quite small, collisions can fragment but still not destroy it. Most of the fragments must be ejected from the collision at greater than the escape velocity from the target body; otherwise they will remain bound in orbit around the center of mass and will probably coalesce back into a single body again (Fig. 9). Or if a rapidly spinning target body were involved in an appropriate off-center collision, the combined angular momentum might be too great for a single body to reform, and a binary or multiple system would result.

A "supercatastrophic" collision might provide sufficient energy and momentum not only to break up an asteroid but to disperse the fragments as well. In that case, the fragments would depart on independent, but still similar, heliocentric orbits. The fragments would rarely or never meet again, and the asteroid population would have gained a family of smaller asteroids at the expense of a larger body (or two bodies, counting the projectile, which is most likely to be small compared with the target body). Such families of asteroids in similar orbits were discovered by the Japanese astronomer Kiyotsugu Hirayama about 60 years ago and are called *Hirayama families*. In principle, study of the spectra of members of a single Hirayama family would help us learn about the interior composition of a precursor asteroid, destroyed by a supercatastrophic collision. Some families have members with similar spectra, implying a homogeneous precursor. Others contain a variety of spectral types, but they are not always easy to "put back together" into a single precursor body that makes physical and geochemical sense.

Quantitative models of asteroids' collisional evolution have been developed. We have confidence in our estimates of the frequencies and energies with which asteroids of different sizes collide with each other. But we do not know how the energy and momentum are partitioned into fracturing the material, heating it up, ejecting fragments into

space, and so on. Lacking experimental evidence on collisions or explosions even remotely approaching the magnitude of asteroidal collisions, we must combine theory with uncertain extrapolation from laboratory-scale experiments to infer the outcomes of asteroidal collisions.

By incorporating these parameters into the model, we can draw a number of interesting — if uncertain — conclusions about the collisional evolution of asteroids. The correct story almost certainly lies somewhere between the following two scenarios. First, it may be that asteroids have been broken up into generations of successively smaller fragments, gradually grinding themselves down to meteoroids and eventually to dust which is swept out of the asteroid belt, out of the solar system, or into the Sun. In that case (which assumes efficient conversion of impact energy into ejecta velocities), the present asteroid population might represent a small remnant of a much larger earlier population. The larger asteroids still existing would be considered the lucky few that have by chance escaped destruction.

The second alternative (if ejecta velocities are generally low) is that the number of asteroids larger than 50 km in diameter hasn't changed much over the aeons, but individual asteroids have been damaged repeatedly, as depicted schematically in Fig. 10. When such asteroids are smashed to bits by repeated collisions, their fragments often reaccumulate into a gravitationally bound collection of rubble. Some of these asteroids will be spun up by repeated collisions and may become binary or multiple systems following an energetic off-center impact.

The latter scenario is now preferred by some scientists, since it seems to explain features of the asteroid size distribution. But it means that asteroidal materials may have been badly shocked and rearranged during repeated impacts, making it more difficult to read the evidence about primordial events.

The asteroids continue to collide with each other today. The smaller impacts crater and crack their surfaces, which are gradually covered over again by regional or even global ejecta blankets from the larger cratering events. Rare large impacts destroy smaller asteroids or reassemble the configurations of larger ones. At least several percent of the asteroids must be binary or multiple systems in which each component separately is more susceptible to collisional destruction than a single body of the same mass. Fragments from rare supercatastrophic collisions yield new Hirayama families; some fragments may be sprayed into resonant orbits that are quickly perturbed into elliptical orbits crossing the orbits of other planets. Such small asteroids and meteoroids lead a transitory existence of only about 10 million years before they impact a planet or become gravitationally yanked into a radically different trajectory.

This collisional and dynamical evolution is just the tail-end of the accretionary processes that gave rise to the terrestrial planets earlier in the solar system's history. As some of these long-protected remnant planetesimal fragments finally enter our atmosphere to crash down as meteorites, they produce a spectacular flash in the nighttime sky before coming to rest on the Earth's surface. Ultimately a few of them are found and dissected in laboratories for clues about the earliest history of the solar system.

The Voyager Encounters

Bradford A. Smith

"Our sense of novelty could not have been greater had we explored a different solar system." — Anon.

This reflection on Voyager 1's reconnaissance of the Jupiter system was expressed by a member of the imaging-science team shortly after our historic encounter with the giant planet on March 5, 1979. It conveyed for all of us our astonishment with the thousands of scenes we had witnessed of this assemblage of bizarre, yet beautiful worlds — worlds which had previously belonged only to the astronomer or to the occasional watcher of the night skies. Changed forever was our distant view of Jupiter and the Galilean satellites; the pale yellow disk surrounded by its starlike points of light had suddenly become real, and even if we failed to understand all that we saw, we had nonetheless become curiously familiar with a distant planetary system. Twenty months later this intimacy would extend to Saturn.

JUPITER: THE PRE-VOYAGER VIEW

Jupiter is the largest of the planets, possessing more than the combined mass of everything else in the solar system, excepting the Sun. Yet, at a mean distance of nearly 800 million km from Earth, it is a difficult object for telescopic observation. The disk of Jupiter displays a series of alternating light and dark bands, parallel to the equator and shaded in subtle tones of blue, brown and orange (Fig. 1). In

Fig. 2. Four hours before its brush with Jupiter in 1974, Pioneer 11 obtained this close-up of the Great Red Spot and neighboring features in the Jovian southern hemisphere. At that point, the spacecraft was as close to Jupiter as the Moon is to the Earth.

Fig. 1. One of the most detailed photographs of Jupiter ever taken from Earth, this image from the Lunar and Planetary Laboratory's 154-cm telescope in Arizona hints at the dynamic turbulence present in the planet's upper atmosphere.

addition to these features, there are numerous bright and dark spots, the largest and best known being the Great Red Spot. By carefully tracking the more conspicuous of such localized features, astronomers — both professional and amateur — were able to map the global circulation patterns of Jupiter's tropospheric winds. The meridional profile showed a pattern of alternating easterly and westerly (zonal) winds that seemed to be related to the banded cloud structure. At the equator was a 30,000-km-wide jet traveling toward the east at speeds as high as 400 km per hour. The only nonzonal winds seen were in the anticyclonic flow around the Great Red Spot, first noted by Elmer Reese and the author in 1966. Centuries of visual and, more recently, photographic records show that the contrast and color of the Jovian cloud system is in a continuous state of change; dramatic increases or decreases in the reflectivity of a single band have been observed over only a few months and the entire planet's appearance can change in just a few years.

Fig. 3. Voyager 1 recorded this view of Jupiter from a range of nearly 33 million km. From that distance, cloud features as small as 600 km across can be resolved. The Great Red Spot and whorls of turbulence to one side dominate the planet's southern hemisphere. Farther to the south are several white ovals that first appeared about 40 years ago. The white puff in the equatorial zone north of the red spot marks one area where bright clouds originate, then stream westward. Voyager color photographs are actually composites of several black-and-white frames taken through colored filters.

Among Jupiter's satellites (13 were known prior to Voyager), the four Galilean satellites stand in a class by themselves; Io, Europa, Ganymede and Callisto are each planet-size bodies in their own right and can actually be seen as tiny disks in a large telescope. Vague, dusky markings have been reported by visual observers, and when combined with records of periodic brightness variations during the course of an orbit, these observations show that each Galilean satellite always keeps the same face toward Jupiter.

Although other types of telescopic instrumentation have told us much about Jupiter and its satellites, information based on direct observation has left much to be desired. Observers attempting to make systematic observations of Jupiter's clouds have been hampered by unpredictable weather and other difficulties; variable observing conditions give rise to sporadic data of nonuniform quality. When the objectives involve time-variable phenomena, such as the morphology and motions of atmospheric features, the inability to obtain uniform and consistent records is a severe impediment to an observer.

The space age came to the outer solar system with the arrival of Pioneer 10 at Jupiter on December 3, 1973. This extraordinary accomplishment was repeated one year later when its sister spacecraft Pioneer 11 flew by the giant planet on December 2, 1974. The colorful images transmitted back by Pioneers 10 and 11 (Fig. 2) gave us our first appreciation of the complex nature of the Jovian cloud system. Much of this intricate structure was located at the interface between light and dark bands where telescopic observations had suggested large wind shears. At high latitudes the morphology of individual features, some as small as 500 km across, suggested convective cloud systems. Structure seen within the Great Red Spot and elsewhere implied complex atmospheric motion; unfortunately, because of the brief duration of the Pioneer encounters, repetitive coverage was seldom obtained and few cloud motions could be measured. Thus, although Pioneer gave us a new and captivating insight into the detailed structure of the global cloud system on Jupiter, it provided little in the way of dynamic information. Pioneer also recorded several images of the Galilean satellites, and while the resolution was superior to that obtained from telescopic observations, it was insufficient to give even a hint of any geologic sculpturing of their surfaces.

VOYAGER: THE APPROACH

The Voyager 1 approach phase at Jupiter began on January 4, 1979, just 60 days before encounter. At this time the narrow-angle (telephoto) camera began systematic imaging

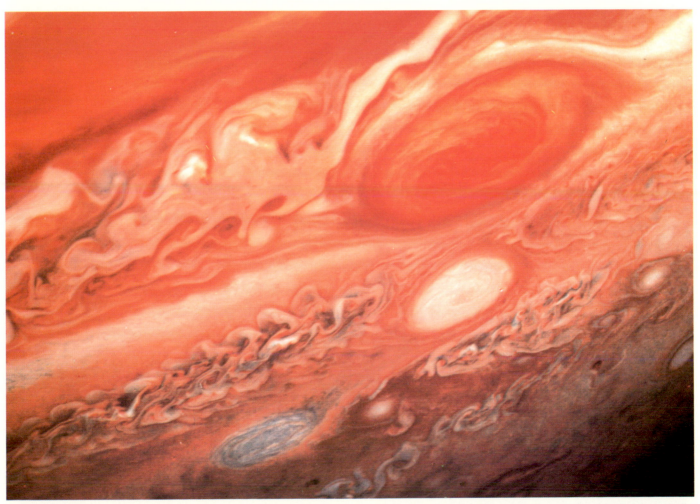

Fig. 4. By the time Voyager 2 reached Jupiter in mid-1979, the turbulence west of the Great Red Spot viewed by Voyager 1 had broken into smaller cloud parcels and vortices. A region of white clouds cap the GRS's northern boundary, preventing smaller puffs from circling the immense cyclonic feature as they had done four months before.

Fig. 5. This is one of the most striking images from Voyager 1, which captures ruddy Io and pearl-like Europa against the vast backdrop of Jupiter, here about 20 million km from the spacecraft. Circulation patterns are evident in the planet's atmosphere, especially the intriguing eddies located within and between cloud bands. Even from this distance, Io shows color variations on its surface.

of Jupiter, taking a multicolor sequence every two hours. More than a month earlier Voyager test images had already exceeded the resolution of the best telescopic pictures taken from Earth; now we were beginning a systematic series of observations designed to collect data on the structure and dynamics of the Jovian atmosphere. Even at that great distance of 60 million km, the pictures were spectacular and getting better every day (Fig. 3). From the very earliest images, it became clear that the structure of the Jovian clouds was even more complicated than we had thought. As the data continued to accumulate and cloud motions became apparent, we could see that the detailed dynamical picture was going to be equally complex. The model that we had developed from telescopic observations was wrong — rarely did we see simple zonal motion, but instead there was a planet-wide pattern of small-scale vortices. The interaction of zonal flow with the anticyclonic vortex of the Great Red Spot (Fig. 4) produced a complicated, time-variable maelstrom in which clouds were torn apart and often swept into the swirling anticyclone. Elsewhere, interacting currents caused clouds to consume their neighbors, only to disgorge them a few tens of hours later. Others engaged briefly, circled each other in a *pas de deux,* then separated and continued on their way. Later we would learn that the vorticity, drawing its energy from the interior of Jupiter, was the source of momentum for the well-known zonal flow.

During the approach phase, we watched certain cloud systems with a special interest. These were the features selected from a ground-based photographic patrol, operated by team member Reta Beebe, as potential targets for close scrutiny at encounter. The continued monitoring of these atmospheric systems by Voyager was essential if we were to predict their precise location and extent at the time of encounter. Final coordinates had to be provided one month before the high-resolution pictures were taken, thereby placing a requirement on the imaging team to make a 30-day forecast of Jovian weather. That we were successful with nearly every target is testimony to the fact that not all Jovian phenomena are unpredictable.

Some 30 days before encounter we exceeded the highest spatial resolution achieved by Pioneer; we had entered *terra incognita*. Jupiter loomed before us in a spectacle of bizarre cloud formations and dazzling variegated color that went beyond the sum of our collective imaginations. The immensity of what we were seeing was staggering: the vortex of the Great Red Spot alone could swallow several Earths. As we approached encounter, Jupiter grew larger than our field of view (Fig. 5). Features barely visible from Earth filled an entire Voyager picture. Later would come the analysis and the scientific significance (see Chapter 12); at the moment we were caught up in the sheer wonder of it all.

Throughout the approach the images of the Galilean satellites had grown in size as they passed back and forth across our television monitors. Among all of the planetary bodies visited thus far by exploratory spacecraft, none elicited less distinct images in our minds than these four

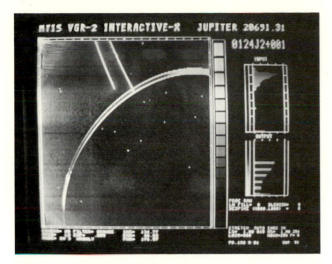

Fig. 6. Soon after passing Jupiter, Voyager 2 turned its television eyes back on the planet, now backlit by the Sun, from within its shadow. Monitors at the Jet Propulsion Laboratory in California flashed with images of the tenuous Jovian ring, which gleamed with unexpected brightness when viewed from that perspective.

large moons of Jupiter. Now they were becoming individually recognizable bodies. Within a few days we would see them as clearly as astronomical telescopes view our own Moon — and we were about to make one of the most important discoveries of the space program.

THE ENCOUNTER

While Jupiter was the star of the show during the approach, we were actually *too* close to the huge, cloud covered planet at encounter. The images were low contrast and fuzzy. As we passed behind on the night side, however, our cameras recorded a number of interesting phenomena such as lightning, a meteor, and auroras.

Just as we passed through Jupiter's equatorial plane, about 17 hours before closest approach, our cameras turned to the side for an 11-minute-long time exposure. In that picture was a faint ring. The discovery was not accidental. Everything we understood about the formation and stability of planetary ring systems at that time had told us that Jupiter could not have a ring. Nevertheless, one does not travel nearly 1 billion km into space without at least a quick look. Even though the image had been a part of the encounter sequence (primarily at the insistence of team member Tobias Owen and experiment representative Candy Hansen), most of us considered the chance of actually finding anything as nil and had actually forgotten about it. The ring, it turns out, was seen much better by Voyager 2 in back-illumination (Fig. 6), but it was a major discovery of the first encounter. More is said about Jupiter's ring (and those of other planets) in Chapter 13.

Even with our camera's 1,500-mm focal length (granted us by NASA to our everlasting gratitude), only five of Jupiter's satellites would show significant disks: Amalthea and the four Galilean satellites. Amalthea, so small that it is seldom seen even with large telescopes, was certain to present a serious targeting problem; we did not know exactly where it would be. Fortunately, it was identified by Voyager navigation team member Stephen Synnott in the approach images. This made it possible to aim our cameras accurately and

during encounter several high-resolution images were obtained. As team member Joseph Veverka pointed out, Amalthea turned out to look a little like a dark red potato, complete with eyes. Synnott would later discover two, very small, new satellites of Jupiter in the Voyager images. A third would be found by David Jewitt, a graduate student working for team member Edward Danielson. Those satellites and Amalthea are described further in Chapter 19.

Our preconceptions of what the surfaces of the Galilean satellites would be like varied from one team member to the next. Many of us thought that Ganymede and Callisto, with their thick ice mantles, would be smooth; ice would flow as a plastic, reducing all impact or tectonic structures to something approaching a geopotential surface. Io, being without ice, would look like a reddish twin of our own moon, covered with sulfur-coated impact craters. Io, in fact, would be the Rosetta Stone for meteorite fluxes in the outer solar system, allowing us to translate the impact records, and therefore the ages, of all other satellite surfaces. Europa probably had a thin coating of ice and, therefore, a hybrid surface — but then, we considered Europa to be the least interesting of the four, anyway. I don't think we could have been more wrong in predicting what we would see on the Galilean satellites. What we had failed to consider is that ice becomes as rigid as steel in frigid space five astronomical units from the Sun; also, we could not have known that there were processes occurring within Io and Europa that had not yet been encountered in planetary exploration.

Fig. 7. This full-disk image of Io from Voyager 1 reveals a variety of features that appear to be linked with the satellite's intense volcanic activity. For example, the circular "doughnut" in the center has been matched with an erupting plume seen in other images. Despite the wildly variegated colors on its surface, Io seems to be covered entirely by only sulfur and sulfur dioxide. No impact craters have been found at all, leading scientists to conclude that the eruptive vents constantly deposit new material on the surface, thus burying the cratering evidence.

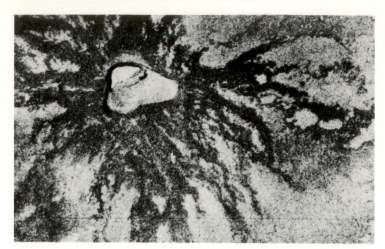

Fig. 8. Voyager 1 passed Io at a distance of only about 30,000 km and recorded intricate details like the volcanic feature seen here. Dark flows of basaltic or perhaps sulfurous lava radiate from a complex caldera which measures about 50 km in diameter.

The first of the ice-mantled satellites to be seen at close range was Ganymede. We had survived the radiation environment at our close pass by Jupiter and the whole team was now standing around, drinking champagne and watching Ganymede on the television monitors. Would there be any topography at all on these icy bodies? The geologists were nervous. Then deputy team leader Laurence Soderblom saw a well-defined impact crater, jumped up, and pointed at the screen; the team cheered, but the expression on Soderblom's face conveyed more relief than jubilation. Several hours later, we were looking at Callisto's battered surface, and craters of all sizes were everywhere.

On both Ganymede and Callisto there was ghostly evidence of impact features that had lost all of their relief, presumably a result of plastic flow in the warm ice of an earlier time. But these were relatively rare; the surfaces of both satellites were sculpted by topography which had held its form over aeons in the 120° K ice. The geologists were ecstatic — an evolutionary record had been preserved. We soon realized that the surface of Callisto is very old, perhaps dating back to the late torrential bombardment period more than 4 billion years ago. It is a world long dead and its aspect vaguely familiar to geologists experienced in studying the silicate bodies of the inner solar system. Some regions on Ganymede also date back to that very early epoch, but others are younger and show a strangely grooved landscape that is suggestive of internal activity. This was *not* familiar terrain, and discussion of the responsible mechanisms would go on for months.

Europa was not seen well by Voyager 1, although what we did see was fascinating. No craters were evident, but the satellite's surface was interlaced with a remarkable pattern of linear features, looking curiously like a Lowellian drawing of Mars. We suspected global tectonics, and this greatly enhanced the anticipation of our more favorable Voyager 2 encounter with Europa — this "least interesting" Galilean moon.

At a distance of 8 million km, one week before encounter, Io was beginning to show circular features which we assumed were craters. Each passing day, however, brought increasing doubts about our Rosetta Stone; the "craters"

Fig. 9. Before moving on to Saturn, Voyager 2 captured this dramatic view of Jupiter's receding crescent. The Great Red Spot appears near the limb.

didn't really look like craters. As we approached, the surface of Io began to take on the most bizarre appearance of any object yet seen in the solar system. Great splotchy regions of yellow, orange, white, and black gave its surface a grotesque "diseased" appearance (Fig. 7). The "craters" were actually albedo features; no true impact craters were to be found. Their absence had profound implications: since Io could hardly have escaped meteoritic bombardment, some very active process had to be destroying or burying the craters. The surface of Io was, therefore, relatively young. Our first estimates of 100 million years seemed to be incredible, but as resolution improved and impact features continued to be absent, our estimates of the surface age became even younger. At 10 million years it was obvious that volcanism must be responsible; in fact, features which looked like volcanic landforms were already becoming recognizable. At one million years we realized that volcanism must be as extensive as on Earth and that it was only bad luck that had prevented us from seeing an actual volcanic eruption taking place as Voyager flew by.

The nature of the volcanism was debated within the team. Red, orange, and yellow sulfurous material seemed to be associated with the volcanic features, but there were closely related black deposits as well. Silicate volcanism, indicated by the black deposits, was proposed by Harold Masursky and Michael Carr. Sulfur was suggested by Eugene Shoemaker and myself. Carl Sagan then pointed out that allotropes of sulfur can take on many colors, including white and black. Long sinuous flows could be seen emanating from volcanic calderas in images that showed features less than 1 km across (Fig. 8). Still, no impact craters could be seen. Voyager had made an important and surprising discovery, but the biggest surprise was three days away.

On March 8th, navigation team member Linda Morabito made her historic discovery. On an image of Io taken for spacecraft trajectory evaluation was a bright volcanic plume rising hundreds of kilometers above the limb. Encounter-weary imaging team members, home for a weekend rest, quickly returned to JPL. Within hours several more volcanic plumes were found in the encounter images. The active volcanoes had been there all along, but special contrast enhancements of the pictures were required to make the relatively faint plumes stand out against the black sky. A final count showed that eight volcanic eruptions were occurring during the flyby of Voyager 1. Our guess that Io might be as volcanically active as the Earth was far too conservative. Team member Torrence Johnson had calculated that the volcanic resurfacing rate on Io could be as much as 10 mm per year. Io's surface was as young as yesterday.

Voyager 2 arrived at Jupiter on July 9, 1979, four months after Voyager 1, giving us a different view of the planet's global wind patterns and a new look at phenomena found by the first spacecraft: the ring system, Europa's smooth surface and the Io volcanoes. As Garry Hunt, Verner Suomi, and Jimmy Mitchell had predicted, Jupiter's weather had undergone a number of changes. Europa was found to have the smoothest surface in the solar system; no topography greater than a few hundred meters was evident. The enigmatic linear features continued to puzzle us although they were obviously a result of some active process in Europa's thin ice crust. Team member David Morrison remarked that the pattern looked like a cracked egg shell and Shoemaker thought it reminded him of packed sea ice. Crater experts Robert Strom and Joseph Boyce remained silent; not a single well-defined impact crater was seen, although a dozen or so circular features may actually turn out to be badly eroded craters. The surface of Europa was relatively young, although not nearly as young as Io's.

One volcano on Io had died in the four months between the encounters, one had turned out of our view, and two new eruptive vents had formed. From these crude statistics it appeared that eruptions may last for several years. The driving volatile for the plumes seemed to be sulfur dioxide or sulfur. There was disagreement among team members as to whether we were witnessing sulfur-enriched silicate volcanism or a new type based entirely on sulfur and sulfur compounds. The issue remains unresolved to this day.

Even as our cameras looked back at the receding crescent of Jupiter (Fig. 9), we had all become aware that this giant planet would never again be the mystical, wandering star of the ancients — or even the mysterious, banded disk of the modern astronomer. The secrets of Jupiter and its satellites had been revealed; these astronomical bodies belonged now to the atmospheric physicist, the geophysicist and the geologist. Meanwhile, ahead of us lay Saturn, and it too would soon become known.

NASA public affairs officer Frank Bristow moderates a well-attended formal press conference at JPL's Von Karman Auditorium. Hundreds of media representatives from around the world flocked to JPL during the Voyager encounters for firsthand reports from project scientists.

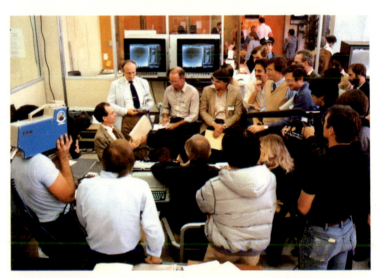

Press conferences during the encounters were often spur-of-the-moment events prompted by the appearance of knowledgeable project personnel. Here chief scientist Edward Stone offers fresh insights to eager reporters in the press room.

Albert Hibbs readies for another broadcast from JPL's "Blue Room," a self-contained television studio that broadcast live and taped results from the Voyager spacecraft during each encounter.

Fig. 10. Before Voyager, this was the best existing color photograph of ringed Saturn, taken by astronomer Stephen Larson in March, 1974. At that time the planet's axis was tipped 27° toward Earth, allowing terrestrial observers to view the rings fully open. Despite this exquisite detail, (the result of combining 16 separate transparencies), Larson's photograph gives no hint of the hundreds of fine ring divisions revealed by Voyager imagery.

Fig. 11. Pioneer 11 saw Saturn's rings from their unlit side, which created a kind of negative effect because sunlight filtered through each ring differently, according to the density of particles within it. Seen from left to right are the C ring (very bright), the nearly opaque B ring (almost black), Cassini's division (thin and bright), the A ring (edged in brown), and a faint, detached arc of the previously unknown F ring. At far right is a small satellite.

SATURN: THE PRE-VOYAGER VIEW

Saturn is the most remote and the dimmest of those planets known to the ancients. Moving slowly about the Sun in the cold, faintly lit fringe of our solar system, it is nearly twice as far from Earth as Jupiter and receives scarcely more than 1 percent of the solar energy that falls on our own planet. Saturn's great distance, more than 1.4 billion km, and low surface brightness placed severe limitations on our pre-Voyager knowledge of the planet, its satellites, and its rings. If the surprises found at Jupiter were an inevitable consequence of our prior ignorance of the Jovian system, Voyager was certain to treat us to a host of unexpected and exciting revelations as we approached the Saturn system.

We could say very little about our prospective target from viewing it through ground-based telescopes. A century of photographic records and more than three and a half centuries of visual observations had produced only a smattering of knowledge. Saturn's disk is marked by a series of light and dusky bands, similar to those on Jupiter, but with far less contrast in both brightness and color (Fig. 10). This surface is unblemished by the light and dark spots that give Jupiter its characteristic appearance, yet it was known that this generally featureless appearance could not be due entirely to Saturn's greater distance. In all, less than a dozen well-documented spots had ever been observed in Saturn's atmosphere, but those few that could be followed gave us an approximate value for the planet's rotation period and showed us that the equatorial zonal jet was considerably stronger than the one on Jupiter; there was little more that could be said about the dynamics of Saturn's atmosphere.

Among Saturn's satellites, only Titan can be seen as a disk, even through large telescopes. Several visual observers reported dusky shadings that seemed to vary with time, but such observations of the tiny disk (1.0 arc second across) were suspect. The other satellites are seen only as points, although they have been noted to display variations in brightness with their orbital longitude; the most striking example is Iapetus, whose brightness during one orbit varies by a factor of six. In 1966 Audouin Dollfus discovered a

Fig. 12. In bringing out faint details, color exaggeration during processing of a Voyager 1 image has transformed Saturn's disk into a psychedelic assortment of colored bands. The largest violet band is actually a faint brown marking, visible from Earth, called the North Equatorial Belt.

small satellite orbiting just outside the bright ring system; in a reanalysis of the 1966 observations, John Fountain and Stephen Larson reported still another satellite in the vicinity of the Dollfus object. Both satellites were confirmed in 1980 and found to be co-orbital. In March, 1980, J. Lecacheux reported the discovery of yet another faint satellite near the Lagrangian point 60° ahead of Dione in its orbit. This brought to twelve the total number of confirmed Saturn satellites prior to the arrival of Voyager 1 in November, 1980.

Saturn is observed, telescopically, to have three bright rings. The more conspicuous A and B rings are separated by the easily seen Cassini Division; the "crepe" or C ring is too faint to be recorded by the limited dynamic range of most photographic emulsions, but is not difficult for a visual observer. Several brightness minima in the A and B rings, most notably the Encke division, were reported by various visual observers, but these features had never been obvious photographically. The existence of the Encke division, however, was confirmed photoelectrically in 1977 when the satellite Iapetus was eclipsed by the Saturnian rings. In addition to the three bright rings, two fainter ones have been reported by ground-based observers: a faint band (subsequently designated D) interior to ring C was reported in 1969 by Pierre Guerin, and an even fainter ring extending

Fig. 13. Three weeks away from its encounter with Saturn, Voyager 2 had this view of its majestic target, 21 million km distant. The planet's surface is crossed by light and dark atmospheric bands. Within the rings are the now-familiar "spokes" (whose multiple images in this color composite show their motion between exposures). The icy moon Tethys is just below the planet; Dione is at bottom. Tiny Mimas appears to the upper left of Tethys, inside Saturn's disk.

outward from the bright rings to beyond the orbit of Dione was photographed in 1966 by Walter Feibelman. During the passage of the Earth through the Saturn ring plane in 1979-80, this faint outer ring (designated ring E) was confirmed by several observing groups and was actually found to extend nearly to the orbit of Rhea.

The Pioneer 11 encounter with Saturn on September 1, 1979, contributed greatly to our knowledge of the magnetic and charged-particle environment of Saturn and of the thermal, photometric, and polarimetric properties of the atmospheres of both Saturn and Titan. In most respects the imaging results, however, were disappointing. No clearly identifiable discrete clouds were seen on Saturn, dashing our hopes that a Jupiter-like spottiness might lie just below the resolution of the better telescopic photographs; the cloud-tops of Saturn in the Pioneer images showed the same banded pattern seen from Earth, but little else. Titan, the only Saturnian satellite to be seen by Pioneer as anything more than a point of light, was frustratingly featureless. However, several new and intriguing features were evident within the ring system (Fig. 11): the Encke division was photographed clearly for the first time and structure was seen within the Cassini division. A narrow (less than 500 km) ringlet, designated the F ring, was discovered approximately 4,000 km outside the outer edge of A; although the actual width of the new ring appeared to be below the resolution of the Pioneer images, a hint of clumpiness was suspected. Not all of Pioneer's contributions to rings and satellites involved direct imaging, however, as the instruments that measured charged-particle fluxes noted a number of minima that implied the existence of several as yet undetected satellites or rings located between the new F ring and Mimas, the innermost of Saturn's brighter satellites.

THE APPROACH TO SATURN

With this and little more to prepare us, we awaited the Voyager 1 encounter with Saturn during the late summer and autumn of 1980. The approach phase of systematic imaging was to begin on August 25th, 80 days before the November 12th encounter, but test images taken in early summer had already exceeded the best images obtained with telescopes. Those of us interested in atmospheric dynamics were greatly concerned over the lack of discrete cloud features in the telescopic, Pioneer, and early Voyager images (Fig. 12); without such cloud systems to track, it would not be possible to map the global circulation of Saturn's atmosphere. Fortunately, our fears were put to rest at the start of the approach phase; contrast enhancement done on our interactive computer terminals revealed a low-contrast bright spot in Saturn's North Tropical Zone. Within a few weeks several more were found, including a relatively large oval feature in the southern hemisphere. This southern hemisphere feature, first seen by experiment representative Anne Bunker, was observed throughout the approach and bore certain similarities to the Great Red Spot on Jupiter. As resolution improved, more atmospheric features — bright and dark — became visible. The morphology of these cloud patterns was encouragingly familiar. We had seen such clouds dozens of times only 16 months earlier at Jupiter; these were very similar, but extremely low in contrast. Our atmospheric enthusiasts relaxed; the data needed to define Saturn's global circulation were being safely recorded by computers on the ground.

Preliminary measurements, made within a few weeks of encounter, confirmed the extraordinary equatorial current that had been inferred from ground-based observations; the

Fig. 14. An enormous impact crater on Mimas (*right*), one of Saturn's inner satellites, gives the moon a surprising resemblance to the Death Star spaceship from *Star Wars* (*left*).

maximum wind speed was 1,800 km per hour, four times greater than Jupiter's equatorial jet. But this was only the first hint that Saturn's zonal wind system is strangely dissimilar to that of Jupiter. Saturn's equatorial jet (if "jet" is still the proper term) is very wide, more than 80,000 km, extending to $\pm 40°$ latitude. Poleward of 40° latitude there appear to be several alternating easterly and westerly zonal currents, similar to those on Jupiter; but the relationship of maximum velocities to the locations of bright and dusky zones is decidedly different. Why should the zonal wind system at the cloud tops on Saturn be so different from that on Jupiter? To a first approximation both planets have the same dimensions, the same bulk composition, and similar cloud morphology; both have internal heat sources. The answer may be found in the physical structure of Saturn's interior as suggested by team member Andrew Ingersoll, or in the strong seasonal effects that are geometrically amplified by the presence of the ring system.

There are, of course, other atmospheric questions to be answered. Why are the clouds of Saturn so low in contrast and unvariegated in color? Saturn is, after all, a colorful planet (Fig. 13); it's just that it is nearly all the *same* color — the *chromophores,* or coloring pigments, are well mixed. Is the low contrast of the clouds intrinsic or due to an overlying haze? The evidence seems to point toward intrinsically low contrast, consistent with the idea that the chemistry within most of the convective cells is similar.

During the Jupiter approach phase, we were content to do little more than wait patiently for the accumulation of atmospheric dynamical data. With two Jupiter encounters behind us, the imaging team was even more relaxed as the spectacle of the ringed planet grew ever larger on our television monitors. It was team member Richard Terrile who broke the quiet routine on October 6th with the discovery of dark, spokelike features extending radially outward across ring B and revolving around Saturn with the ring particles. A hastily prepared "movie" from early October's time-lapse sequence of Saturn's rotation gave a dramatic visual rendition of Terrile's findings. Working quickly with Voyager project personnel, we were able to modify a similar, higher resolution sequence scheduled for October 25th, pointing the camera off Saturn and toward one ansa (or tip) of ring B. The sequence was carried out flawlessly, and the results were spectacular. Although the data were everything that we could have hoped for, the interpretation of the

spokes was to become a headache that would continue to persist even during the Voyager 2 encounter. Spokes were readily apparent in two specially designed time-lapse sequences made during the approach. Commented one frustrated team member, "I wish they'd just go away."

This was not the only surprise to come out of the approach phase; team member Stewart Collins and Terrile, while tracking the motion of the co-orbital satellites, found two new satellites just outside and inside the F ring. Although very small (some 100 km across), it was immediately evident that these "shepherding" satellites play a crucial role in the Saturn ring system, stabilizing ring F against disruptive non-gravitational forces. Just before encounter, Terrile found still another small satellite (20 by 40 km) orbiting only 800 km beyond the outer edge of ring A. This tiny object might be considered insignificant anywhere else, but it appears that in its unique location the satellite provides stability against outward diffusion for the entire bright ring system of Saturn.

As the final days of the approach phase were upon us, anticipation grew. The satellites were finally being seen as individual disks and there was something very peculiar about the bright rings — they were daily showing more and more structure, far more than could be accounted for by simple satellite resonance theory. We were prepared for an exciting encounter, but even bigger surprises were to come.

THE ENCOUNTER

The Saturn encounter would be more intense than either of those at Jupiter; the compact system of satellites and rings and the rapidly changing viewing geometry meant that the most interesting data would all be transmitted within just a few hours of Voyager's closest approach to Saturn. Eighteen hours before passing Saturn, Voyager 1 would encounter Titan and cross over to the unlit side of the rings. Five hours after closest approach, the spacecraft would cross back to the illuminated side of the rings, now viewing a back-illuminated planet with a phase angle of 150°. During this brief interval would occur the closest approaches to Titan, Dione, Mimas, the co-orbital satellites, and Rhea. Our best views of Tethys, Enceladus, and the outer satellites would have to await the encounter by Voyager 2.

Our first brush with a Saturnian satellite was with Titan, considered by many to be one of the most important targets of the entire Voyager program. After so many years of expectation, the imaging results were disappointing to even the most loyal Titan enthusiasts. On the evening before encounter, team members watched silently as dozens of images of the cloud enshrouded satellite were displayed with monotonous, featureless repetition. The only details visible at the interactive terminal were a slight hemispheric difference in cloud brightness, displaying a curiously sharp demarcation at Titan's equator, and a global haze layer that thickened near the north pole to produce a polar hood; 10 months later Voyager 2 would find that the polar hood had evolved to a polar collar. None of the hoped-for holes in the clouds, no surface features to be seen. James Pollack wore the only smile — he was interested in Titan's atmosphere. (Further enhancement of the images revealed a series of low-contrast bands parallel to Titan's equator, giving the satellite an appearance similar to that of Jupiter and Saturn). Although the imaging results were a disappointment, other Voyager instruments were reassuringly successful (see Chapter 15) and Titan still retains its high position on any list of interesting bodies in the solar system.

From the very beginning we knew that we might not see the surface of Titan, but now we were approaching the bright, icy satellites on whose surfaces we would surely see features, in some cases as minuscule as 1 km across. This group of bodies represents a class of objects never before encountered. Intermediate in size between the Galilean satellites and Amalthea, they are composed largely (in some cases, perhaps entirely) of water ice. After our experience at Jupiter, we were prepared for almost anything. What we saw turned out, at first glance, to be rather commonplace: at high resolution the icy surfaces of Dione and Rhea looked very much like the Moon or Mercury, although the morphology of the individual craters seemed slightly different. There were puzzling features, however. The poorly resolved hemispheres of both satellites showed bright, wispy streaks that might have been caused by ejecta from impacts, but some of the team members believed they could have been created by the condensation of volatiles leaking from the interior. Were these small bodies exhibiting internal activity? Furthermore, both Dione and Rhea had regions in which craters seemed to be too few in number, as though many had been obliterated or covered up. Was this still more evidence for internal activity at some time in the distant past? Mimas seems to be more normal — except for the giant impact crater that is more than a third the diameter of Mimas itself. As the image of Mimas and this absurd crater first appeared on our television monitors, there was a sense of *deja vu*. Of course! It was George Lucas' Death Star. The resemblance (Fig. 14) is uncanny. Aside from its strange appearance, the very size of the crater seemed improbable. It is difficult to understand how Mimas could have received such an impact and survived.

Iapetus, Hyperion, and Tethys, seen poorly by Voyager 1, were imaged at much higher resolution by the second spacecraft. Icy Tethys displayed an enormous impact feature, proportional in size to the large crater on Mimas, and a battered, irregular Hyperion was revealed with a spin orientation and rate that continue to defy analysis. Iapetus was seen well enough to show one heavily cratered, bright hemisphere and another very dark, apparently featureless hemisphere. Unfortunately, the images did not show enough detail for us to rule out any of several competing hypotheses offered to explain the satellite's dichotomous character. Exogenous mechanisms proposed by astronomers Sagan and Morrison, endogenous processes cham-

Fig. 15. Four days after its dazzling and flawless encounter with Saturn, Voyager 1 looked back on the planet from a distance of more than 5 million km. This view of a crescent Saturn — unobtainable from Earth — is but one of 17,000 pictures by the spacecraft.

pioned by geologists Masursky and Shoemaker, and hybrid models suggested by Soderblom, all remain viable. (More discussion appears in Chapter 16.)

Enceladus also was poorly seen by Voyager 1, but not so poorly that its highly anomalous appearance escaped notice. At the limit of resolution (10 km) the observed surface appeared to lack topography of any kind! Since there is no way that Enceladus could have escaped the meteoritic or cometary bombardment suffered by its neighboring satellites, some mechanism must be destroying its craters at an astonishing rate. Enceladus is in an eccentric orbit forced by Dione, suggesting to team members Allan Cook and Terrile that tidally dissipated energy, similar to that experienced by Io and Europa, is also responsible for the rapid resurfacing of this Saturnian satellite. Voyager 2, however, revealed a heterogeneous surface, cratered in some places but devoid of impact features elsewhere. The very existence of topography was a great relief to the team geodesist, Merton Davies, and the complex nature of the surface has delighted (if not confounded) the geologists.

Several observations were planned to observe the two co-orbital satellites, the Lagrangian companion to Dione, and two new moons recognized from the ground after the Voyager 1 flyby — tiny bodies occupying the Lagrangian points preceding and following Tethys. The irregular shapes of the co-orbital satellites and the proximity of their orbits implies that they are now the remaining pieces of a single satellite which received an impact having rather more serious consequences than the one that formed the big crater on Mimas.

Meanwhile the rings continued to draw our attention. Even when Voyager was still several weeks away from Saturn, the three bright rings were starting to show far more structure than could be accounted for by a simple application of resonance theory. Now, as resolution improved, the three major rings seemed to be breaking down into hundreds of individual ringlets. Some indeed were located where major resonances with satellites should occur; but they were everywhere else as well. In some regions, the structure appears to have organization, with the spacing of individual ringlets following some sort of arithmetic or geometric progression. In other regions the spacing and widths of ringlets is apparently random. In all, there are over a thousand ringlets at the resolution of the Voyager cameras and probably still more that we are unable to see. The Cassini division alone contains more than a hundred individual ringlets and some of them are eccentric. Others have been found in ring C, and even the outer edge of ring B is out of round. Eccentric ringlets within the bright ring system were not high on our list of expected phenomena; in fact, such a suggestion prior to their discovery would probably have been met with verbal abuse. Although the dynamic mechanisms responsible for some of the ring structure have so far been elusive, there now exists an enormous data base in which we hope to find the answers. It was team member Jeffrey Cuzzi who summed it up when he said that we now have far more information about the Cassini division alone than previously for the entire ring system.

In the final hours before encounter with Saturn our attention was fixed on the F ring. Resolution was improving rapidly and the apparent decreasing width of the ring kept pace. Already we knew that it could not be more than 100 km wide and, furthermore, it was exhibiting a curious non-uniformity in brightness which we thought might be due either to variations in width or particle accumulation. The most detailed images of the F ring to be obtained by Voyager 1 were received just 10 hours before closest approach to Saturn; those of us watching the monitors were stunned. If, by our third planetary encounter, we had become somewhat jaded to the unpredictability of the outer solar system, our sense of astonishment was brought back in an instant. In some tabulation of ring phenomena that we least expected to see, the observed structure of the F ring would have been somewhere off the top. In the pictures were three individual strands, each approximately 20 km wide and separated by a few tens of km; they appeared to be knotted, kinked, and braided. To me, it was the most improbable picture yet sent back by either Voyager spacecraft.

The F ring's (or rings') apparent deviations from simple Keplerian motion must be due to complex gravitational interactions with the two shepherding satellites or to interactions with nongravitational forces such as those generated by charged particles moving in Saturn's magnetic field. Non-gravitational forces become important when particles are small, and the enhanced brightness of the F-ring when viewed under forward-scattering conditions suggests that a large fraction of its particles have diameters of only a few times the wavelength of light. Additional Voyager images have shown that the F ring's structure varies along the circumference, and probably changes with time. At the moment, however, all of our explanations are likely to be little more than an exercise in vigorous armwaving.

Without doubt the biggest surprises of the Voyager encounters with Saturn have come from the rings. Ironically, prior to the Voyager 1 encounter, it was with the rings alone that we really felt comfortable; the astonishing discoveries, we thought, were certain to be found elsewhere.

As somewhat of an anticlimax we found the D ring. But it was extremely faint, too faint. This ring could not possibly have been seen from the Earth. Observers (myself included) had reported it, Voyager had photographed it, but it was all just a coincidence — another of Voyager's ironies.

EPILOGUE

As the Voyager spacecraft left the Saturn system, their cameras gave us memorable views (Fig. 15). On December 19, 1980, the two cameras on Voyager 1 were turned off, probably forever. But even as Voyager 1 sails outward toward interstellar space, Voyager 2 has begun its lengthy journey to Uranus, an odyssey that will take longer than the entire flight to Saturn. The spacecraft will pass Uranus on January 24, 1986, and, if all is well, it will be sent onward to Neptune, arriving at this now most distant planet in late August, 1989.

As we have explored ever outward, the solar system has become stranger and stranger. Who could doubt but that Uranus and Neptune hold secrets that will continue to amaze us. Voyager has dramatically extended the eyes of the scientific world and, by so doing, has given all of us — scientist and layman alike — a sense of "being there."

Jupiter and Saturn

Andrew P. Ingersoll

Jupiter and Saturn are the Sun's principal companions. Together they account for 92 percent of the extrasolar mass of the solar system. The most basic statistics about them emphasize their enormity: Jupiter possesses 318 times the mass of the Earth, or about 0.1 percent that of the Sun; Saturn is 95 times more massive than our planet. Both are some 10 times Earth's diameter and about one tenth that of the Sun. Clearly, these bulk physical properties place them in a class midway between the Earth and the Sun.

At wavelengths of the visible spectrum, Jupiter and Saturn shine by reflected sunlight (Figs. 1 and 2) and are among the brightest objects in the night sky despite their great distance from us. They also glow brightly at infrared wavelengths, partly because like all planets they reradiate to space a fraction of the energy received from the Sun. In addition, however, Jupiter and Saturn emit their own stored energy, generated long ago during the gravitational contraction of the nebulae from which they condensed. Had these planets been 10 times more massive, their internal temperatures might have risen high enough to trigger nuclear fusion. Then our solar system would have contained a multiple star, and conditions on all planets would have been considerably different than observed today. Even though the gas giants never reached that stage, enough energy was stored to last until the present.

Jupiter and Saturn are of great interest to planetary scientists for several reasons. By virtue of their size and mass, they occupy a distinct place among objects in the solar system. They and their satellite families constitute two miniature "planetary" systems whose character and evolution can be compared with our perceived history of the solar system proper. Finally, these giants exhibit a wide range of atmospheric phenomena that includes multicolored clouds arranged in parallel bands, huge circulating storms which can last for decades or centuries, lightning and giant auroras (Fig. 3), and swift-moving cloud currents that approach supersonic speeds relative to the planets' interiors (Fig. 4).

Fig. 1. Jupiter, as viewed by the Voyager 1 spacecraft from a distance of 54 million km. The overall color is reasonably accurate, although color contrast has been enhanced.

Fig. 2. Voyager 1 recorded Saturn and its rings from a distance of 18 million km. The color enhancement is similar to that in Fig. 1. Note that cloud features on Saturn are fewer and lower in contrast than on Jupiter.

Fig. 3. The night hemisphere of Jupiter displayed surprising evidence of lightning (bright patches below center) and auroras (curved arcs at top) in this Voyager 1 image. The planet's north pole lies roughly midway along the auroral arc.

Molecule	Sun	Jupiter	Saturn
H_2	0.89	0.90	> 0.94
He	0.11	0.10	< 0.06
H_2O	1.0×10^{-3}	1×10^{-6}	—
CH_4	6.0×10^{-4}	7×10^{-4}	5×10^{-4}
NH_3	1.5×10^{-4}	2×10^{-4}	2×10^{-4}
Ne	1.4×10^{-4}	—	—
H_2S	2.5×10^{-5}	—	—

Table 1. Fractional abundances (by number of molecules) for a solar-composition atmosphere and for Jupiter's and Saturn's atmospheres near the cloud tops. The list contains seven of the 10 most abundant elements in the Sun (and the universe). The other three — Si, Mg, and Fe — are believed to reside in the cores of the giant planets. Dashes indicate unobserved compounds. All numbers are uncertain in the least significant figure.

Thus, at one extreme, we can compare Jupiter and Saturn with the Sun and stars, and their satellites with our own solar system. At the other extreme we can compare the atmospheric phenomena observed on these planets with those on Earth. Our understanding of terrestrial atmospheric phenomena was inadequate to prepare us for what we have found recently about the gas giants, but by testing our theories against observations of these planets, we can understand the Earth in a broader context. This chapter, therefore, delves into their atmospheres, interiors, and possible internal histories.

OBSERVATIONS

Our knowledge of Jupiter and Saturn is derived from earth-based telescopic observations begun 300 years ago, modern observations from high-altitude aircraft and orbiting satellites, and the wealth of recent findings from interplanetary spacecraft. Between 1973 and 1981, four unmanned probes flew by Jupiter (Pioneers 10 and 11, Voyagers 1 and 2), and three of those by Saturn (Voyager 2's encounter: August, 1981).

Basic characteristics like mass, radius, density, and rotational flattening were determined during the first era of telescopic observation. Galileo's early views revealed the Galilean satellites which bear his name. Newton estimated the mass and density of Jupiter from observations of those satellites' orbits. Other observers, using improved optics, began to perceive atmospheric features on the planet. The most prominent of these, the Great Red Spot (Fig. 5), can be traced back 300 years with near certainty, and it may be older still. Beginning in the late 19th century, astronomers made systematic measurements of Jovian winds by tracking features visually with small telescopes. Photographic observations with larger telescopes later augmented these early

efforts. Most recently, tens of thousands of features on Jupiter and Saturn have been tracked with great precision using the Voyager imaging system. The constancy of these currents over the long time intervals spanned by classical and modern observations is one of the truly remarkable aspects of Jovian and Saturnian meteorology.

Determination of chemical composition in the atmospheres began in the 1930's with the identification of methane (CH_4) and ammonia (NH_3) in the spectra of sunlight reflected from their clouds. About 1960, molecular hydrogen (H_2) was detected. Because hydrogen is a simple, symmetrical molecule, its vibrational and rotational absorptions are weak. Fortunately, hydrogen occurs above the clouds in such abundance that its absorption lines are nevertheless detectable in the spectra of both planets. It was quickly verified that the proportions of hydrogen, carbon, and nitrogen in Jupiter's atmosphere were consistent with a mixture of solar composition. Similar inferences have been made for Saturn, although actual observations of these compounds (especially ammonia) are extremely difficult. During the 1970's, as infrared detectors improved, observers recorded absorptions by other gases present in extremely minute amounts (one part per million or even per billion by volume). Ethane (C_2H_6), acetylene (C_2H_2), water, phosphine (PH_3), hydrogen cyanide (HCN), carbon monoxide (CO), germane (GeH_4), and compounds with deuterium (^2H) and isotopic carbon (^{13}C) all were detected in Jupiter's atmosphere in this way (see Table 1). The list for Saturn is shorter than for Jupiter, partly because Saturn is a fainter object and partly because it is colder; many of these compounds freeze at the level of its cloud tops and become harder to detect.

Helium, the second most abundant element in the Sun, is presumably an important constituent of Jupiter and Saturn as well. Unfortunately, it has no detectable absorptions to make its presence known. However, instruments aboard the Pioneer and Voyager spacecraft noticed the effect of collisions between helium and hydrogen molecules on the latter's infrared absorptions, giving us a kind of "back-door" identification of helium. And we can also combine infrared measurements with radio occultation data (obtained as the spacecraft passed behind the planets) to determine the molecular weight of the mixtures in their atmospheres. Voyager experimenters refined both of these methods to yield helium abundances with an uncertainty of only 2–3 percent by volume. The values they derived (10 percent for Jupiter and less than 6 percent for Saturn) are roughly consistent with solar composition, but the evidence suggests that some helium has been depleted from the upper atmosphere of Saturn, a fact which provides intriguing clues to its internal structure and history.

The temperatures and energy budgets of the gas giants have been studied from Earth for decades, but the best determinations have once again come from spacecraft. Both infrared and radio-occultation experiments have probed the cloud tops (in the upper troposphere) up to where the temperature increases with height (the lower stratosphere). Ultraviolet instruments probe the much greater altitudes of the thermosphere and ionosphere. We find that pressures near the cloud tops range from hundreds of millibars (mb) to 1 bar, as on Earth, but the temperatures there are much colder — about 125° and 95°K based on infrared emissions from Jupiter and Saturn, respectively (Fig. 6). From these observations, we infer that Jupiter radiates between 1.5 and 2.0 times the amount of heat it absorbs from the Sun, and that Saturn radiates 1.5-2.5 times its absorbed heat. These excesses provide us with additional insight into internal structures and histories.

The latitudinal distributions of temperatures and heat flux for Jupiter and Saturn (Fig. 7) indicate how effective the winds are in redistributing heat from equator to pole. We would expect to see thermal gradients if this took place near the cloud tops. However, the planets exhibit almost complete lack of thermal contrast, when averages are taken with respect to longitude, time of day, and season. This fact, combined with their latitudinal banding, suggests that the heat transport occurs at deeper levels.

We have inferred the upper atmospheric structure of these worlds by observing the effect of free electrons on spacecraft radio signals occulted by the planets, by comparing ultraviolet emissions on the day and night side, and from watching the Sun in the ultraviolet as it sets behind the planet as seen from the Voyager spacecraft. Temperatures can be deduced from changes in electron density with altitude. The distribution of various emissions over the disk and limb are clues to composition and excitation mechanisms. Excitation stems mainly from charged particles (causing auroras) and energetic solar photons (airglow) striking the atmosphere from above. These processes can alter the chemical composition of deeper layers by causing stable compounds to form that become mixed downwards.

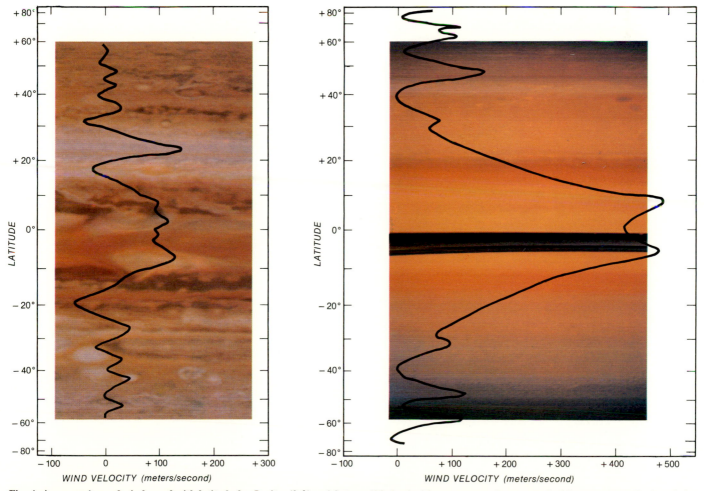

Fig. 4. A comparison of wind speed with latitude for Jupiter (*left*) and Saturn. Wind velocities are eastward, measured with respect to each planet's internal rotation period (9 hours 55.5 minutes for Jupiter, and 10 hours 39.4 minutes for Saturn), which are based on observations of the planets' periodic radio emissions. Negative velocities are, therefore, winds moving westward with respect to these reference frames. The Voyager imaging team has combined these results with images of each planet projected so that the latitude scales match the data and longitude lines run vertically. (The black band across Saturn's equator is the rings and their shadow on the planet.)

ATMOSPHERIC STRUCTURE AND COMPOSITION

Temperatures, pressures, gas abundances, cloud compositions — all as functions of altitude and horizontal position — are the principal components of atmospheric structure. Atmospheric dynamics, which is discussed in the next section, concerns the wind and its causes and effects. However, the distinction between structure and dynamics is not always clear. Atmospheric circulation alters the structure by carrying heat, mass, and chemical species from place to place. The structure, in turn, affects wind patterns by controlling the absorption of sunlight, emission of infrared radiation, and release of latent heat, all of which lead to heating, expansion of the gas, and pressure gradients which drive the wind.

The vertical temperature structure is perhaps most fundamental to understanding a planetary atmosphere. This is normally given as temperature versus pressure (Fig. 6), the latter derived from the altitude using a relationship known as hydrostatic equilibrium. The thickness, in kilometers, of a given pressure interval is less for Jupiter than for Saturn, owing to the more intense Jovian gravity. Temperatures at the deepest levels measured, which correspond to nearly 1

bar of pressure, tend to follow an adiabatic gradient (this is how temperature would change with pressure for a parcel moving vertically at a rate such that there is no heat exchange with the surroundings). An adiabatic gradient is usually a sign that the atmosphere is well mixed by convec-

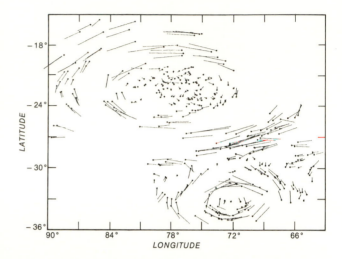

Fig. 5. Extreme color exaggeration has been used to bring out details in this high-resolution mosaic of the Great Red Spot and a white oval (known as BC since its formation more than 40 years ago). The accompanying diagram shows wind vectors in this region, determined from changes observed over a 10-hour period. Each dot marks the position of the cloud feature measured; an attached line indicates the direction of flow, and its length is proportional to the wind velocity.

tive currents. We would expect an adiabatic gradient to extend down almost indefinitely because heat from the interior cannot be carried off by infrared radiation, which is blocked by the opacity of the gases there. Thus convection must carry the internal heat up to levels where radiation to space occurs. For Jupiter and Saturn this occurs near pressure levels of 100–300 mb. At altitudes higher than this, the temperature increases with height due to the absorption of sunlight.

The rise in temperature with altitude above the 100-mb level probably requires more heating that can be accounted for by sunlight striking a gaseous atmosphere. Some scientists attribute this to a layer of fine black dust, produced photochemically, but other evidence of its existence has been hard to find. Farther up, the atmosphere is so thin that temperature is extremely sensitive to small energy inputs. Voyager discovered auroral emissions from Jupiter, especially intense at latitudes where magnetic field lines from the orbit of the satellite Io intersect the planet. The energy producing these auroral emissions is equivalent to about 0.1 percent of the total incident sunlight (a large input at these

altitudes). Thus the charged-particle flux from Io seems capable of causing the high temperatures (1,000–1,300° K) observed near the top of Jupiter's atmosphere, since solar photons produce temperatures only near 200° K. The dissipation of upward-propagating waves could probably also account for the high temperatures, but little evidence for these waves exists.

The vertical cloud structure is inferred by several means. First, one can compute the altitudes above which each atmospheric constituent can condense (Fig. 6). The procedure is to take a uniform solar composition mixture (Table 1), or whatever mixture is appropriate, and compute the partial pressures of each gas at each level. We compare these with the saturation vapor pressures, which are determined from the temperatures at each level. Condensation occurs where the computed partial pressure exceeds the saturation vapor pressure, and since the latter falls rapidly with temperature, clouds will form at the coldest layers. Above the cloud-forming levels, particle fallout may reduce the abundance of condensates and prevent clouds from extending above the top of the convective zone. When these calculations are worked for Jupiter and Saturn, we find three distinct cloud layers (Fig. 6). The lowest is composed of water ice or possibly liquid water droplets. Next are crystals of ammonium hydrosulfide (NH$_4$SH), which is basically a compound of ammonia (NH$_3$) and hydrogen sulfide (H$_2$S). At the top we expect an ammonia ice cloud.

These cloud structures assume solar composition throughout. But the real situation is more complex. Vertical mixing could carry particles from the lower clouds upward, thereby changing the composition of the upper clouds. Moreover,

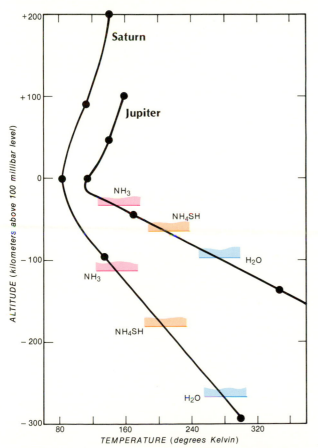

Fig. 6. Pressure-temperature profiles for the upper atmospheres of Jupiter and Saturn, as determined by both ground-based and spacecraft measurements at radio and infrared wavelengths. Each dot marks the point where atmospheric pressure is 10 times greater than the next dot above it; the total range indicated runs from 1 millibar at top to 10 bars near the bottom. Colored bands show the altitudes at which various clouds should form, based on a gaseous mixture of solar composition. Temperatures are lower on Saturn because it is farther from the Sun, and the range of cloud altitudes is broader because Saturn's weaker gravity allows a more distended atmosphere than on Jupiter.

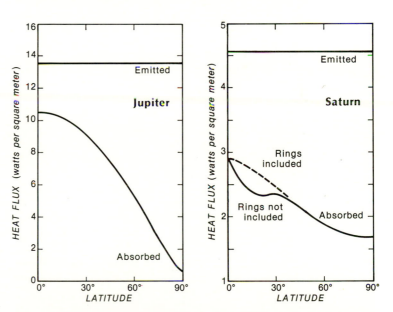

Fig. 7. A comparison of absorbed solar energy and emitted infrared radiation for both planets, averaged with respect to longitude, season, and time of day. Saturn is farther from the Sun and colder, so its heat fluxes are lower (the rings also shadow a large band around Saturn's equator). However, Saturn's seasonal cycle introduces a large uncertainty in the curves shown. Both planets emit more infrared radiation than the amount of sunlight they absorb, implying an internal heat source. The emitted radiation is also more uniform, which suggests heat transport across latitude circles at some depth within the planet. These results are derived largely from Pioneer and Voyager infrared observations, as well as from those in visible light. Irregularities in the energy flux associated with belts and zones are not shown.

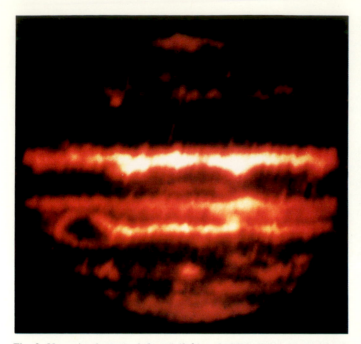

Fig. 8. Near-simultaneous infrared (*left*) and visible-light views of Jupiter. Holes in the upper cloud deck allow radiation to escape from warmer layers underneath; these spots appear bright in the infrared. The infrared image was taken at Palomar Observatory by Richard Terrile.

hydrogen sulfide has not been detected on either Jupiter or Saturn, possibly because it precipitates out below the cloud tops, and possibly because it changes to sulfur when exposed to sunlight. The existence of water clouds is also uncertain. Water vapor apparently occurs only as one part per million by volume, even though solar composition would imply abundances 1,000 times greater. The paradox is that we can see through holes in the clouds (Fig. 8) to warm levels with temperatures of at least 300° K, where both H_2O and H_2S should be abundant. Perhaps these "hot spots" are the deserts of Jupiter, where dry air stripped of its condensable gases is descending into the well-mixed interior. The observations nevertheless cast some doubt on the solar composition model.

The abundance ratios of various gases provide clues to chemical processes that are occurring inside the two planets. If the atmospheres are in chemical equilibrium, with hydrogen the dominant species, virtually all the carbon should be tied up in methane, all the nitrogen as ammonia, all the oxygen as water, and so on. But in Jupiter's case, we have identified gases such as C_2H_6, C_2H_2 and CO, which imply disequilibrium. Both C_2H_6 and C_2H_2 are relatively easy to account for, since they are formed in the upper atmosphere as by-products of methane photodissociation. In these reactions, solar ultraviolet photons at wavelengths less than 1600 angstroms knock hydrogen atoms off the methane (CH_4) molecule, leaving free radicals that can react with other CH_4 molecules to form C_2H_6 and C_2H_2. The existence of CO is more interesting, since oxygen is not readily available in the upper atmosphere — water, the principal oxygen-bearing molecule, tends to condense in the lower clouds. Several explanations have been proposed. One is that oxygen ions, injected into the magnetosphere by the SO_2 volcanoes on Io, enter the upper atmosphere from above. Another explanation is that CO is created at depth in Jupiter's atmosphere, and is mixed upwards before it is destroyed. But phosphorus

should also react with oxygen at depth in this way, so the existence of phosphine (PH_3) is something of a mystery. A critical unknown in these theories is the vertical mixing rate as a function of altitude.

The spectacular colors of Jupiter and more muted colors of Saturn provide further evidence of active chemistry in the atmospheres. Every observable feature in the Voyager photographs corresponds to clouds of various colors and brightness. Infrared images show that cloud color correlates with altitude (Fig. 8). Blues have the highest brightness temperatures (calculated from emissions at specific wavelengths), so they must lie at the deepest levels and are only visible through holes in the upper clouds. Browns are next highest, followed by white clouds, and finally red clouds, such as those in the Great Red Spot, which is a very cold feature judging from its infrared brightness. The trouble is that all cloud species predicted for equilibrium conditions are white. Color must come when chemical equilibrium is disturbed, either by charged particles, energetic photons, lightning, or rapid vertical motion through different temperature regimes. The most likely coloring agent is probably elemental sulfur, which forms a variety of colors depending on its molecular structure. Sulfur is definitely present on Io, which has many of the same colors as Jupiter (Fig. 9). Some scientists believe phosphorus explains the Great Red Spot's color, and organic (carbon) compounds have been proposed by others to explain almost all of the colors.

Thus coloration is a subtle process, involving disequilibrium conditions and trace constituents. The correlation with altitude presumably reflects the processes that cause chemical reactions to occur. For example, higher altitudes receive more sunlight and a higher charged-particle flux. Certain regions may contain more lightning (Fig. 3). And other regions may be sites of intense vertical motion. A different question is why these processes should be organized into large-scale patterns that last for years and sometimes for

Fig. 9. Io transits the southern hemisphere of Jupiter in this Voyager 2 image with a resolution of about 200 km. Note the similarity in color between the planet's clouds and Io's sulfur-covered surface.

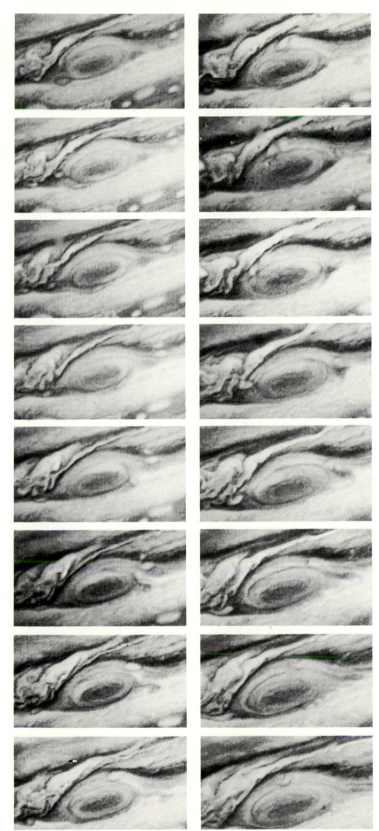

Fig. 10. A blue-light image of the Great Red Spot was taken every other rotation of Jupiter over a period of about two weeks in this sequence, which begins at upper left, continues down each column, and ends at lower right. Note the small, bright clouds that encounter the Red Spot from the east, circle counterclockwise around it, and partially merge with it along the southeast boundary.

centuries. This question involves the dynamics of the atmospheres, to which we now turn.

ATMOSPHERIC DYNAMICS

The dominant observable dynamical features in Jupiter and Saturn's atmospheres are the counterflowing eastward and westward winds (Fig. 4). Instead of one westward current at low latitudes (the trade winds) and one eastward current at high mid-latitudes (the jet stream) as on Earth, Jupiter has five or six of each kind of current in each hemisphere. Saturn seems to have fewer of such currents than Jupiter, but more than our planet; its currents are also stronger than Jupiter's. In fact, the eastward wind speed at Saturn's equator is about 500 m per second, about two-thirds the speed of sound there. As on Earth, these winds are measured with respect to the rapidly rotating interiors. For Jupiter and Saturn, which have no solid surfaces, the rotation of the interior is deduced from that of the magnetic field, which is generated in the metallic core. Even supersonic wind speeds would be small compared to the equatorial velocity due to the planet's rotation.

These currents, the zonal jets, apparently have not changed their latitudinal positions during 80 years of modern telescopic observations. During the four months between the two Voyager encounters of Jupiter, the zonal velocities changed by less than the measurement error, in this case about 1.5 percent. This is remarkable for several reasons. First, although the zonal jets on Jupiter correlate with the latitudinal positions of the colored bands, the bands often change their appearance dramatically in a few years, while the jets do not change. Second, as revealed in Voyager photographs at 30 times the resolution of earth-based images, there is an enormous amount of eddy activity in addition to the zonal jets. The time required for structures to be sheared apart by the eastward and westward jets is only about 1 to 2 days. This is also the lifetime of small eddies that suddenly appear in the zonal shear zones. Larger eddies, including the long-lived white ovals and the Great Red Spot (Fig. 5), manage to survive by rolling with the currents. Voyager repeatedly observed smaller spots encounter the Red Spot from the east, circulate around it in about 6 days and then partially merge with it (Fig. 10). How the zonal jets and the large eddies can exist amidst such activity is something of a mystery.

An important fact about the eddy motions was learned from statistical analysis of Jupiter's winds. During one 30-hour period of Voyager 1 atmospheric observations and another 30-hour period for Voyager 2, many thousands of 100-km features were tracked, usually at 10-hour intervals. Averaged over the planet, the typical (root mean squared) zonal wind was found to be 50 m per second. Departures from the zonal mean wind (eddy winds) at each latitude were moving from 10 to 15 m per second. And a positive correlation was seen between the northward and eastward velocity components at those latitudes where the zonal wind increases with latitude. A negative relation was found in the opposite situation. This correlation is consistent with the idea that the zonal winds are continuously pulling the eddies apart (Fig. 11). The kinetic energy of the eddies is not lost, however, but goes into the zonal jets. Thus the eddies help maintain the zonal jets, not vice versa. The eddies presumably get their energy from buoyancy, and are driven either by heat from Jupiter's interior or by solar heat absorbed at low latitudes.

A similar process helps maintain the Earth's jet streams. However, on Earth this energy transfer averaged over the globe is only one-thousandth of the total energy flowing through the atmosphere as sunlight and infrared radiation. On Jupiter the transfer of energy from eddies to mean zonal flow is more than one-tenth of the total energy flow. In other words, Jupiter is able to harness the thermal energy flow 100 times more efficiently than our own atmosphere. One possible explanation is that the Earth's eddies — the mid-latitude cyclones and anticyclones — get their energy from the sideways transfer of heat due to temperature differences from equator to pole. Jovian eddies get half their energy by vertical convection of heat from the interior. The details have not been worked out, but the latter process is likely to be more efficient than the former.

The constancy of Jupiter's zonal jets still must be explained. If the eddies and zonal jets occupy the same thickness of atmosphere, the eddies could double the kinetic energy of the jets in about 75 days at the observed rate of energy transfer. If the zonal jets extend much deeper than the eddies, the doubling time will be much longer. One explanation for the more than 80-year lifetime of the zonal jets is that the mass involved in the jet motions is at least several hundred times greater than that of the eddies. If the eddies were confined to cloudy layers with pressures no greater than about 5 bars, the jets would have to extend down to pressures of 1,000 bars or more. Such behavior is not as unlikely as it first seems. In a rapidly rotating sphere with an adiabatic fluid interior, the small-amplitude motions (relative to the basic rotation) are of two types. The first are waves with periods near the planet's rotation period. The second are steady zonal motions on coaxial cylinders (Fig. 12), each cylinder moving about the planet's rotation axis at

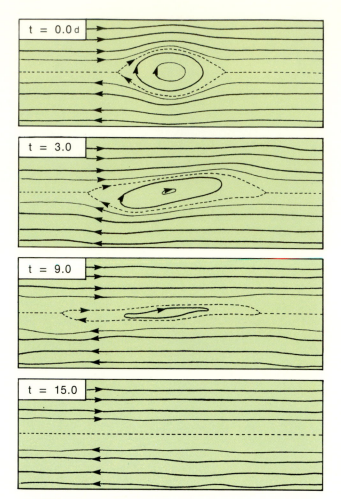

Fig. 11. A computer model of an unstable vortex in a zonal shear flow. Here the initial vortex is spinning too slowly to remain intact and becomes pulled apart. Its energy is transferred into the zonal jets, since total energy is conserved in this model. On Jupiter and Saturn, as on Earth, vortices like this one get their energy from buoyancy. This figure and Fig. 14 are from work by the author and P. G. Cuong.

a unique rate. The observed zonal jets, as far as we know, could be the surface manifestation of these cylindrical patterns.

The problem with such speculation is that we lack information about winds below the visible cloud tops. A totally different approach is to treat Jupiter simply as a larger version of the Earth. Computer models designed for the Earth's atmosphere have given realistic zonal wind patterns when applied to Jupiter (Fig. 13). This is somewhat surprising, since the models assume an atmosphere less than 100 km thick with a rigid lower boundary (the Earth's surface) and no internal heat flow. On the other hand, the process of eddies driving zonal flows is a very general one and seems to occur in a wide range of situations on rotating planets. In fact, several years before Voyager measured it, atmospheric scientists proposed eddy-to-mean-flow energy conversion as an explanation of Jupiter's zonal flow, largely on the basis of these meteorological models.

Clearly, the lack of data on vertical structure has allowed a wide range of models to develop explaining the zonal jets of both Jupiter and Saturn. Some progress may be made by

Fig. 12. The possible large-scale flow within the molecular fluid envelopes (see Fig. 15) of Jupiter and Saturn. Each cylinder has a unique rotation rate; zonal winds (Fig. 4) may be the surface manifestation of these rotations. The tendency of fluids in a rotating body to align with the rotational axis was observed by Geoffrey Taylor during laboratory experiments in the 1920's and was applied to Jupiter and Saturn by F. H. Busse in the 1970's. Such behavior seems reasonable for Jupiter and Saturn if their interiors follow an adiabatic temperature gradient.

examining the secondary predictions of each model, and comparing these with actual observations.

THE GREAT RED SPOT

There is a similarly wide range of theories concerning the long-lived oval structures, well known on Jupiter and discovered by Voyager on Saturn. As already mentioned, Jupiter's Great Red Spot (GRS) is probably more than 300 years old. It covers 10° of latitude, and so is about as wide as the Earth. The three white ovals slightly to the south of the GRS first appeared in 1938; other spots have lasted for months or years. These long-lived ovals, which drift in longitude but remain fixed in latitude, tend to roll between opposing zonal jets. The circulation around their edges is almost always counterclockwise in the southern hemisphere and clockwise in the northern hemisphere, indicating that they are high-pressure centers (Fig. 5). They often have cusped tips at their east and west ends. And at the time of the Voyager encounters, both the GRS and the white ovals had intensely turbulent regions extending off to the west.

Any theory of the GRS and white ovals must explain their longevity and their isolated nature. Longevity involves two problems. The first concerns the hydrodynamic stability of rotating oval flows. If these spots were unstable they would last only a few days, which is roughly the circulation time around their edges (or the lifetime of smaller eddies that are pulled apart by the zonal jets). The second problem concerns energy sources. A stable eddy without an energy source will eventually run down, although on Jupiter this dissipative time scale could be several years.

How do theorists account for the long lifetimes of Ju-

piter's ovals? The "hurricane" model postulates that these structures are giant convective cells extracting energy from condensing gases below (latent heat release). The "shear-instability" model postulates that they draw energy from the zonal currents in which they sit. Still another model postulates that they gain energy from the smaller, buoyancy-driven eddies, much as the zonal jets gain energy. It is also possible that the large ovals gain energy by absorbing smaller eddies.

One model that explains most features of the observed flow fields is the *solitary-wave theory*. Assuming certain conditions, notably a very stable density stratification, there is a class of waves in the visible clouds in which the east-west flow lines would be displaced slightly to the north and south, creating a wave. But unlike ordinary waves in which the displacement is periodic, this wave has just one crest, and is therefore called a solitary wave, or *soliton*. When two solitons collide, they pass through each other unscathed, suffering only a net displacement during the event. Thus the lifetime of a soliton at a given latitude band is limited only by the lifetime of the zonal currents that define the band. One criticism of the solitary-wave theory is that it only describes structures that have a long east-west dimension; short wave disturbances are neglected, yet in reality these can grow to large amplitude and perhaps affect the behavior of the long-wavelength solitons.

An alternate view of Jovian vortices assumes that the effective density stratification is small. This assumption rules out the kinds of waves discussed above. But if there is a basic zonal flow that extends from the top of the clouds well into the interior, the flow lines can be permanently displaced when there is a discontinuity of a certain quantity called *potential vorticity* (related to angular momentum). Regions of anomalous potential vorticity can exist as closed, circulating spots. These vortices or spots exist in the slightly less dense cloud zone and do not affect the zonal flow of the deeper atmosphere. When two such vortices collide they merge, unlike the solitary waves. After merging there is a short transient phase during which the new vortex ejects material, usually westward and equatorward. The new larger vortex finally settles down to a steady state until it encounters another vortex. A computer model can expose this behavior (Fig. 14), but it is also typical of Jovian spots. The merging provides a natural explanation for how large Jovian spots maintain themselves against dissipation — they devour smaller transient spots produced by buoyancy.

The above might be called a *modon theory* of Jovian spots, after a concept proposed by oceanographers to explain circulating rings of water that break off from the Gulf Stream and other major currents. A criticism of the modon theory is that it requires a zonal jet flow in the deep atmosphere beneath the clouds, and we have no knowledge of whether such a flow exists or not. The cylindrical pattern in Fig. 12 is theoretically possible and would support the modon theory. We are now studying how this flow might be maintained against dissipation and determining whether it is hydrodynamically stable. The observed zonal velocity profiles on Jupiter and Saturn (Fig. 4) exhibit a fair degree of north-south symmetry but there are also significant departures from symmetry, especially in the peak amplitudes of zonal jet pairs in the north and south. Differences in internal structure (Fig. 15) could account for the difference between Jupiter's and Saturn's wind profiles. Further theoretical work and further analysis of Voyager data are needed; our data may ultimately limit the unknown parameters so that the correct model will emerge.

HEAT BUDGET AND INTERNAL STRUCTURE

As mentioned earlier, both Jupiter and Saturn radiate more heat than they absorb. This fact leads to several questions: What is the source of the internal heat? Can it be accounted for in any simple way, such as by the slow cooling of an initially hot object created 4.6 billion years ago? What is the effect of this excess heat on the atmospheric structures and circulation? What is its effect on the internal structure?

We consider the internal structure first. A crucial step in modeling the interior is knowing that temperature increases with depth along a given adiabatic profile. Here we assume that only by convection could the quantities of heat observed be carried from the interior to the cloud tops, and convection leads to an adiabatic temperature gradient. Knowing that the interior is warm, we can deduce that there is essentially no solid surface, only a gradual transition from gas to liquid followed by an abrupt transition from molecular liquid to metallic liquid (in which the molecules have been stripped of their outer electrons), and finally to an ice-rock core at the center (Fig. 15 and Table 2). Also, knowing the effect of temperature on density, we can derive the overall hydrogen-helium ratio from the bulk density and moment of inertia. These calculations are generally consistent with the solar composition models for both planets. Finally, knowing the temperature distribution we can estimate the amount of stored heat both now and in the past, and see if it could have lasted 4.6 billion years.

Attempts to model this gradual cooling have yielded an interesting difference between Jupiter and Saturn. One starts with an initially warm planet. How warm is not important, since very little time is spent in the warm stage. As the planet cools, it contracts. The rate of energy loss is followed to the present, where the model result is compared with observation. For Jupiter the calculated and observed heat fluxes are consistent; for Saturn, they are not quite consistent: the observed internal heat flux is greater than the calculated value. Either Saturn is only half as old as Jupiter, an unlikely possibility, or else another energy source besides cooling and contraction is contributing to Saturn's heat output.

A source which would operate on Saturn but not on Jupiter, as required by the observations, has been proposed. Because Saturn is smaller than Jupiter and farther from the Sun, its surface and interior are colder. Calculations suggest that Saturn's interior temperatures are too low for helium to be uniformly mixed with hydrogen throughout the metallic zone. Instead, helium should be condensing at the top of this zone, and raindrops of helium should be falling toward the center of Saturn, converting their gravitational energy into heat. This process began about 2 billion years ago when temperatures first dropped to the helium condensation point. On Jupiter this point can only have been reached recently, and the process is not yet generating significant amounts of heat.

This hypothesis requires that the helium in Saturn should be concentrated toward the center. The outer edge of the metallic zone lies 45 percent of the distance from the center to the cloud tops (Fig. 15). Helium that condenses at the top

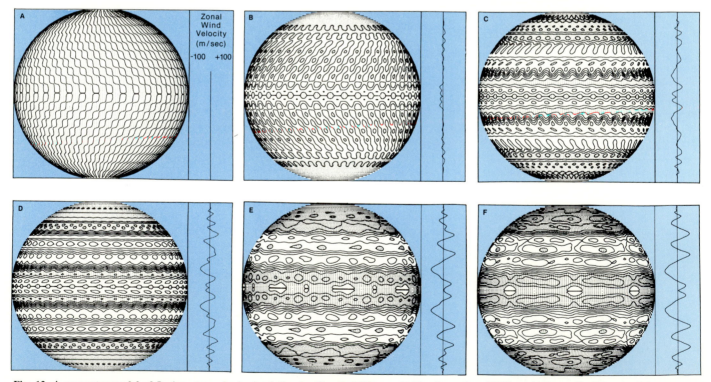

Fig. 13. A computer model of Jovian atmospheric circulation by Gareth Williams. Unlike the author, Williams believes all of the clouds' energy exchanges occur within a narrow layer of the atmosphere (as is the case on Earth); the eddies get their energy from sunlight. His model first produces small-scale eddies (*a*), which become unstable (*b*) and give up their energy to zonal jets (*c,d*) that eventually dominate the flow (*e,f*). Eddies driven by internal heat demonstrate the same behavior, so remaining questions center on the depths of both the eddies and the zonal flow. Adapted from *J. Atmos. Sci.*, 35, 1978, 1399-1426.

of this zone is derived from the entire gaseous envelope which surrounds the metallic zone. This is due to the fact that the envelope is a convecting fluid and its compositional differences become homogenized quickly compared to the age of the solar system. In order to explain Saturn's excess heat flux, the envelope should have lost about one-half its original helium. Two types of observation support this hypothesis.

First, as already mentioned, the helium-to-hydrogen ratio in the atmospheres of both planets has been estimated from spacecraft observations. The Voyager results are more accurate, and they imply just about the right amount of helium depletion if Saturn started with a uniformly mixed atmosphere of solar composition. Second, the mass of the envelope relative to Saturn's mass has been determined from measurements of the gravity field. The envelope responds more to centrifugal forces than the core does, so the amount of gravitationally induced oblateness, or flattening, is a measure of the relative mass of the envelope. Both satellite

orbits and spacecraft trajectories have been used to fix this oblateness, and we find that the envelope has less mass than expected if the planet were well mixed. These observations are consistent with the idea that one-half of the heavier helium has settled out toward the core. In similar studies performed for Jupiter, no significant helium depletion is found. Jupiter's helium-to-hydrogen ratio could be equal to the solar value throughout the planet. Thus it appears that we can account for both Jupiter's and Saturn's internal heat by postulating that both planets started hot 4.6 billion years ago with essentially uniform solar abundance mixtures of hydrogen and helium.

Resolving these questions is important because it helps us acquire confidence in our models of the interiors and early histories of these planets. These models say that large bodies like the Sun, Jupiter, and Saturn give off large amounts of excess energy shortly after forming. This early high-luminosity phase is the basis of our current understanding of the differences between the inner and outer solar system, and between the inner and outer Galilean satellites (see Chapters 14 and 20). According to our scenario, volatile material orbiting the Sun in the inner solar system was either lost by evaporation or otherwise failed to be incorporated into the planets as they were forming. The result is a general increase of volatile content and decrease of density as we move away from the Sun. The same reasoning seems to account for the density differences and water-to-rock ratios of the Galilean satellites. The innermost satellite Io apparently has no water, whereas the outer ones, Ganymede and Callisto, seem to have approximately solar ratios of water to rock.

However, preliminary Voyager results on the satellites of Saturn do not show the same trend. In fact, the Saturnian satellites are all low density objects, showing little evidence of volatile loss (Chapters 15 and 16). The most likely explanation seems to be that Saturn's early high-luminosity phase was weaker than Jupiter's, owing to the smaller mass of Saturn. Thus no water, even close to the planet, was

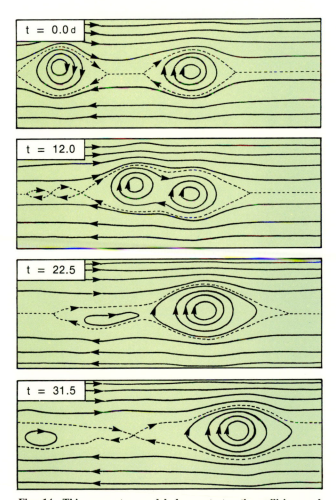

Fig. 14. This computer model demonstrates the collision and merging of two stable vortices (first two frames), followed for a short time by the ejection of material (last two frames). Each vortex by itself, including the resulting larger vortex, could last indefinitely. This merging behavior agrees with observations (Fig. 10), and contrasts with the solitary-wave behavior, discussed in the text, in which disturbances pass through each other without interacting. Merging provides a means by which large oval spots could maintain themselves, given a supply of smaller buoyancy-driven eddies.

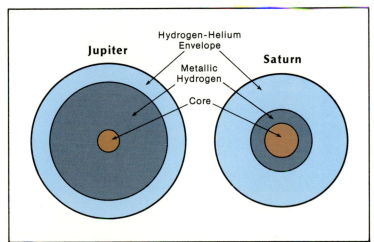

Fig. 15. The interiors of Jupiter and Saturn are shown schematically. The cloud-zone thickness is only 0.1-0.3 percent of the radius. Beneath that, clear atmosphere extends downward and gradually transforms into the liquid of the molecular fluid envelope. An abrupt transition occurs where this liquid becomes metallic. A second abrupt transition marks the interface of the hydrogen-helium zone with an ice-silicate core. Saturn's metallic zone is smaller than Jupiter's — a consequence of Saturn's lower mass, gravitational field strength, and internal pressures.

	(1)	(2)	(3)	(4)
JUPITER				
Fractional radius	1.0	0.76	0.15	0.0
Radius (10^3 km)	71	54	10	0
Pressure (10^6 bar)	10^{-6}	3.0	42	80
Density (g/cm³)	2×10^{-4}	1.1/1.2	4.4/15	20
Temperature (Kelvin)	165°	10,000°	20,000°	25,000°
Fractional mass	1.0	0.77	0.04	0.0
Mass (10^{30} g)	1.90	1.46	0.08	0.0
SATURN				
Fractional radius	1.0	0.46	0.27	0.0
Radius (10^3 km)	60	28	16	0
Pressure (10^6 bar)	10^{-6}	3.0	8	50
Density (g/cm³)	2×10^{-4}	1.1/1.2	1.9/7	15
Temperature (Kelvin)	140°	9,000°	12,000°	20,000°
Fractional mass	1.0	0.43	0.26	0.0
Mass (10^{30} g)	0.57	0.24	0.15	0.0

Table 2. Properties of Jupiter's and Saturn's interiors are given at four distinct levels: (*1*) the 1-bar level in the atmosphere, (*2*) the transition from molecular to metallic hydrogen, (*3*) the outer boundary of the (hypothetical) rock-ice core, and (*4*) the center of the planet.

driven out of the Saturnian system. (This explanation does not account for the small, irregular variations of density in the Saturn system, or the irregular distribution of surface ages as revealed by crater statistics.)

In closing this discussion of Jupiter and Saturn, let us speculate on their internal heat flux and its possible role in atmospheric dynamics. This heat is carried to the surface from deep in the interior by convective currents. Since this process is normally very effective in transporting heat, one does not expect large convective velocities or large departures from an adiabatic temperature distribution. As mentioned earlier, an adiabatic interior is consistent with the cylindrical pattern of zonal flow (Fig. 12) whereby the measured zonal velocities (Fig. 4) are the surface manifestation of much deeper motions. If that were the correct pattern, the slow convective motions driven by the internal heat source would maintain the rapid zonal motion in the interior. Current theoretical research seems to indicate that such processes can occur in the fluid interiors of giant planets.

There are no direct measurements of the zonal velocity profile with depth. However, there are limits to the rate at which velocity can change with depth. According to the thermal wind equation of meteorology, this change is proportional to the temperature gradient with latitude. However, no temperature gradients were found on either Jupiter or Saturn by the Pioneer and Voyager instruments

at the deepest measurable levels. Therefore, the velocity change with depth must be extremely small, and consequently the winds measured at the cloud tops must persist below. A rough estimate is that the eastward wind speeds cannot fall to zero (relative to the internal rotation rate) at pressures less than about 10,000 bars — about 1,000 times the pressure at the cloud base (Fig. 6). This is far too deep to be affected by solar energy, which is mostly absorbed in the clouds. This leaves internal heat as the obvious source for these deep motions.

The internal heat flux may also provide the mechanism for maintaining the poles and equator at the same temperature. Convection is normally very effective in producing an adiabatic temperature distribution, especially when the heat sources are located below the heat sinks. This is the arrangement for Jupiter's and Saturn's interiors. Once an adiabatic state is reached, it is maintained by the buffering effect of the internal circulation. For Jupiter and Saturn the heat is derived from the cooling of the planet, and heat sources are distributed in a spherically symmetric manner throughout the interior. The heat sinks are in the atmosphere; their total influence is defined by the infrared emission to space minus the sunlight absorbed. The net heat loss is therefore greatest at the poles and least at the equator, although it is positive at all latitudes (Fig. 7). We expect that small departures from adiabaticity have arisen in the interior to drive the internal heat flow poleward. These departures are too small to be observed or to drive significant lateral heat fluxes in the atmosphere. The atmosphere is effectively short-circuited by the interior; there is no large-scale temperature gradient to disrupt the banding in the atmosphere.

The situation is very different on Earth. Here the poleward heat transport must take place in the atmosphere and oceans. Absorption of sunlight by the oceans and subsequent latent heat release from water in the atmosphere leads to a net heat gain in the tropical atmosphere. Radiation to space leads to a net heat loss in the polar atmosphere. These combine to create appreciable temperature gradients with latitude, making the atmosphere decidedly nonadiabatic. Our atmosphere and oceans show obvious evidence of mixing across latitude circles.

Again we find ourselves speculating on reasons for a pronounced difference between the gas giant planets and the Earth. And again the cause of the difference — in this case the bandedness of Jupiter and Saturn — may lie in the fluid interiors beyond our reach. Some of our speculations may turn out to be wrong. The entire circulation may be taking place in the atmosphere. Further analysis of Voyager data and theoretical work should resolve many of our questions.

Planetary Rings

Joseph A. Burns

"I do not know what to say in a case so surprising, so unlooked for, and so novel," confessed Galileo Galilei when Saturn's rings apparently vanished in 1612, only two years after he had discovered them. At the time of their disappearance, the rings were actually just presenting a narrow edge-on view to Earth (as they do every fifteen years or so when we pass through the ring plane) but, since their nature as a thin flat disk encircling Saturn was not yet realized, Galileo was confounded. The great Italian scientist was actually never sure of their precise form — indeed, in 1610, the year of his discovery of the four Jovian moons that now bear his name, he believed Saturn to be three bodies (Fig. 1a) only to think six years later that the rings were two great arms or handles stretched toward Saturn. Many alternative

Fig. 1. Our view of Saturn's rings has improved during the past three and a half centuries. *a*, drawing by Galileo (1610); *b*, a sketch by Gassendi (1634); *c*, Fontana's view (1646); *d*, a drawing by Riccioli (1648); *e*, Huygens' sketch (1655); *f*, J. D. Cassini's drawing (1676) showing his division; *g*, first photograph of the rings by A. Common (1883); *h*, Lyot's diagram (1943) from Pic du Midi Observatory; *i*, photograph (1974) from the 1.5-m telescope of Catalina Observatory in Arizona; *j*, Pioneer 11's (1979) view of light scattered through the rings; *k*, Voyager 1's (1980) contrast-enhanced image of backscattered light.

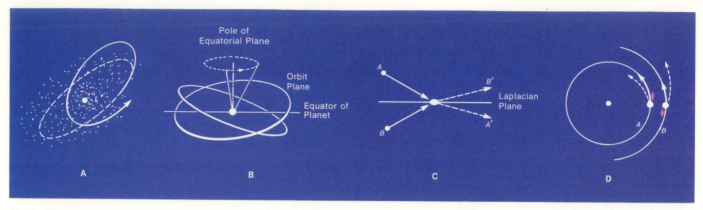

Fig. 2. The gross evolution of a circumplanetary cloud. *a,* In a swarm of particles shown surrounding a central planet, the dotted line illustrates an inclined, elliptical orbit of one of the particles if it were undisturbed. The solid line depicts the same particle as its nearly planar path is gradually modified by perturbations. *b,* As seen in a side view, the pole of an inclined orbit precesses about the pole of an oblate planet's equator; the inclination remains constant but the orbit's orientation shifts. *c,* Becuase of relative drifts, two nearby particles *A* and *B* can intersect to allow collisions, which preferentially occur near the Laplacian plane (see text). These collisions reduce out-of-plane velocities so that the cloud slowly flattens (dashed lines). *d,* As seen in this plan view, particle *A,* nearer the planet, moves faster than *B* and therefore passes it. In any such collision, equal and opposite forces (colored arrows) push on the particles; *A* becomes slowed, causing it to drop on the (dashed) path closer to the planet, while *B* moves farther away.

interpretations were put forth by other skywatchers in the early 17th century (see Fig. 1) as the poorly viewed Saturn system varied its configuration due to changing Sun-planet-Earth orientations. It was not until 1659, nearly one-half century after Saturn appeared in Galileo's crude telescope, that Christiaan Huygens first correctly solved the puzzle about the nature of Saturn's "appendages."

Today we know the gross morphology of planetary rings. Nevertheless, Voyager 1's discovery in 1979 of a faint band circling Jupiter, and the detection of nine narrow rings about Uranus from stellar occultation measurements in the late 1970's, have each brought forth new mysteries. In addition, detailed revelations concerning Saturn's elaborate ring system from Voyager observations have raised at least as many problems as they have solved.

The vast growth in our knowledge of planetary ring systems during the past few years has allowed us to begin addressing the questions of which properties are fundamental to all planetary rings and which are specific to a particular set of rings. In this chapter I will first outline some dynamical concepts that govern the overall structure of planetary rings, then describe the ring systems about Jupiter, Uranus, and Saturn. Finally, I will speculate on the possible origin of planetary rings and their long-time stability, raising the issue of which other planets have had, or will have, rings. Throughout we will find that these planetary ornaments continue to be just as surprising and novel as they were to Galileo, the first human to view them.

DYNAMICAL CONCEPTS

Through a modest telescope Saturn's rings appear to be a continuous sheet of matter, broken by at most a single division. The smoothness of their appearance — and their nearly opaque character — motivated Pierre Laplace and Jean Cassini, as well as other early astronomers and mathematicians, to wonder whether the rings were solid, liquid, or particulate. This question was answered when James Clerk Maxwell demonstrated mathematically that the rings had to be "comprised of an indefinite number of unconnected particles," for which he was awarded the Adams Prize in 1857.

Four decades later James Keeler confirmed Maxwell's idea by noting that the sunlight reflected off the rings was Doppler-shifted such that the individual ring particles must each be in its own orbit about Saturn, with the innermost ones moving faster than their outer counterparts by some tens of percent.

How might such a system have evolved dynamically? A cloud of particles moving about a central mass will develop through three stages (Fig. 2). First, any distribution of objects placed in orbit about a planet will swiftly flatten to a thin disk in a specific plane. This end state can be understood by considering the fate of two nearby particles which, if in orbit about an isolated spherical planet, would each move on separate planar elliptical paths. Slight additional forces, caused, for example, by the planet's oblateness (nonsphericity) or by a perturbing satellite, will force the orbital planes of the two particles to drift (or *precess*) gradually relative to some mean plane called the *Laplacian plane*. For the first of these perturbations, the Laplacian plane lies in the planet's equatorial plane (Fig. 2b) and, in the other case, it is in the satellite's orbital plane. The gradual drift between the two particles allows them to collide occasionally, reducing the relative velocity between them (Fig. 2c). In particular, the bodies' motions out of the Laplacian plane are reduced until they lie essentially in that plane.

For a system containing many particles, damping of the orbital motions into a plane happens rapidly and continually since any renegade particle not moving "in step" with the remaining particles will pass through their mean orbital plane twice every orbit; in doing so, it will sometimes strike another member of the planetary ring. The frequency of such damping collisions depends on what areal fraction of the disk is filled with particles: for a nearly opaque ring like Saturn's, each particle collides with another during almost every passage, or on the average once every few hours. This process is very effective, for the vertical thickness of Saturn's ring system (apparently just hundreds of meters) is but a tiny fraction of its breadth — more than 200,000 km.

Because the strength of each gravitational perturbation (due to planetary oblateness, a satellite, the Sun) depends on the distance to the perturber, the mean plane mentioned above is actually a slightly warped surface, looking like a snapped-down hat brim extending out from the planet. The warp in this surface off the equatorial plane is less than a kilometer for known planetary rings since the disturbing satellites lie near their planets' equatorial planes; if Neptune were to have a ring, it could be pulled dramatically out of planar configuration by massive Triton, because of this satellite's highly inclined orbit.

Even after the first stage of dynamical evolution ends and the cloud of particles has collapsed to a flat disk, individual ring members will continue to interact with one another in a systematic manner (Fig. 2d). Particles close to a planet always move faster than those further away to counteract the increased tug of gravity. This creates a *differential shear motion*, in which inner particles gradually overtake and sporadically rub against their slower-moving, outer compatriots; the resulting collisions tend to retard the inner particles, causing them to drop nearer the planet, while their jostled partners are driven outward. So during the second dynamical phase, planetary rings slowly spread unless their boundaries are restrained in some way. The final stage in a ring's history is reached when the ring particles are so diffusely spaced that they no longer collide with one another; thereafter, the ring does not change dynamically.

The out-of-plane damping that occurs in the initial formative phase never produces a perfectly thin layer. A monolayer state is not achieved because interparticle collisions always scatter objects somewhat out of any plane. In particular, the differential shear motion brings particles of radius r together with a typical velocity of ωr, where ω is the orbital angular velocity about the planet. Any collision reorients the relative velocities of the particles involved as well as damping them slightly. Therefore, particles in a ring should have random velocities relative to one another on the order of ωr; the vertical component of this relative velocity produces a ring at least several particles thick. The scattering of particles by one another's gravitational fields can also inflate the ring if the objects are large enough (at least 10 m across).

Collisions that slightly thicken a ring also cause it to spread radially inward and outward, if the ring's edges are unbounded. The rate of this spreading is directly proportional to the square of the particles' radii. As the ring smears out, the ring particles also drift gradually toward the planet under the influence of both *Poynting-Robertson drag* (due to impacts with photons of light) and *plasma drag* (due to collisions with magnetospheric plasma). But the rate of orbital decay caused by these processes is *inversely* proportional to particle size (the bigger they are, the less they're affected). Consequently, if rings we observe today formed with the planet, the particles comprising them can be neither too small nor too large.

The model of a flattened, nearly opaque disk of particles orbiting a central object pertains to more than just planetary rings. Probably it also is a good representation for one phase in the evolution of the protoplanetary nebula (see Chapter 20), from which the planets supposedly developed. But unlike the planets' accumulation, large objects cannot grow within planetary rings because the latter almost always lie close to planets, within what we term the *Roche limit*. The Roche limit is that distance at which tidal stresses on an

object exceed the self-gravity of the object. In other words, anywhere inside this limit a fluid satellite can no longer remain intact but instead gets torn apart by planetary tides. This distance is $2.446\,R(\varrho_p/\varrho)^{1/3}$, where ϱ_p is the density of the planet and ϱ that of the orbiting object, while R is the planet's radius. Solid satellites, because of their material strength, can approach closer than the Roche distance, but, if they exceed a certain size, will also be broken apart by tidal forces. If they do rupture, the resulting fragments will continually reimpact one another, so that the largest member is gradually ground down in size to some tens of kilometers.

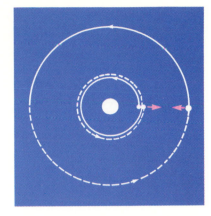

Fig. 3. In a 2:1 resonant orbit, the inner satellite orbits the planet in half the time the outer satellite takes. Thus, during every other passage the two objects repeat precisely the same configuration, so that perturbative effects are amplified.

The observed reality that all planetary rings lie within the Roche limits of their planets has naturally prompted the popular speculation that ring systems arise when comets or satellites stray so close to a planet that they are tidally disrupted. However, these arguments are incomplete because the process works both ways: the tides that tear objects apart just as well prevent material from accreting when located within the Roche limit. In other words, if a primordial disk of debris were extended out from a planet's surface to a great distance, material far from the planet could accumulate into satellites, whereas that within the Roche limit would not grow so easily. In addition, any material very close to a planet might even be lost to its surface by various drag effects. Thus a primordial circumplanetary disk should ultimately develop into a system with exterior satellites, rings, and then a gap just above the planet's surface; that is much like what we see today around Saturn and Uranus.

Another orbital distance — the *synchronous position* R_s, at which a satellite orbits in the same time as its planet rotates — may be important in the development of planetary rings. It can separate the evolutionary paths of particles in two ways. First, planetary tides push satellite orbits outward when they are located beyond R_s, but inward when they are closer; thus large nearby objects can gradually drift toward planets. Second, *electromagnetic forces*, which may dominate gravity for small particles — say, typically less than a micron across — reverse direction in a relative sense at R_s and, as such, may cause material to accumulate there.

Satellites undoubtedly play a pair of fundamental roles in sculpting the planetary rings we see today. First, they gravitationally perturb the orbits of ring particles, a process that probably accounts for the gross form of Saturn's rings, confines the narrow Uranian rings, and perhaps influences the structure of the Jovian ring. Second, the largest objects (which could be called satellites, although we will coin the name *mooms*) embedded in the rings can serve as either sources or sinks for ring particles. Collisions between members of a planetary ring may liberate new ring material or, instead, the objects may coalesce; the outcome depends upon their precise locations, densities, shapes, and the physical characteristics of their surfaces. Moreover, impacts by interplanetary projectiles may excavate debris off larger ring members to supply ring material and to abrade away ring objects. A combination of these effects (which tend to determine the particles' total surface area) may account for observed brightness variations within rings.

Resonant orbits are conspicuous features of the solar system's present makeup since matter is often either absent from, or packed into, such locations. Resonances between two objects occur at those positions where the orbital period of one object is an integer fraction of (or "commensurate with") the other's period (see Fig. 3).

Numerous examples of resonances exist. For instance, at *Kirkwood gaps* in the asteroid belt, the number of minor planets is substantially less than at adjacent locations; these gaps are located where asteroids would have orbital periods that are simple fractions (like ⅗, ½, ³⁄₇, ⅖, ⅓, and so on) of the

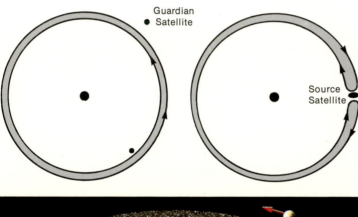

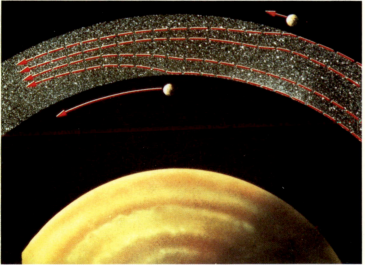

Fig. 4a (top left). The Goldreich-Tremaine model for constraining narrow rings employs "guardian satellites" that act on individual particles to confine them to narrow paths. Moving slower than the ring, the outer satellite exerts a slight drag on particles trying to move outward, causing them to lose energy and "fall" closer to the planet. Conversely, the inner satellite, moving faster, adds energy to nearby particles and kicks them into higher orbits. Together, these forces herd the particles into a narrow ring, as illustrated in the lower diagram.   Fig. 4b (top right). The Dermott-Gold-Murray model envisions a small satellite orbiting within the ring. Each time a particle passes the satellite, mutual gravitation causes them to exchange orbital positions, but since the satellite is the much more massive object, it is not affected as obviously. To an observer riding on the satellite, ring particles would appear to follow horseshoe-shaped paths bringing them first toward, then away from, the satellite.

orbital period of Jupiter, the primary perturber of the asteroid belt. Such resonant positions are not always vacant and indeed sometimes they may be unusually populated: the Hilda asteroids have orbital periods two-thirds that of Jupiter's and there are literally a thousand Trojan asteroids moving with the Jovian period.

The same resonance pattern is seen in satellite systems, where oftentimes moons have commensurate orbital periods; the most notable example being the Laplace resonance among the three inner Galilean satellites (where the orbital periods of Io, Europa, and Ganymede have a ratio of 1:2:4). This resonance means that the satellites repeat any particular configuration periodically and their alignment continually reinforces (through gravitational forces) the eccentricity of Io's orbit. Since the tidal flexing of Io requires that the satellite move along an eccentric orbit, the resonant condition is fundamental to the tidal heating of its interior and therefore to its volcanism (see Chapter 14). Resonances are common among the satellites of Saturn, and near-resonances are seen in the Uranian satellite system.

The *Cassini division,* which separates the two major classical rings of Saturn, is like a Kirkwood gap, since a particle there would have a period about one-half that of the Saturnian moon Mimas. Many gaps in Saturn's A ring are located at resonant positions with the co-orbital satellites 1980 S 1 and 1980 S 3, and the shepherd satellites 1980 S 26 and 1980 S 27; the boundaries of both the A and B rings are near resonant locations. Particle deficiencies are also noted in the rings at those positions where Saturn's oblateness would force ring particles to precess with periods that match the orbital periods of Titan and Iapetus. However, Voyager pictures of Saturn's rings show that the situation is more complicated: there is *not* a simple one-to-one correspondence of resonant locations with ring structure; even Cassini's division, while generally containing less material than its surroundings, is filled with many ringlets.

The standard explanation for the low particle densities in resonant gaps is that, at such positions, the effects of perturbations can build up because the involved particles and satellites continually repeat their relative configurations. Particle orbits are most disturbed in such regions, so particles encounter one another more frequently and more violently than average, and this interaction inhibits the accumulation of material. But the argument may be flawed since nearby particles, except in a very narrow band close by the exact resonance location, are equally perturbed and march side-by-side along similarly affected orbits.

An alternative viewpoint — put forth by Peter Goldreich and Scott Tremaine in the late 1970's — claims that spiral density waves, like those in galaxies, are initiated at resonant locations in circumplanetary disks. Through gravity, a satellite can alter a ring particle's orbit, making it elliptical. The overall effect of this on a populous disk of particles is to cause bunching in spots that give rise to a *spiral density wave.* If the perturbing satellite is exterior to the ring, this wave moves outward, carrying negative energy and angular momentum. An entrained ring particle, if allowed to proceed without interference, would eventually return to its original position. But the rings are dense enough to make collisions probable, and particles involved in such collisions ultimately move inward toward the planet. This process clears a wide gap just outside the resonance position.

The same physics can be invoked to demonstrate that a satellite near a planetary ring can push that ring away; simultaneously, the moon is repulsed. Their model is a prime contender to account for narrow rings, such as those in the Uranian system. Small (about 10 km across) undetected satellites on each side of a ring could constrain its edges and prevent ring spreading (see Fig. 4a); such moons can also generate elliptical rings. The Goldreich-Tremaine mechanism received substantial vindication when Voyager 1 discovered two new satellites shepherding the narrow outer F ring of Saturn, and pinpointed yet another guarding the outer boundary of the A ring. (Density wave structures in the Cassini division and in the outer part of the A ring may have been glimpsed by Voyager 1.)

However, according to Stanley Dermott, Thomas Gold, and Carl Murray, satellites may account for the narrow Uranian rings in a totally different way. They maintain that a moon lies hidden within each ring, and that particles in the ring move along elongated, horseshoe-shaped orbits (as seen by an observer riding on the satellite; Fig. 4b). Since the satellites circle inside the synchronous distance R_s, they are gradually dragged by tides toward the planet; tidal forces also draw material off the satellites' surfaces, which becomes part of the ring material. Ring particles follow trajectories which are essentially locked to the satellites' path, and thus planetary rings can be either elliptical or inclined. The Dermott-Gold-Murray model has been verified to the extent that Saturn's two co-orbital satellites (S10 and S11) undergo the horseshoe dance.

THE GENERAL CHARACTER OF KNOWN RINGS

We can gain considerable insight into the nature of planetary ring systems by matching actual ring dimensions against the dynamical models described above, and also by comparing the three known systems. Fig. 5 shows the overall dimensions of the various rings. The Saturn system is obviously much more prominent than the other two. Essentially all rings are situated within the Roche limit and, except for the exterior portion of Saturn's ring, all lie within the synchronous orbit. Perturbing satellites border the rings of Jupiter and Saturn, with some actually interlacing the outer edges of their respective rings. We presume as-yet unseen satellites also circle contiguously to Uranus' rings. The main band about Jupiter grades continuously into a fainter inner sheet (see Fig. 7) which seems to extend down

Table 1. RING PROPERTIES

	Jupiter	Saturn	Uranus
Primary ring radius (R_p)	1.72-1.81	C: 1.23-1.52 B: 1.52-1.95 A: 2.02-2.27 F: 2.33	1.60-1.95
Secondary ring radius	1 (?)-1.72	D: 1.11-1.23 G: 2.8 E: 3-8 (?)	
Width of main and narrowest structure (km)	6,000; ≤600	20,000; ≤0.1	100; ≤5
Thickness (km)	<30 (halo 10^4)	<0.2	?
Optical depth	10^{-5}-10^{-6}	≈0.1-large	≈0.1-1
Typical particle size	micron (μm)	cm-10 m	cm-m (?)
Surface mass density (g/cm²)	10^{-9}-10	10-100	10-100 (?)
Total mass (kg)	10^7-10^{17}	10^{17}-10^{19}	10^{14}-10^{16}(?)
Albedo	low?	0.2-0.6	0.03

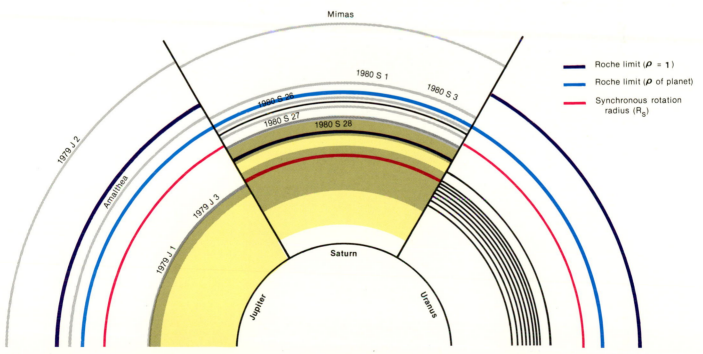

Fig. 5. The three known planetary ring systems are compared by setting each planet's radius equal to one unit of measurement. Illustrated here are the distribution of ring material, nearby satellite locations, the synchronous orbit radius R_S (for Uranus a 16-hour rotation period is assumed), and the Roche distances for satellites having either the planet's density or that of water.

to Jupiter's cloud tops. In contrast, the other ring systems are separated from their planets and seem to be comprised of discrete ribbons.

Table 1 lists a few other physical properties of the three ring systems. The rings about Jupiter and Saturn are very thin, although Jupiter's ring has a halo component that extends 10,000 km out of the orbital plane. The *optical depth* is a measure of the amount of light that can penetrate a layer; for small values, it is essentially the fraction of photons that do not pass through the layer. The Jovian ring, especially its halo, is diaphanous. Most components of the other systems are almost opaque in places, although the inner Uranian rings and the outermost (G and E) rings of Saturn are relatively ethereal.

The *typical particle size* should not be taken too literally since it depends on how the averaging is done, and it may come from biased observations or from faulty reasoning. Almost certainly, a range of particle sizes exists in each ring. The *surface mass density* for Jupiter and Uranus is estimated by multiplying the optical depth times the average particle's size and density; the wide limits for Jupiter's ring arise from the assumption that the typical particle could be either like those seen best in forward-scattered light (a few microns across) or one just below the resolution limit of the Voyager cameras (a few km). Spiral density waves seen within Saturn's rings allow us to estimate their surface mass density. *Total ring mass* is the surface mass density times the ring area. Jupiter's ring is distinguished by its meagerness.

The *albedo* characterizes the fraction of light that is reflected from a surface. The particles in Saturn's rings are bright, returning most of the sunlight that strikes them, as are most of Saturn's icy moons. The ring material in the other systems is dark, presumably indicating not icy compositions but ones of silicon or carbon compounds.

THE RING OF JUPITER: FAINT YET DISTINCT

As Pioneer 11 traversed Jupiter's magnetosphere in 1974, approaching to within 1.6 Jovian radii (R_J) of the planet's center, counts of high-energy particles dropped whenever a known satellite was passed. Thus, when the number of energetic particles decreased at about 1.7-1.8 R_J, investigators Mario Acuna and Norman Ness proposed that the giant planet might have a satellite or a ring system at that position. Most scientists dismissed this possibility since terrestrial observations had not revealed anything there, and other explanations of the reduced counts were available. Never-

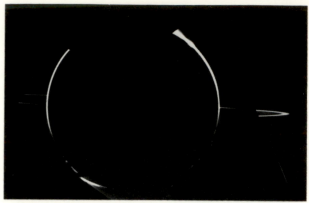

Fig. 6. The ring of Jupiter is shown in this mosaic of wide-angle images from Voyager 2, taken as the spacecraft looked back from within the planet's shadow from 1.45 million km away. Highlighting of the ring and Jupiter's limb is due to strong forward-scattering by small particles in the ring and upper Jovian atmosphere, respectively. On either side of the planet, the arms of the ring curve back toward the spacecraft but are cut off by Jupiter's shadow as they approach its bright-rimmed disk.

theless the Voyager 1 spacecraft was programmed to look along Jupiter's equatorial plane at the instant it was crossed. In this way the spacecraft detected a faint disk of material extending inward from about 1.81 R_J.

Voyager 2 observations (Figs. 6, 7, and 8) distinguished three components for the Jovian ring. The main band starts abruptly at 1.81 R_J and ends more gradually 6,000 km closer to the planet, at 1.72 R_J. Even on the brightest part of the ring, matter covers a very small fraction of space (the ring's optical depth τ is about 3×10^{-5}). Compared to the rings about Uranus and Saturn, the Jovian version is unusually smooth; its brightness is almost uniform but for a slight 600-km-wide enhancement centered at approximately 1.79 R_J. A more tenuous sheet, with τ equal to about 7×10^{-6} merges continuously with the brighter outer belt and then stretches smoothly down to Jupiter's upper atmosphere. A halo (with an optical depth no greater than 5×10^{-6}) enshrouds the ring and sheet; it has a lenticular or lens-shaped cross-section, thinnest at the bright ring's edge but extending as much as 10,000 km above the equator when close to the planet.

The rings are more than 20 times brighter in forward-scattered light (which is mainly diffracted) than in reflected light, implying that a significant number of the ring particles

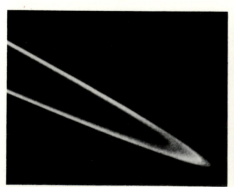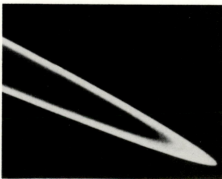

Fig. 7. Different processing of a Voyager 2 image of Jupiter's ring has brought out specific details: *a,* the bright ring between 1.72 and 1.81 Jovian radii; *b,* the faint sheet interior to 1.72 radii; and *c,* the fainter halo that envelops the other two components.

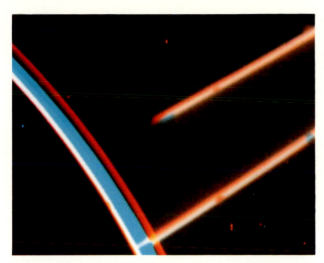

Fig. 8. Jupiter's ring is an orange line in this composite of two images taken through orange and violet filters while Voyager 2 was in Jupiter's shadow. Since the images were recorded at different times and registered on the rings, Jupiter's limb appears as slightly displaced, colored images.

are only a few microns across. Such small objects have correspondingly short lifetimes because (a) they can dynamically evolve out of the system in a thousand years or so due to Poynting-Robertson drag and plasma drag; and (b) in a comparable time, they are eliminated through catastrophic micrometeoroid impacts or through more gradual erosion by the abundant energetic particles that exist in Jupiter's environment.

An immediate implication of the brief grain lifetimes is that the small particle population must be regenerated continually if the ring is to be a permanent feature. One way to supply ring material is to bombard large bodies in the ring with high-velocity particles from outside the system. Apparently, a population of boulder-sized objects surrounds Jupiter and supplies fine material to the visible ring in just this way. The impacting particles could be interplanetary micrometeoroids or even grains of volcanic dust drawn off the surface of Io.

Objects in a plasma like the one that surrounds Jupiter maintain a negative electric potential on their surfaces so that the number of protons striking the surface is the same as the number of swifter-moving electrons. Due to this potential, Jovian ring particles feel an electromagnetic force which is about 1 percent of the planet's gravity for 2-micron-sized grains but which is comparable to the Jovian gravity for objects one-tenth that size. These forces cause ring particles to move relative to one another and might also scatter typical particles.

Plasma drag would eventually force particles from the outer portion of the ring toward the inner, more tenuous sheet. Here the ring material, already greatly reduced in size, would be subject to interactions with energetic ions in the magnetosphere and be ground down even further. They ultimately become small enough that electromagnetic forces acting on them overpower the effects of gravity; once this point is reached, the particles are swiftly yanked out of the ring plane and into the extended halo. Because the Jovian ring is relatively uncrowded, this fine ring debris gets pumped into the halo before interparticle collisions can

dampen its motion. Thus, the halo does not flatten as Saturn's does.

Jupiter's main ring probably contains a variety of particle sizes. These may be pieces from the catastrophic breakup of a small satellite, or may just represent uncompleted accretion of a satellite located within the Roche limit. Whatever their origin, they have gradually spread from differential shear and drag, filling the observed region. The presence of unusually big mooms, serving either as sources or sinks of material, may generate some minor brightness variations in the ring; the satellites J14 (an object 25 km across that skirts the ring's outer boundary) and J16 (a comparable-sized object a little further into the ring, whose reality some astronomers doubt) may be merely largest of these.

The Jovian ring particles, like those in Uranus' rings, are believed to be dark, presumably siliceous. Their near-infrared spectrum resembles that of the nearby satellite Amalthea: it is flat, showing no absorption peaks.

Despite the insubstantial nature of Jupiter's ring system, it has illuminated our understanding of planetary ring systems because certain effects are highlighted by its low optical depth and the smallness of its component particles. In particular, it is apparent that relatively large mooms must surely be present in Jupiter's ring. Accordingly, we wonder whether the other ring systems are also fed from as-yet unseen reservoirs.

URANUS' RINGS: NINE, NARROW AND NEAT

The way in which a star's light is extinguished as a planet passes in front of it tells much about the planet's atmospheric properties and also permits measurement of a chord across the planet's disk. With this in mind, several teams of astronomers trekked to observatories surrounding the Indian Ocean in order to witness in March, 1977, the occultation of the star SAO 158687 by Uranus. They re-

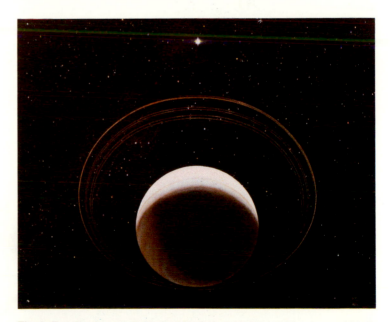

Fig. 9. Don Davis created this backlit view of Uranus and its rings, which lie in the planet's equatorial plane. Because Uranus' polar axis points nearly in the ecliptic plane, the rings can appear almost fully open to observers on the Earth. They were discovered this way in 1977, as the system passed in front of a bright star (see text). The rings are actually much narrower and darker than portrayed here.

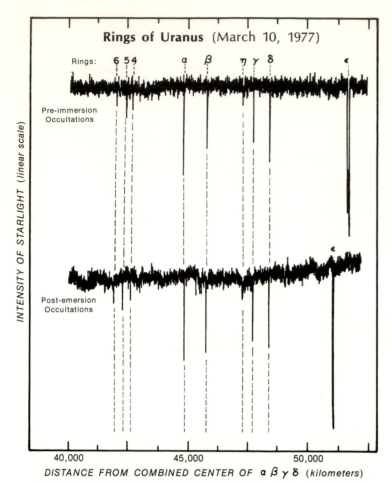

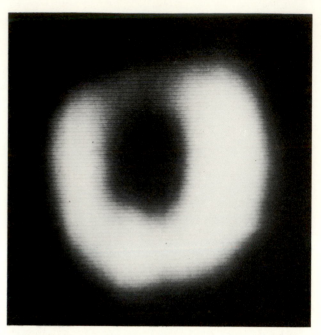

Fig. 11. This first picture of the rings of Uranus — minus the planet — was obtained by scientists using the 5-m telescope at Palomar Mountain. They scanned the planet at two infrared wavelengths, one at which the planet appeared brighter than the rings and one at which it appeared darker. By subtracting one scan from another, they effectively made the planet "disappear" altogether, leaving the image of the rings. The rings are actually extremely narrow; the width of the single wide band shown here corresponds to the brightness of the reflected light. The apparent gap occurs at the narrowest point of the ε ring.

Fig. 10. The starlight from SAO 158687 was recorded before (top) and after (bottom) its occultation by Uranus' disk. Dips in the tracing are due to each of the nine Uranian rings (labeled individually). Note that the star's light is diminished at nearly the same distances on either side of the planet, revealing the rings' circular nature. Most rings obscure the star very well and rather abruptly; this indicates that they are nearly opaque and have well-defined edges, the latter suggesting the presence of nearby satellites to prevent the rings from diffusing. The variations in opacity for ring ε may be due from a higher density of particles near its edges. (Small fluctuations in the star's brightness are caused by system "noise".) From James L. Elliot, *Annual Reviews of Astronomy and Astrophysics, 17, 1979, 445-75.*

ceived an unexpected bonus when a set of narrow rings was discovered around the planet (Fig. 9). The best observations came from James Elliot's group, who watched the occultation high above the Indian Ocean through the Kuiper Airborne Observatory's 91-cm telescope.

As Fig. 10 shows, the rings substantially attenuate the starlight, interrupting it at approximately symmetric points about the planet; thus the Uranian rings are nearly filled with material and are almost circular. Subsequent stellar occultations have identified a total of nine rings, extending between 1.60 and 1.95 planetary radii, and identified in order of increasing planetary distance (6, 5, 4, α, β, η, γ, δ, ϵ). Compared to their circumferences (some 250,000 km), the rings are remarkably narrow — most do not exceed 10 km in width and only one, the outermost ring, spans as much as 100 km.

An unexpected revelation is that at least six Uranian rings are somewhat out-of-round (having eccentricities of 0.001-0.01) and have variable widths. Both of these charac-

teristics are best illustrated by the ε ring; its distance from Uranus varies by about 800 km and its width changes from 20 to 100 km, in a linear fashion with its distance from Uranus. The latter observation can be produced by a model in which two aligned, elliptical rings (of slightly different eccentricities) delimit the ring boundaries.

The elliptical bulges of the Uranian rings precess slowly about the planet, just as they should due to the planet's oblateness. However, we would expect the rings to precess differentially, since normally the rate of precession would depend on distance from the planet: the inner edge of a ring of any width would precess more rapidly than its outer boundary. This process would shear each Uranian ring into a circular band in no more than a few hundred years. But this has not happened; in fact, each ring retains its form, precessing as a rigid whole. The eccentric nature of the rings seems to be continually reinforced, either by satellites or by the ring material itself.

An additional property of the ε ring is visible to some extent in Fig. 10 and has been seen better in subsequent occultations: its opacity (and therefore its density) decreases towards the ring's center. This irregular mass distribution appears across any radial slice through the ring, regardless of where it is made, even though the ring is both elliptical and markedly variable in width. Perhaps the particles flow along discrete "tubes" that do not interact over the times of our observations.

The total amount of light reflected from the Uranian system implies that spaces between the rings are relatively free of debris — all material lies concentrated in the observed bands. Uranian ring boundaries are crisp, suggesting the ineffectiveness of differential shear that should cause interparticle collisions and thereby generate diffuse boundaries. Sharply defined rings can only be explained if satellites bound the edges in some way; in either the guardian-satellite model or the horseshoe-orbit model, about 10 unidentified satellites are needed. Such moons could be as small as 10 km across and, as such, would be imperceptible through ground-based telescopes.

At a variety of wavelengths, the rings of Uranus reflect less than a few percent of the light that strikes them. This low albedo implies that the ring particles lack coatings of water, ammonia, or methane frosts, materials common among so many of the outer-planet satellites (including the Uranian satellites, which seem to be covered mainly with dirty water ice). Only one well-defined absorption feature is seen in the ring spectra, near 2.2 microns; it could be caused either by ammonia frost covering a small fraction of the surface, or by certain hydrated silicate minerals.

Because of their low albedo, intrinsically small cross-section, and proximity to Uranus, the rings are extremely difficult to see. In fact few observations — apart from those precisely defining their geometric configuration — have been successful. Observers have obtained low-resolution maps of the ring system by subtracting sets of scans made at a wavelength of 1.6 microns (where the planet dominates the rings) from corresponding scans at 2.2 microns (where the planet absorbs virtually all the light that strikes it). Such "images" (Fig. 11) show the system to be azimuthally asymmetric, that is, brighter on the side of the planet where the ϵ ring is widest.

SATURN'S SYSTEM: RINGS GALORE

Ground-based observations of the Saturnian system at infrared, visible, and radio wavelengths have revealed considerable information on the properties and overall form of the rings. But these results have been all but eclipsed by recent findings from the Voyager spacecraft, whose better resolution has allowed previously unimaginable structure to be discerned. Furthermore, Voyager observations of variations in the rings' optical properties with changing Sun-ring-spacecraft aspect will ultimately disclose much about the size, shape, and distribution of particles in the system.

Saturn's rings are much more elaborate and complex than the other two known sets of rings. Among their distinct, classically recognized components are the bright A and B bands, which are separated by Cassini's division (first noticed in 1675). Optical depths average 0.3-0.5 for the A ring and greater than 1 for B. Interior to these is the crepe or C ring, recognized in 1850, with an optical depth of roughly 0.1. Pierre Guerin claimed in 1969 that he had found a very faint D ring, closer to Saturn, but isolated from the C ring. Voyager 1 confirmed that some material, including a few narrow, widely spaced bands, extends from the C ring down at least halfway to the planet; however, the D ring is essentially undetectable from Earth.

Exterior to the traditional system lies the E ring, which has such a low optical depth (about 10^{-6}-10^{-7}) that it is scarcely more than a slight concentration of debris in the satellite orbital plane. It becomes visible only when the ring system is viewed approximately edge-on. The highest-quality observations of the E ring, made in 1980, show that material extends out to at least eight planetary radii, thickening toward the outer edge. Some findings suggest that its density may peak near Enceladus' orbit. Voyager detected only the E ring's forward-scattered light, so many of its particles must be micron-sized, like those in Jupiter's ring.

In 1979, Pioneer 11's imaging photopolarimeter located a new, slender F ring (see Fig. 12), just beyond the A ring, and separated from it by about 4,000 km. Apparent absorptions of charged particles measured during the flyby hinted that undiscovered satellites or rings might also lurk beyond the classical rings' limit. In 1980 and 1981, the Voyagers found several new satellites and confirmed the existence of another pair that had just been properly identified by terrestrial astronomers. Also seen was the F ring's shadow falling on one small satellite, as well as the faint G ring at a distance of about 2.8 Saturn radii.

Clues about the vertical structure of Saturn's rings have come from telescopic work. The rings exhibit an "opposition effect" in which their brightness surges nonlinearly as the phase angle (the Sun-Saturn-Earth angle) approaches zero. This indicates that some reflecting regions are in shadow at phase angles other than zero degrees: either the particles have intricate surface texture or, more probably, some of the various particles making up the ring occasionally shadow one another (so that the ring is "many-particles thick"). In the latter case the particles are separated by distances 5-10 times their size.

Random velocities induced by collisions, or by gravitational scattering off large ring members, can account for a thickened ring. Scattering might also be responsible for azimuthal brightness variations which have been noticed from Earth only in the denser parts of the A ring. When the rings are fully open (that is, when one of Saturn's poles is seen tipped 26° toward us), they appear about 10 percent brighter than normal in those quadrants immediately preceding the line between Earth and Saturn. This brightness variation doubles when the rings have intermediate tilts (about 10°), only to decrease as the rings approach their edge-on configuration. Such variations could be explained by synchronously aligned, nonspherical (or asymmetrically mottled) particles. But this mechanism is unlikely since

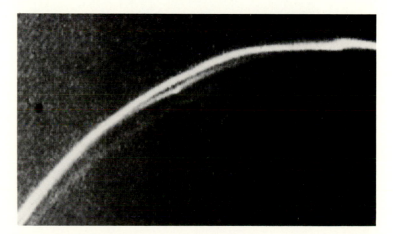

Fig. 12. The F ring of Saturn, as photographed from its unilluminated side, shows two knotted and braided components, plus a more diffuse third ring. The narrow rings are about 10 km across and have occasional bright knots. These clumps retained their integrity for 15 orbits and perhaps mark the positions of large ring members.

Fig. 13. The rings of Saturn, in a computer-processed image from Voyager 1. The threadlike, outermost F ring can be seen along with one of the two guardian satellites that skirt its edges. Interior to this lies the bright A ring (the darkest gap within it is called the Encke division). As seen here, the Cassini division consists of four ringlets, each about 500 km across. The broad B ring does not exhibit as much structure because it is more uniformly opaque. The inner C ring shows many narrow ringlets up until the point where it ends abruptly. All told, some 95 rings are visible in this picture, although hundreds more appear in other images.

particle sizes are too small for tides to be effective in locking rotations and, in any event, collisions should disrupt any well-ordered configuration. Instead, the preferred explanation is that large objects engender spiral density wakes, acting to cluster particles. When viewed from above the ring, the pattern would contain a series of short arcs, like the blades of a turbine. These structures appear "broadside" to one's view in the morning and evening quadrants, and therefore reflect more light; such a mechanism is most effective when the rings are only moderately filled, as the A ring is.

The composition of a typical particle in the A or B ring is known from ground-based work in the 1970's. Using reflection spectroscopy in the near-infrared, Gerard Kuiper, Carl Pilcher, and their colleagues identified water ice as the primary constituent of the particles' outer layers. A slight reddening of rings in the visible portion of the spectrum indicates some surface contamination, perhaps from trace impurities, micrometeoroid debris, or radiation-induced modification of the water-ice lattice. We could even be observing clathrate compounds (where methane or ammonia molecules are packed into the lattice), which are indistinguishable from water ice in our existing near-infrared and visible spectra.

Agreeing with the identification of ice are the high albedos computed for the particles from multiple-scattering calculations. These computations match variations in ring brightness with phase and tilt angles at several wavelengths, and predict ring optical depth and particle brightness with changing phase angle. Voyager observations, because of their vastly extended phase-angle coverage, are sure to yield stronger constraints on the nature of the rings.

Although both infrared and microwave observations are consistent with the ring particles' interiors (as well as their surfaces) being composed of ice, metallic ring particles would also satisfy these measurements. But cosmochemical models point to water ice as the primary solid compound in the outer solar system, and when taken with the low densities of most Saturnian satellites, they support the interpretation that Saturn's ring particles are entirely snowballs.

The particles in the C ring and within Cassini's division have noticeably different properties than the rest of the ring. They are somewhat bluer and larger, generally darker, and not as forward-scattering as the average material in the A and B rings. The eccentric C ringlets have distinct optical properties as well. Such segregation of particle compositions — and its maintenance over geologic time scales — hint

that diffusive processes are less effective than once believed. Abrupt changes in properties across ring boundaries tell of differences among ring components that may have survived from primordial time.

We can estimate particle sizes using several lines of evidence. Earth-based observations of the cooling of ring particles when they are in eclipse imply that, if solid, the particles are bigger than about 1-2 cm; more likely, they are even larger than this but are covered with low-conductivity water frost. Moreover, much of the ring population must exceed a few centimeters in size because Saturn's rings are remarkably efficient radar reflectors at 3.5 cm and 12.6 cm. Their low radio brightnesses and strong radar reflectivities have led Jeffrey Cuzzi and James Pollack to conclude that a many-particle-thick layer with a "power-law" size distribution (with many more centimeter-sized particles for every meter-sized one) would satisfy the observations. Radio signals transmitted through the rings by Voyager 1 hint that the particles in the C ring have a mean size of about 1 m, and that there are at least some objects greater than 10 m in the

Fig. 15. An elliptical ringlet (arrow) in the C ring appears in this figure, which was made by splicing together two Voyager images of opposite ansae. Except for the ring located in the dark band, all other areas have the same radial position whether seen on the east or the west side of the planet. Like the Uranian eccentric rings, this noncircular ring about Saturn is broader when it is farther from the planet. The picture was taken from a range of 3 million km.

Fig. 14. A view from 1.5 million km above and behind Saturn's rings illustrating the phonograph-record appearance of the rings. The C ring is relatively faint as it crosses the planet's limb. The ringlets in ring B are so closely packed that in places they entirely screen the planet; most of the observed radial structure arises from variations in optical thickness. The illuminated crescent of Saturn is overexposed.

Cassini division. Objects in this size range could explain the variations in ring brightness and the many-particle-thick nature of the rings, since the large objects would scatter vertically a broader range (and therefore greater numbers) of ring members.

Several caveats should be given about this model. First, the results pertain only to the *average* ring member. Undoubtedly some particles in the A and B rings are quite small, such as those continually being generated by micrometeoroid impacts. Very likely, a few large objects (tens of meters in size to perhaps even a kilometer or so) are present, if the intricate ring structure is to be explained. Furthermore some scientists, buoyed by the cool temperatures (about 55° K) of the unilluminated sides of the rings, would argue that a single layer of material is better because it would shadow portions of particles more effectively. And others still maintain that, although unlikely, metallic objects larger than a few centimeters cannot be ruled out on the basis of present observations.

Close-up photos by Voyager disclosed Saturn's rings to be more finely divided than anticipated. Rings lay within rings:

upwards of a thousand (Fig. 13) have been counted, and many more are suggested from a high-precision Voyager 2 record of optical thickness obtained during a stellar occultation by the ring system. Even the so-called "divisions" are crammed with ringlets: Cassini's division contains perhaps 100 (!) rings, including an eccentric one, and other material is seemingly clumped nonuniformly in the narrow Encke gap. The classical divisions have therefore been revealed as just *relative* absences of matter, in comparison to adjacent parts of the A and B rings.

Brightness variations in the A and B rings are due to the spacing of ringlets and to optical-depth variations; Saturn's disk can be seen through most of the "opaque" B ring (see Fig. 14) and only a few places are truly impenetrable to light. Surprisingly, the C ring (Figs. 13 and 15) exhibits a more pronounced structure than either A or B. This arises in part from its lower optical depth, which makes multiple-scattering insignificant and image enhancement more effective.

As mentioned earlier, a few prominent features of the rings appear to be associated with satellite resonances. Nevertheless, much of the radial ring structure seems to be

Fig. 16. Eight hours after its close approach to Saturn, Voyager 1 looked back upon the ring system and recorded this image. Clearly visible is the F ring, seen as a thin, detached band along the outer edge. Its brightness is probably due to minute particles that preferentially scatter light in the forward direction. The broad, bright B ring contains light-colored spokes (also strongly forward scattering) toward lower right. At left, considerable structure is also apparent in the dark gray C ring.

Fig. 17. A false-color image of the inner B ring and C ring taken by Voyager 2 from a distance of 2.7 million km. Note the sharp color difference between C (blue) and B (gold), which suggests distinct compositions. However, three ringlets can be seen within C that also exhibit the same pale yellow color as B-ring material.

uncorrelated with the simple perturbations of known satellites: some strongly perturbed locations contain material, others do not; some gaps are at resonant positions, many are not. Presumably the intricate structure of the rings is due to large objects (which I call "mooms") within the disk; these serve as sources and sinks of ring matter, and perturb nearby material. But a systematic search for the satellites presumed to clear gaps within the Cassini division found nothing larger than 5-9 km across. An alternative explanation for the radial ring structure holds that it is a consequence of dynamic instability driven by certain collision situations. The highly organized ring structure seems to conflict with the diffusional spreading of rings, which should produce a more continuous density distribution. The influence of large mooms on ring spreading is not known.

Eccentric rings may give further evidence for large objects residing in the rings. Several such ringlets are in the C ring (Fig. 15), another in the Cassini division, and perhaps a third in the F ring itself. Eccentric rings can be generated by the Goldreich-Tremaine mechanism but require appreciably large satellites nearby.

Two small satellites, 1980 S 26 and 1980 S 27, herd the particles in the narrow and braided F ring and presumably prevent its spreading. These objects are also implicated in creating the F ring's unusual appearance: the clumps and the braids have a typical spacing of about 9,000 km, which is comparable to the relative motion of the ring particles with respect to the satellites over one orbital period. Kinks seen in two Encke gap ringlets must similarly be due to perturbations by unseen moonlets. (Even though the strands making up the F ring appear interwoven, the paths of individual particles in the separate bands do *not* pass through one another.) By comparing the F ring's brightness in Figs. 13 and 16, we realize that particles in the outer ring efficiently forward-scatter visible light just like the Jovian ring particles do, suggesting that the F ring consists principally of micron-sized grains. Electromagnetic forces acting on these small particles might account for the knotted shape of the outer band, but that is uncertain.

Again based on scattering properties, small grains are implicated in the formation of "spokes" visible in the B ring (Figs. 16-18). Spokes occur near the densest part of the B ring, about 104,000 km from Saturn's center, and extend outward to the edge of the Cassini division. In appearance each is a pair of opposing triangles (like an hourglass) with the narrowest point located near the synchronous-orbit radius. The spokes are believed to form swiftly as radial strips. These then differentially shear by relative orbital motion and so are presumably produced by orbiting particles. Since small grains are involved, electromagnetic effects (which, for example, might levitate small particles out of the ring plane) may be the cause. In this regard, the spokes' location near the synchronous orbit (where the effective electric field changes), together with Voyager's discovery of lightning in the ring's vicinity, may be pertinent.

Saturn's ring system is a dynamical place. Spiral density waves have been identified in the outer Cassini division and in parts of the B ring; other wavelike features occur in the outer A ring and near the Encke gap. The boundary between the B ring and Cassini's division (see Fig. 19) moves radially in and out 140 km, apparently forced by Mimas. In the same region, numerous radial features on scales of 20 km seem to

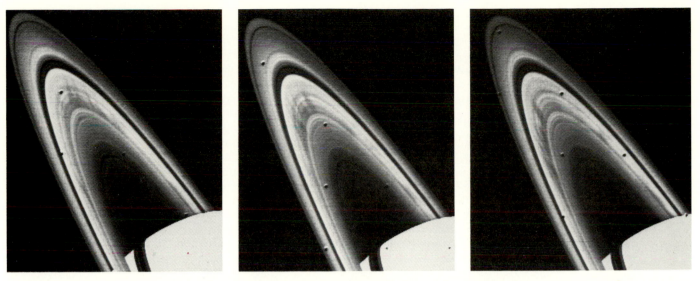

Fig. 18. Taken during a 1½-hour period on October 25, 1980, this series of images from Voyager 1 show dark fingers or spokes moving across the B ring of Saturn. (Dark spots are reference marks introduced by the imaging system.)

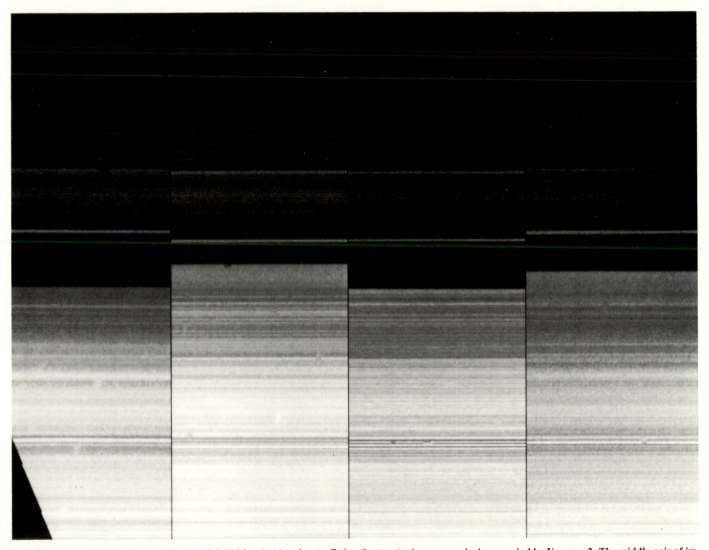

Fig. 19. Four radial slices through the Cassini division (top) and outer B ring (bottom) taken over a six-hour period by Voyager 2. The middle pair of images are of the east ansa, while the outer two are of the west ansa. Not only does the position of the B ring's outer edge change, but fine structures (roughly 20 km wide) within it vary with both time and longitude. Larger-scale features (about 500 km) are circular and remain correlated. An eccentric ringlet occupies the variable gap at the Cassini division's inner edge.

be transient and noncircular; presumably they are caused by waves and instabilities.

It is fitting that Saturn's rings, as the first discovered set, remain the most complex and interesting system. Fortunately they are also the best observed and, once the Voyager findings are fully assimilated, are sure to provide many fundamental insights into the nature of all planetary rings.

THE ORIGIN OF PLANETARY RINGS

Two possible modes for the origin of planetary rings — tidal breakup and halted accretion — were mentioned briefly at the outset of this chapter. The first of these was espoused in the mid-19th century by Edward Roche, who suggested that Saturn's rings might have been generated when an object was tidally torn apart. If applied to a satellite drawn in by planetary tides, then this mechanism could explain only ring systems located within the synchronous orbit but would not apply to the outer reaches of Saturn's rings. However, the breakup could also be of an interplanetary body (a comet?) that strayed inside the Roche limit and was torn apart. Whether the precursor object was a satellite or a comet, some fragments — even after subsequent collisional grinding — would still be as large as tens of kilometers. These might cause the elaborate structure in Saturn's rings or could be the satellites that fashion the narrow Uranian rings.

New planetary ring systems may soon develop elsewhere in the solar system. Phobos, Mars' closest satellite, lies at 2.76 Martian radii, well inside the synchronous orbit, and has been noticed to be evolving inward due to tides. If Phobos remains intact, it should strike Mars in about 100 million years; more likely, it will fracture or, at the very least, be denuded of its loose surface covering and thereby produce another faint ring system. Neptune's giant satellite Triton also has a tidally collapsing orbit and should reach its planet's vicinity at about the same time as Phobos' demise. Planetary rings then will be commonplace and not at all the rarity they once were thought to be. Indeed, following the discovery of rings about Uranus and Jupiter, some ring enthusiasts have maintained that there may have been, or even may be, rings also about Venus, Earth, and the Sun itself!

The alternate hypothesis for the origin of planetary rings was first propounded by Laplace and the metaphysicist Immanuel Kant at the end of the 18th century. Their view was that Saturn's rings formed from the primordial circumplanetary nebula out of which the planet's satellites grew. This occurred much in the manner that the planets themselves accumulated in the primordial solar nebula; in fact, essential properties in their nebular model of the solar system's origin were derived by observing Saturn's resplendent entourage.

The modern version of the nebular theory (see Chapter 20) has giant planets forming in the outer parts of the circumsolar cloud. A local density enhancement develops in the solar nebula about a core and, when pressures are high enough, an instability grows. The end result is a large gaseous protoplanet at the center of a flattened disk of gas and dust.

Large objects can aggregate, at points beyond the Roche limit, out of any matter that condenses in the circumplanetary nebula. The composition of these accumulated objects will depend upon distance from the protoplanet since, in its early stages, the protoplanet is an appreciable heat source and thereby governs the locations in which specific volatile compounds can condense out of the nebula. Investigations by James Pollack and his colleagues show that temperatures in the neighborhood of today's planetary rings were above the triple point of water for several million years after the systems first formed. Hence in these regions, only less-prevalent, less-volatile materials, like silicates, could condense. Apparently, before temperatures about Jupiter and Uranus dropped low enough, the gas disk dissipated, perhaps due to an ancient episode of violent solar wind called the "T-Tauri" phase. In contrast to Jupiter and Uranus, Saturn cooled earlier and allowed water vapor to condense into the magnificent rings we see today. The remaining two ringed planets have only modest silicate rings.

Particles that condense very close to a planet will not survive long. Gas drag causes orbits to decay and any particles forming in a gaseous disk are rapidly lost to their planets. Moreover, the planets themselves were originally more distended than seen today and so the gaps between the observed rings and the planets are not surprising.

Material that accretes within the Roche limit cannot grow without bound. The particle size is limited by the balance of disruptive tidal stresses against the attraction of gravity and interparticle "stickiness." In this regard, the apparent absence of a significant fraction of large particles in Saturn's rings could be meaningful. Indeed the identification of objects larger than a kilometer or so in Saturn's rings would be very damaging to the nebula model.

A RING RECAP

The austerely beautiful rings encircling Jupiter, Saturn, and Uranus remain the "most extraordinary marvel" that Galileo called Saturn's in announcing its discovery. As our knowledge of planetary rings has improved, the three systems have been seen as perhaps more individualistic than the planets they surround. Uranus' nine narrow bands are made of small rocky chunks and display an intriguing dynamical structure. Comprised of icy snowballs, Saturn's rings are baroque in their organization and variety. Jupiter's ring is a mere wisp that must be continually generated from unseen moons.

The dynamical processes occurring presently in planetary rings may provide an appropriate analog for events in the early solar system or in the galaxies. The detailed structure visible even today in the ring systems could be a fossil record of an intermediate stage in the accretion of orbiting bodies. Planetary rings are not only exquisite; in a very fundamental way they may represent the solar system's distant past.

The Galilean Satellites

Torrence V. Johnson

For over three and a half centuries following their discovery, the four largest satellites of Jupiter remained tantalizing points of light in astronomers' telescopes, tiny disks barely discernible even under the best atmospheric seeing conditions. The discovery of Io, Europa, Ganymede, and Callisto, announced in 1610 by Galileo, provided strong support for the Copernican solar system. Observations of their relatively rapid motions around Jupiter have fascinated the amateur and intrigued the professional astronomer ever since. Ole Roemer used their eclipses to determine the speed of light, Albert Michelson measured their diameters with his stellar interferometer, and mathematical analyses of the satellites' motions in the earlier parts of this century emphasized the importance of resonant phenomena in celestial mechanics.

In the decade of the 1970's, new observations and techniques resulted in a renewed awareness of the importance of the Galilean satellites to solar system studies. Telescopic observations of the satellites' spectra led to knowledge of their surface compositions; more accurate measurements of their diameters were made; clouds of neutral sodium and ionized sulfur related to Io were discovered; the Pioneer 10 and 11 flybys provided the first *in situ* measurements of the satellites magnetospheric environment. And finally, in 1979, the Voyager spacecraft transformed our view of these objects from dots of light into places, new worlds seen clearly for the first time.

GENERAL CHARACTERISTICS

The Galilean satellites, together with tiny Amalthea, a trio of moons discovered by Voyager (1979 J 1, 1979 J 2, and 1979 J 3), and Jupiter's tenuous ring, form one of the three known regular satellite systems, the others being those of Saturn and Uranus. These systems are characterized by satellites and rings with circular, coplanar orbits in the planet's equatorial plane (orbital characteristics of the Jovian satellites are given in the Appendix). All of these systems thus resemble the Sun's planetary system and are frequently referred to as "mini solar systems." Fig. 1 conveys this impression of a planetary system; it shows, moving outward from Jupiter, the orbits of Io, Europa, Ganymede and Callisto.

The Galilean satellites lie deep within the Jovian magneto-sphere. Instead of being continually exposed to the solar wind, as our Moon is, they are immersed in the Jovian plasma environment and are bombarded by high-energy, trapped charged particles. The satellites, particularly Io, appear to interact strongly with this environment in several ways. Io, at least, supplies significant amounts of material,

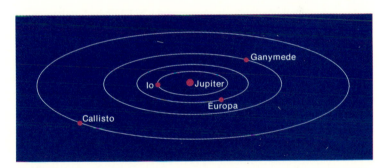

Fig. 1. The orbits of Jupiter's Galilean satellites as viewed from a point about 20° above their orbital planes and some 9,000,000 km from Jupiter.

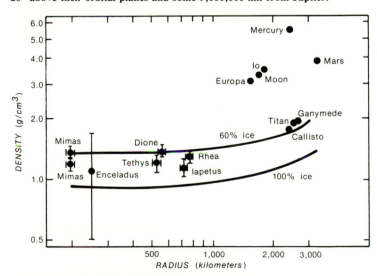

Fig. 2. The radii and densities of the Galilean satellites are compared with those of other bodies; solid lines denote models calculated by Mark Lupo and John Lewis. Both density values for Mimas are based on classical (mass/volume) determinations. The higher one, however, is suggested by a new, greater mass found for nearby Tethys by analyzing the Voyagers' Doppler-shifted radio signals.

— atoms and ions — to the magnetosphere; and charged-particle impacts on the satellite surfaces have been suggested as a cause of coloration and even significant erosion over geologic time.

The satellites are all relatively massive, roughly comparable to the Moon. They thus perturb each others' orbits significantly, and this situation, together with a resonant relationship in the orbital motions of the inner three satellites, permitted deSitter and Sampson to make reasonably good determination of the satellites' masses in the 1920's. Much more accurate values are now available from the tracking of spacecraft, but the old values were in general accurate to 20 percent or better. The resonant relation in the orbital motions of Io, Europa and Ganymede (studied by Laplace) also appears to play a major role in powering the volcanoes on Io, which will be discussed later.

Table 1. RADII AND DENSITIES (Voyager data)		
	Radius (km)	*Density* (g/cm^3)
Io	1,816 ±5	3.55
Europa	1,563 ±10	3.04
Ganymede	2,638 ±10	1.93
Callisto	2,410 ±10	1.81

The diameters of the satellites were very uncertain prior to 1970 due to their small angular size as seen from the Earth, but stellar occultations and spacecraft images have improved this situation considerably. Table 1 gives current values for the satellites' radii, masses and derived densities. These densities provide us with the most direct evidence of basic differences in the bulk compositions of the satellites. The lunarlike sizes and densities of Io and Europa mark them as essentially rocky, silicate-rich bodies, Io being slightly more and Europa slightly less dense than the Moon. On the other hand, the low densities of Ganymede and Callisto (less than 2 g per cubic centimeter) strongly suggest a large non-rock component in their composition, most likely water in some form. Fig. 2 compares the sizes of and densities of the Galilean satellites with those of the terrestrial planets, the Moon and some of the satellites of Saturn. Io and Europa are quite similar to the Moon in bulk properties but differ greatly in surface appearance and evolution (Fig. 3). Ganymede, Callisto, and Titan form a new class of large (Mercury-size) objects with much lower density than the more familiar terrestrial planets.

Fig. 3. What vistas await would-be travellers to the Galilean satellites? Don Davis has rendered these landscapes based on Voyager results. From left

John Lewis, in pioneering work done in the early 1970's, pointed out that densities like those of Ganymede and Callisto are precisely what one would expect from condensation of solar-composition gas at temperatures where water ice is stable. Such a body would have approximate equal weights of silicates and water (see Chapter 20). He also argued that such a composition would almost inevitably lead to early melting and differentiation as a result of accretional, radiogenic, and possibly tidal heating, combined with the relatively low melting temperature of water. Other scientists then considered the possible interior structures of these differentiated, water-rock planets, and their models emphasize the possible role of convection within warm, slowly deforming ice in transporting heat from the satellites' interiors. These models suggest either that complete melting of the water may never have occurred or, if it did, that convection in an ice crust could have frozen the remaining water in a short time. Fig. 4 illustrates current interior models for the satellites. Note that Ganymede and Callisto both have similar structures of ice crust, convecting water or ice mantle, and silicate-rich core.

Why are there two classes of Galilean satellites? The fact that the higher density objects Io and Europa are closer to Jupiter than the low density satellites leads naturally to a comparison with the solar system itself. In the early 1950's, Gerard P. Kuiper suggested that heat from Jupiter early in its history could have prevented lighter elements from condensing or caused them to be "boiled off" the inner satellites. Recent investigations suggest that, indeed, Jupiter's starlike early stages could have produced enough heat to make the environment of the inner satellites significantly warmer than that of the outer satellites. James Pollack, Fraser Fanale, and their co-workers have investigated a number of models of early Jovian history, which use the current value of energy output from Jupiter (over twice the energy received from the Sun) as a boundary condition. They find that most plausible cases result in a period of some 100 million years where conditions would have permitted the formation and retention of water ice at the present position of Ganymede and Callisto while allowing only higher density (although probably water-enriched) silicates to exist at Io and Europa. As with models of solar system formation

to right are the surfaces of Callisto, Ganymede, Europa, and Io. Each scene shows Jupiter at its correct relative size.

the dynamics of accretion may have influenced the satellites' composition and energy balance by mixing materials within the region of satellite formation and by bringing in material from outside the system. A generally accepted detailed model of satellite formation is still lacking, but the striking variation in the amount of condensed volatile material, mostly water ice, with increasing distance from Jupiter is strong circumstantial evidence that the early Jupiter, with its immense gravitational field and central heat source, played a dominant role in determining the nature of its satellite bodies — just as the Sun must have on a vaster scale.

Voyager data provide us with a wealth of new information concerning the satellites' surface processes and evolution. However, most of our knowledge of surface composition comes from analysis of the remarkably diagnostic reflection spectra taken during the last decade using ground-based and airborne telescopes. In the early 1970's several groups began using Michelson interferometer spectrometers to obtain the first relatively high-resolution infrared spectra of the Galilean satellites. Carl Pilcher and his co-workers, and Uwe Fink and Harold Larson independently obtained spectra from 1.0 to 3.0 microns that exhibit strong absorption for Europa and Ganymede and either weak or absent absorptions for Io and Callisto. Later work showed weak absorptions in Callisto's spectrum. (Data suggesting infrared structure in these satellites' spectra were acquired a decade earlier by Kuiper and Soviet planetologist V. I. Moroz.)

Geometric albedo is an object's brightness compared with a perfectly diffusing disk under normal illumination; Fig. 5 shows the satellites' variations of albedo with wavelength from 0.3 to 5.0 microns — effectively the entire range dominated by reflected solar radiation. The deep absorptions near 1.4 and 1.8 microns in Europa and Ganymede's spectra were quickly identified as arising from water ice; the strength of these absorptions and the high albedos suggest large amounts of relatively clean ice. Callisto's weaker absorptions and low albedo, on the other hand, indicate a "dirty" surface, although debate continues on the degree to which frozen hydrated soil can account for the spectra compared to free frost mixed with a dark background.

In Io's case, the combined analysis of the ground-based spectra and evidence from a variety of Voyager observations has given us a much better picture of its surface composition than either data set alone provided. Io's high albedo, very red color, and lack of diagnostic absorption features from 1-3 microns posed a difficult problem to planetary scientists. Many workers suspected that the sharp drop in Io's blue and ultraviolet reflectance was in some way connected with

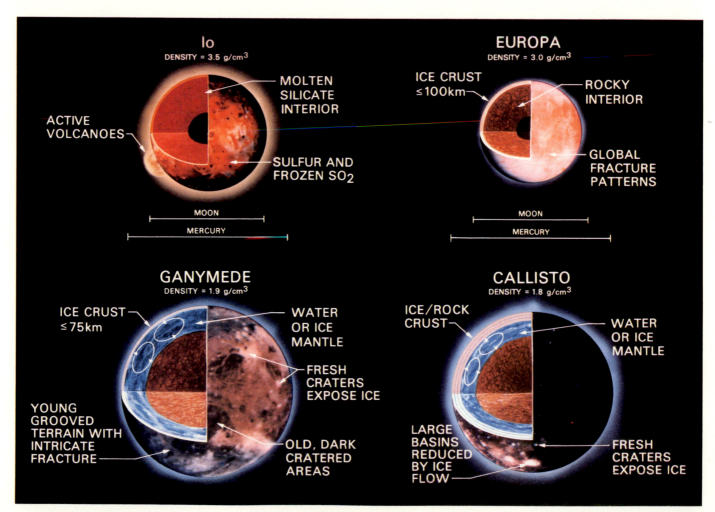

Fig. 4. These schematic illustrations portray each satellite's interior as presently understood. Earlier models, based primarily on telescopic observation, were generally confirmed in light of the Voyager findings, although it had been thought that Io's interior was solid and that Callisto's density was somewhat lower. Horizontal scale bars indicate diameters of the Moon and Mercury.

sulfur or sulfur compounds (a feeling reinforced by the discovery in 1976 of ionized sulfur concentrated along Io's orbit), but the source of the sulfur and the nature of the non-sulfur component, if any, remained subjects of speculation.

In 1978, infrared spectra taken from the high altitude observatory at Mauna Kea and from the Kuiper Airborne Observatory showed that Io's spectrum had a deep absorption near 4.1 microns. This feature was not identified immediately although a number of candidate materials, including most silicates, were ruled out. Spurred on by the prospect of the approaching Voyager encounters, a number of laboratories worked diligently to identify the substance causing the feature and find a consistent explanation for Io's optical properties. As often happens in science, a number of clues fell into place nearly simultaneously. In the weeks following the Voyager 1 Jupiter encounter in March of 1979, the discovery of volcanism on Io, the observation of abundant sulfur and oxygen ions in the Jovian magnetosphere, and the discovery of sulfur dioxide (SO_2) gas in the vicinity of one of Io's volcanoes (Loki) coincided with laboratory data from two independent groups identifying frozen SO_2 as the source of the 4.1 micron absorption feature.

Although there remain many problems in understanding the chemistry of Io, it now appears that the materials dominating the upper surface layers are forms of sulfur and sulfur compounds, including at least SO_2, brought to the surface continually by volcanic activity. What other materials may be there, either mixed with the sulfur-bearing components or buried beneath them, remain to be discovered. This includes the sodium- and potassium-bearing phases inferred from the presence of neutral "clouds" of these elements escaping from Io and from observations of sodium ions in the magnetosphere.

CALLISTO

Impact craters dominate the surface of the outermost Galilean satellite. In fact, Voyager images show that virtually the entire surface is covered with large craters, standing nearly shoulder-to-shoulder. Callisto is unique among the

cratered planets investigated to date in having no "plains" units where craters have been obliterated by more recent processes. The Moon, Mercury and Mars, for instance, all have significant portions of their surfaces covered by material of volcanic origin. The only places on Callisto where the crater density appears to be significantly lower than average are near the centers of the several large ring structures believed to be the scars of even larger impacts (seen dramatically in Fig. 6).

Fig. 7 illustrates the cumulative crater frequency (the number per million square kilometers) as a function of crater size for the Galilean satellites and, for comparison, selected lunar and Martian regions. The curve for Callisto's average surface shows a crater population nearly as dense as that of the lunar highlands. In order to use such data to estimate the age of a satellite's surface, a number of factors must be taken into account, including the differences in the flux of crater-producing objects at different places in the solar system, possible differences in the impacting bodies themselves (for instance, were they asteroids or comets?), the effects of Jupiter's gravity, and the response of ice-rich target surfaces during impacts — all of which await detailed evaluation for the Galilean satellites.

Studies of lunar cratering and radioisotope chronology, the current population of asteroids and comets capable of

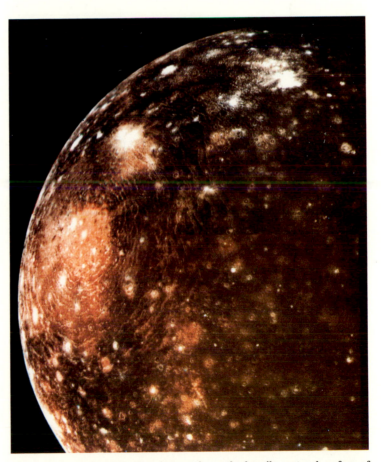

Fig. 6. This mosaic of Voyager images shows the heavily cratered surface of Callisto. Most prominent is an extensive ring structure known as Valhalla, which is similar in many respects to large, circular impact basins that dominate the surfaces of the Moon and Mercury. Valhalla's bright central area is about 300 km across; sets of discontinuous concentric ridges extend out to some 1,500 km from the center.

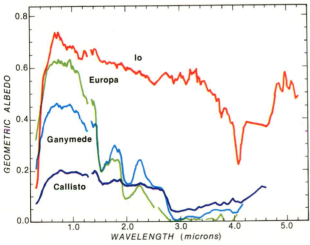

Fig. 5. Studies of Jupiter's largest satellites from Earth show that they have distinctive spectral signatures. Absorptions in the near-infrared reveal the presence of water ice on the surfaces of Europa, Ganymede, and Callisto. The reflectivities change quickly through the visible spectrum. From an article by Roger Clark and Thomas McCord, *Icarus, 41,* 1980, 323-329.

Fig. 8. This portion of Ganymede's surface exhibits typical examples of both dark cratered and light "grooved" terrains. The dark regions, frequently angular or polygonal in shape, are somewhat higher in albedo than Callisto but considerably lower than Ganymede's global average; their color also matches Callisto closely.

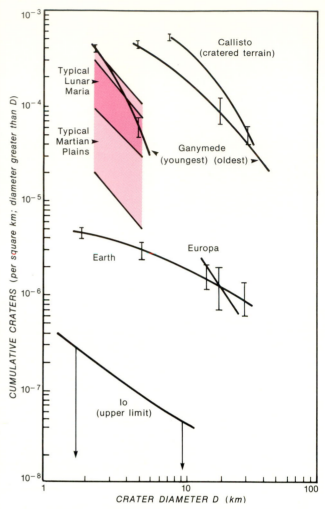

Fig. 7. Crater size-frequency curves for the surfaces of the Galilean satellites are compared with the Earth, Moon and Mars to show relative ages. The oldest (most heavily cratered) terrains on Ganymede and Callisto are comparable in age to the oldest surfaces on the Moon, Mars, and Mercury. The youngest terrain on Ganymede plots within the range covered by the lunar maria and Martian plains. The curve for Europa is very uncertain because of the scarcity of identifiable craters on the Voyager images. Io's surface is so fresh that *no* craters were identified down to 1-km resolution of the Voyager images; a theoretical upper limit for their distribution has been given.

impacting the planets, and comparisons of the cratering records on the terrestrial planets all suggest that the very high crater densities of the lunar uplands, the heavily cratered portions of Mercury, Mars, and Callisto's surfaces cannot have been produced recently in solar system history. Impact specialists generally agree that the cratering flux (at least in the inner solar system) declined very rapidly in the first half billion years following planet formation about 4.6 aeons ago (see Chapter 4). Despite the uncertainties, it seems very likely that Callisto's surface dates back to this early period and has been little modified since (except by a continual but lower intensity "rain" of comets and asteroids).

At first glance, the surface of Callisto shows no obvious evidence of its icy nature. Most of the craters look superficially like those on the rocky terrestrial planets. But examining the details of crater shapes and of the largest

impact features, we find a number of striking differences from similar features on rocky objects. Craters of all sizes on Callisto are much flatter than their counterparts on terrestrial planets. In addition, the largest impact basins lack central depressions (usually partially or completely flooded by volcanic material) and surrounding ring mountains; examples of such basins on the terrestrial planets are the Imbirum and Orientale basins on the Moon and the Caloris basin on Mercury. In place of such structures, Callisto exhibits a number of large high-albedo, circular features with or without concentric rings and radial structure (Fig. 6 shows the largest of these). Such features are similar in scale to traditional basin structures but show little or no topographic relief. Surface temperatures on Callisto range from equatorial noontime highs of perhaps 140-150° K to subsurface averages 100° K or less. Ice this cold acts more as a rock than the relatively volatile substance we are familiar

Fig. 9. A detailed Voyager image along Ganymede's terminator reveals complex ridge and groove systems that crisscross the entire scene. Such terrain presumably results from deformations in the moon's thick, icy crust. The smallest features seen are about 3 km across.

GANYMEDE

The ice giant of the satellite system, Ganymede displays many of the characteristics of Callisto's surface, but also has a baffling array of tectonic features unlike anything previously seen on the terrestrial planets. Seen at low resolution, Ganymede looks deceptively like the Moon, with irregular dark regions on a brighter background (see Fig. 8). These dark regions are probably the basis for Ganymede's mottled appearance in maps drawn from telescopic observations and in low-resolution Pioneer 10 images taken in 1974. Despite the fact that Ganymede's overall reflectance is four to five times higher than the Moon's, these dark regions were occasionally referred to as "mare." High resolution Voyager images showed how fundamentally different Ganymede and the Moon are. The dark regions, far from being younger flows of lavalike material, are instead the oldest parts of Ganymede's surface, very heavily cratered, and resemble Callisto's surface in many respects. The light regions, on the other hand, are clearly younger than the darker areas and exhibit a variety of features suggestive of tectonic activity. Fig. 7 indicates that, while not quite as heavily cratered as Callisto's surface, dark areas on Ganymede are probably also quite ancient. Two of the largest ones now bear the names of Galileo and Simon Marius (who independently discovered the satellites at the same time as Galileo).

The grooves which dominate the lighter regions are parallel sets of ridges and troughs, kilometers to tens of kilometers wide and with perhaps a few hundred meters of vertical relief. They form sets or "bands" of grooves which wander for thousands of kilometers across Ganymede's surface, forming intricate patterns, particularly when different sets intersect or interact with one another. Fig. 9 shows a particularly complex region seen very near the terminator in a Voyager 1 image. These groove systems border the dark terrain everywhere and in some places seem to extend into blocks of the dark terrain. Lower crater densities in these areas suggest that they developed later than dark regions and probably grew at their expense, destroying previously existing dark crust. Although found less frequently than on the dark terrain, craters seen on the grooved areas are still plentiful in absolute terms, occurring in greater numbers than on the lunar maria, for instance.

with. However, even at Callisto's temperatures, we expect ice to "flow" at a glacierlike crawl when subjected to stress over geologic time. Deeper in the crust, tens of hundreds of kilometers below the surface, temperatures will be even higher due to an outward conduction of radioactively produced heat from the deep interior. Thus, large impact structures which "feel" these warmer depths may deform even faster. Furthermore, the material properties of ice probably play an important role in the initial impact event and subsequent deformation of the newly formed crater.

This combination of initial deformation followed by viscous flow is believed to be responsible for the flat crater shapes and the large ring structures on Callisto. Giant impact events — which would have produced classic impact basins on the terrestrial planets — probably never formed a real topographic depression for any length of time on any of the ice satellites, particularly if the cold, brittle crust was thinner and weaker early in the satellite's history. Smaller impact events, involving only the upper few kilometers or so of the crust, formed distinct craters but these probably looked different from ones in rock even from the beginning. Subsequent glacial flow further reduced their topographic profile, again more effectively if the early crust was thinner and "softer" than at present.

Although the history of crater-producing events at the Galilean satellites is not yet firmly established, Eugene Shoemaker believes that crater densities of this order must indicate a very old surface. He suggests that, as with the Moon, the two different terrains are both greater than 3 to 3½ billion years old. We know that the production of grooves was still proceeding when many large craters were being formed, because as shown in Fig. 10, large craters both overlie and are crosscut by groove systems.

Fig. 10. This section of Ganymede, measuring about 300 km on a side shows a group of craters that were formed before, during, and after the grooved terrain around them developed.

Fig. 12. Nearly an entire hemisphere of Ganymede is seen in this Voyager 2 image. The prominent dark area, called Galileo Regio, is about 3,200 km in diameter and contains light-colored, closely spaced bands that resemble those seen on Callisto. Bright spots are relatively recent impact craters, while light brown circular areas are probably the remains of older impacts. Part of Galileo Regio may be covered with a bright frost.

 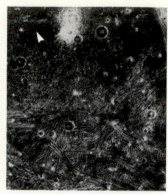

Fig. 11. These images include regions of Ganymede where groove systems or bands appear to change their width or become offset by a linear, faultlike feature. In the picture at right, note the crater (arrow) that has been cleaved in two and spread apart by a furrowed trench running through it.

The exact nature of the immense groove systems criss-crossing Ganymede's surface is not yet known. They appear to result from tensional, not compressional mechanisms, analogous to the formation of graben valleys on Earth. Their high albedo (about 40 percent) indicates that these areas are "less dirty" in some ways than the darker areas. Whether this results from new material supplied from depth during their formation or from reworking of the existing crust is not clear. In addition, there are indications that the grooves' rates of formation may have varied and that many parts of the crust were mobile to a degree (Fig. 11).

On the terrestrial planets, particularly the Moon and Mercury, fresh craters are brighter than their surroundings and exhibit bright, radiating, ray patterns (Tycho is a classic example on the Moon). Bright ray craters result from the exposure of freshly pulverized crustal material which has not had time to develop a dark, impact-"gardened" surface and from reworking of the upper layers of the soil by impact ejecta (see Chapter 4). But *dark* ray craters, which are relatively common on Ganymede, demonstrate that other processes may be important in ice-dominated soils. It is possible, for instance, that contamination of the material thrown out of an impact crater by dark, subsurface units or even by projectile material may be more important on Ganymede than on the Moon or Mercury.

Craters on Ganymede are similar in form to those on Callisto. The large, old craters (particularly on the ancient, dark terrain) are very flat for their size, showing the effects of the icy crust and viscous flow. These scars of ancient impacts have been dubbed "palimpsests," in archaeology a word referring to a parchment which has been scraped clean and written over again. Numerous palimpsests occur on both Ganymede and Callisto with diameters of several hundred kilometers, usually recognized by a light, circular patch (see Figs. 12 and 13). This central patch may represent the limit of continuous ejecta from the original crater, or it may be a more complicated phenomenon marking "ground zero," where cleaner ice was generated either during the impact itself, or by later transport of water from a cleaner sub-surface layer.

The central portion of the Valhalla ring structure on Callisto may be the largest member of the family of palimpsests. No comparable feature is present on Ganymede;

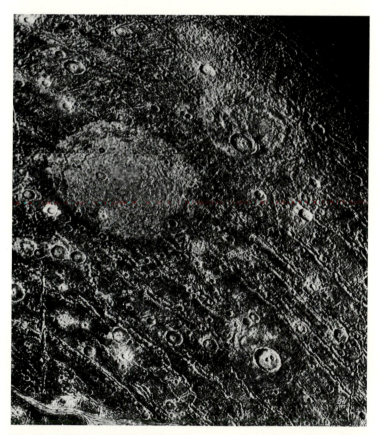

Fig. 13. A higher-resolution view of central and southern Galileo Regio reveals impact craters in various stages of degradation. Nearly all craters appear quite flat, and two prominent light-colored patches show only traces of crater rims, now almost completely erased by slow plastic flow in the icy crust. These features have been dubbed "palimpsests," a word in archaeology referring to a parchment that was scraped clean and written over again.

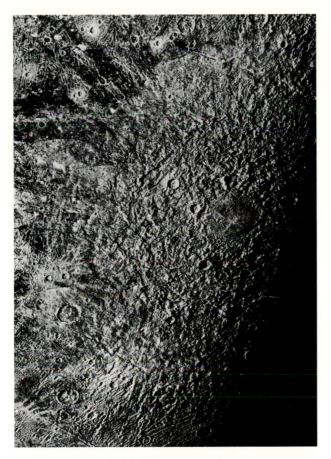

Fig. 15. The best Voyager 1 image of Europa shows a bright, low-contrast surface with darker mottlings and a tantalizing network of lines that crisscrosses much of its globe.

Fig. 14. A fresh impact basin on Ganymede, situated near the terminator in this view, is surrounded by ejecta. Some 175 km across, this basin shows considerable relief and differs markedly from older craters. Evidently it formed when Ganymede's crust became rigid enough to sustain major topographic features.

indeed, there is no piece of undisturbed dark crust large enough to contain such a feature. However, a series of parallel furrows run across Galileo Regio (also discernible on some nearby dark crust) which strongly resemble the ring structures on Callisto. If this is the remnant of a similar structure, the original ring system would have to have been even larger than Valhalla, affecting nearly an entire hemisphere on Ganymede. Unfortunately, the center of this hypothetical structure is no longer apparent; the area where it should lie is extensively modified by grooved terrain. If the furrows on Galileo Regio and the surrounding dark terrain are part of one large ring structure, some movement and rotation of the dark blocks of cratered terrain seem to be necessary to account for its current configuration.

Ganymede's groove systems — indicating the breakup of dark crustal blocks, the offsets of groove systems along transcurrent faults, and the possible rotation of dark areas containing the furrows — are all strongly suggestive of tectonics on a global scale at some early period in the satellite's history. This activity may bear some resemblance to global plate tectonics on Earth, although there are obvious differences such as the composition of crust and mantle and the lack of any identifiable subduction zones (regions of crustal sinking) on Ganymede. The satellite's crustal grinding may not have lasted long, for craters on the younger, grooved terrain show less degraded forms than on

the dark regions, indicating a stiffer crust following groove production. In the southern hemisphere, Voyager 2 images show an impact basin more similar to terrestrial planet basins than to Callisto's ring structures (Fig. 14). This basin has many large craters superimposed on it and must itself be reasonably old. Thus, the combined evidence suggests that Ganymede's crust had stiffened to a point where it retained even large impacts at an early stage in its history, perhaps 3½-4 billion years ago.

This general picture of the early evolution of Ganymede and Callisto fits the available data. Both satellites, in the later stages of accretion, probably possessed heavily cratered crusts darkened by the combination of ice and dark silicates contained within incoming debris. Their interiors were quite likely molten water (or at least warm, convecting ice), and the combination of internal activity and occasional large impacts kept disrupting the dark crusts and bringing fresher, brighter material to the surface. As the satellites cooled, the crusts became thicker and more rigid, less liable to disruption. Callisto's crust may have stiffened somewhat earlier than Ganymede's and thus continued to collect the final dark debris of accretion. Ganymede's crust, on the other hand, probably remained active longer, so that by the time it became stiff, less material remained to rain down on its surface. We do not yet know whether this difference in history resulted from chance (did several large bodies strike Ganymede but not Callisto late in the accretion stage?), or from small differences in early energy sources such as radioactive heating, differences in accretional heating, and tidal effects.

There could easily be variations to this general picture, such as the possibility that Callisto's crust, for some reason,

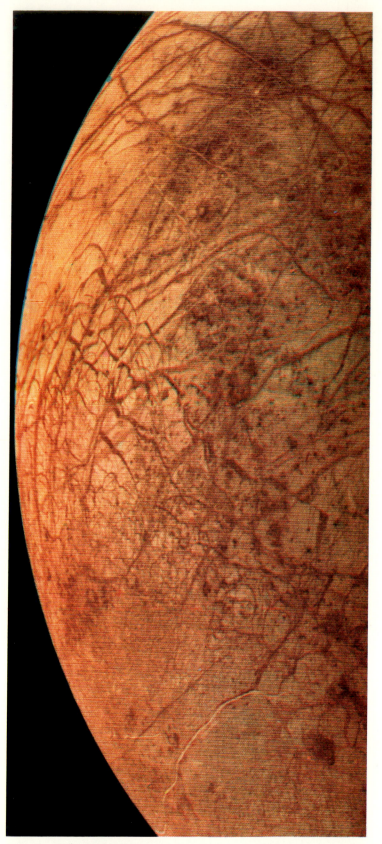

Fig. 16. When Voyager 2 observed Europa from 241,000 km away, the satellite's Lowellian maze of streaks resolved into a vast tangle of light and dark markings, tens to hundreds of kilometers across, that suggest filled-in cracks. Despite their superficial resemblance to fractures, the stripes have very little vertical relief, either positive or negative. They are primarily albedo markings, perhaps controlled by underlying fracture pattern in the crust.

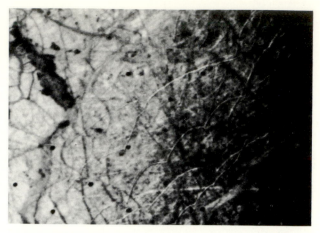

Fig. 17. Only under extremely oblique lighting was it possible to find any topographic relief on Europa. This near-terminator view is an enlarged section of a Voyager 2 image. Many narrow ridges, perhaps a few hundred meters high, loop across the surface in bright, arcuate patterns. Other ridges occasionally run down the center of the dark, broader stripes.

was never thin or soft enough to be disrupted or that subsequent darkening by the infall of meteoritic material has complicated the situation. More detailed studies of the two satellites will be needed to understand more fully the differences in evolution between them.

EUROPA

For many of the planetologists watching early images of Europa arriving from Voyager 1 (Fig. 15), the aspect of this distant satellite was eerily reminiscent of Percival Lowell's perception of canals crossing ruddy Mars. Higher-resolution views from Voyager 2 (Figs. 16 and 17) dispelled any hint of a relationship to the Mars of Lowell's imagination but they revealed instead a surface unlike that of any other planet (although large fields of sea ice in the Earth's polar regions come close). After the Voyager encounters, Europa remains perhaps the most enigmatic of the Galilean satellites. Our best views resolve features no smaller than about 4 km across, and these cover but a small fraction of the globe. This low-resolution, when combined with Europa's low-relief, "billiard-ball" surface, makes normal geologic interpretation very difficult.

Among other points, we'd like to know the age and thickness of Europa's unique crust. A primary observation from Voyager data is the lack of craters larger than about 50-100 km across. At similar resolution, both Ganymede and Callisto exhibit evidence of many impacts in this size range, so Europa's surface must be younger than, say, the grooved terrain on Ganymede. A key issue is *how much* younger, and the data only partially supply us with answers. Several small craters (about 20 km across) can be spotted near Europa's terminator, Fig. 18 shows a few examples. This crater density is quite low but could still indicate a surface as old as the lunar maria (3-3½ billion years old) if objects have struck the satellite only one-tenth as often as they have the Moon. However, recent estimates suggest much younger ages, perhaps as low as 100 million years. If these younger ages are correct, then significant resurfacing must be going on even at present on Europa.

The thickness of the ice covering the satellite is linked to

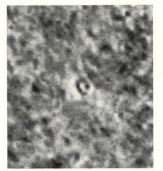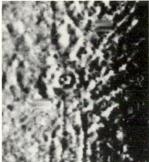

Fig. 18. Isolated craters, about 20 km in diameter, can be seen in this near-terminator Voyager 2 image. Although infrequent, these pits may still suggest that Europa's surface is relatively old.

the formation of the surface stripes and ridges. Europa's relatively low density, 3 g per cubic centimeter, places some loose limits on the ice thickness. If the bulk of the satellite has a density between the Moon (3.3) and Io (3.5), and all its water has been outgassed and now lies frozen on the surface, then an ice crust 75-100 km thick would be needed to reduce the mean density to 3.0. A thicker layer would be possible if the rest of Europa has an unusually high-density composition, for instance one more iron-rich than either the Moon or Io. A lower limit is more difficult to estimate, since even a light frosting of high-albedo ice would be enough to obscure underlying dark material. However, if surface features such as thin ridges result from processes operating in the ice crust, it must at least be comparable in scale to the height and width of these features (at least a few hundred meters).

The stripes themselves suggest tensional forces and, taken together (in the absence of any major compressional zones), they indicate an increase in surface area of perhaps 5 percent. Several models of Europa's crust associate this expansion in some way with a "frozen ocean." The simplest case envisions the outgassing of virtually all Europa's water immediately following accretion, resulting in an initially liquid layer topped by an ice crust. If radioactive decay and tidally generated heat are not sufficient to maintain the liquid layer, the freezing of this "ocean" would significantly expand its volume and increase its surface area — but not enough to explain all the "extra" area in the stripes. This model also suggests that we are looking at a very old surface, contrary to some of the estimates based on cratering. A similar model holds that water may be continually coming out of the interior and resurfacing the satellite.

The other major concern is the stripes' global distribution. Despite the small surface area covered at high resolution, the major linear patterns can be traced around the globe in the lower resolution pictures. Has Europa undergone fracturing on a global scale? Did tidal forces or a wrenching rotational slowdown cause these patterns? Further analysis of Voyager data will certainly yield further constraints to this problem. For instance, we now know of at least two stratigraphically distinct fracture systems, and can therefore infer that they did not occur all at once.

Also awaiting explanation are numerous dark, mottled regions which may have been "dirtied" by the upwelling or mixing of silicate-laden water and ice. Although some pictures suggest that these areas may be somewhat rougher than average, they do not appear to be blocks of older crust

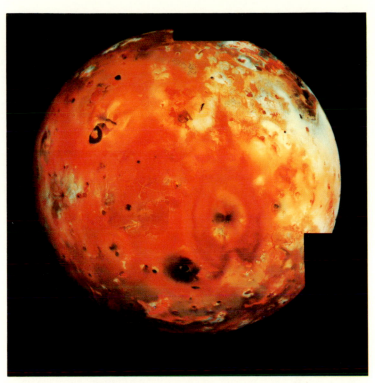

Fig. 19. Many volcanic features are evident in this mosaic of Io's disk, although they were not initially recognized as such. The large, heart-shaped marking just to the right and below center is a ring of volcanic ejecta being thrown out by the eruptive plume Pele inside it. The dark, horseshoe-shaped arc at upper left resembles terrestrial lava lakes and is associated with another eruptive center, Loki, to its immediate upper right. Numerous other dark spots mark volcanic calderas, and flow features appear all over the surface seen here.

as on Ganymede. The Voyagers' color data also indicates that there may be at least two distinct types of this material, with different ultraviolet reflectances. It is possible that external effects (such as the implantation of sulfur ions from the Jovian magnetosphere) may be affecting these color properties as well as the chemistry of the original material, particularly since the region with the lowest ultraviolet reflectance lies on the side of the satellite continually being overtaken by the plasma in Jupiter's rotating magnetic field.

IO

Before the Voyager 1 encounter with Jupiter, Io was already known to be one of the strangest bodies in the solar system. It was, therefore, doubly surprising that Io surpasses even the wildest expectations of planetary scientists. The combination of intense volcanic activity, exotic chemistry, and complex interactions with the Jovian magnetosphere makes Io one of the most active and interesting planetary bodies yet explored.

Most obvious to the eye are the satellite's unusual red-orange color (due primarily to sulfur, as discussed earlier) and the absence of any of the "classical" impact features: ray craters, impact basins, and craters of all sizes. Instead, the landscape is dominated by many volcanic landforms.

The highest resolution images of Io were of the Jupiter-facing hemisphere, that is, regions to the left (west) and beyond the western limb in Fig. 19. The best images achieved resolutions of about 0.5-km and show a wealth of surface detail and volcanic landforms (Fig. 20). Sulfur is apparently responsible for Io's coloration, and some planetary geologists speculate that the flows themselves may be composed of sulfur rather than silicate rock lavas. They suggest that the varying shades of color in and around the flows result from different forms (or allotropes) of sulfur which are stable at different temperatures. Sulfur flows are rare but not unknown even on the Earth. An opposing view is that the flows are more "ordinary" lavas, basaltic or otherwise, but are colored by a high sulfur content and superficial deposits.

Calderas, both with and without obvious flow structures, are abundant on Io's surface. Perhaps 200 with diameters greater than 20 km pock the globe; the Earth, with 3½ times more surface area, has only 15 or so. As opposed to many terrestrial and Martian volcanoes, however, the eruptive centers on Io apparently do not build up large constructs of lava similar to classical shield-type volcanic mountains. This observation, and the very long flows stretching for hundreds of kilometers from some calderas, suggest that the volcanic fluids on Io (whether molten rock or sulfur) have very low viscosities when erupted.

Regions of considerable topographic relief occur on Io (Fig. 21), in contrast to the flat, icy surfaces of the other satellites. The origin of these mountains is still a puzzle. They do not appear to be volcanic constructs; there are, for instance, no summit craters on any of them. Neither are they aligned in chains or ridges in ways that suggest terrestrial

mountain-building or plate tectonics. It may be that such evidence has been covered by the recent volcanic products, making the mountains we see just the tops of an even more varigated topography. In any case, the very existence of these mountains implies significant strength in some parts of Io's upper crust and lighter crustal blocks may even be "floating" like icebergs in a denser mantle.

The dominance of volcanic morphology and the absence of impact craters would probably have qualified Io for the title "Most Volcanic Planet," but the discovery of eruptions in progress during the Voyager 1 flyby (Fig. 22) confirmed the title unequivocally. In all, up to nine eruptions spewed material more or less continuously during Voyager 1's flyby.

The fountainlike plumes of Io are among the most impressive and beautiful sights in the solar system. The satellite's low gravity (about one-sixth of Earth's) and lack of appreciable atmosphere lets volcanic gas and dust rise unimpeded to great heights, then fall slowly back to the surface, frequently forming symmetrical mushroom-shaped plumes. Fig. 23 shows several views of two eruptive plumes, Prometheus and Loki.

Most of the plumes discovered by Voyager 1 were observed several times over a period of days; those seen repeatedly showed little or no change during this time. Voyager 2, arriving four months after Voyager 1, was reprogrammed to make a study of the volcanic activity found by the first spacecraft. It found no new plumes, but several marked

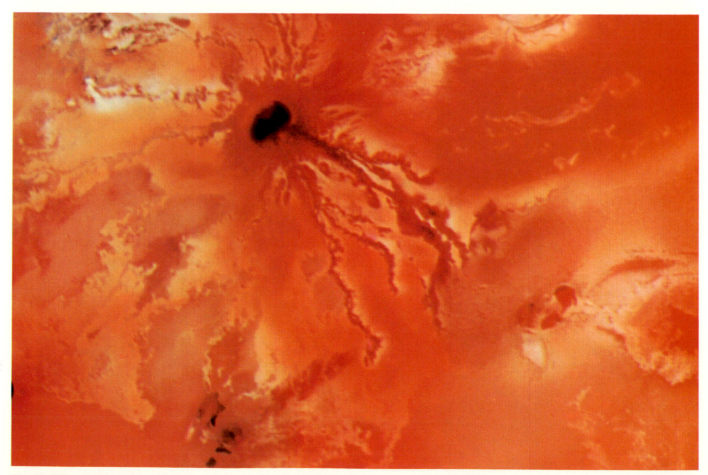

Fig. 20. This volcanic center, with flows radiating from it in many directions, bears strong resemblance to terrestrial volcanoes having fluid, low viscosity lava flows. However, this one (and many others like it) are distinguished by the strong colors they exhibit.

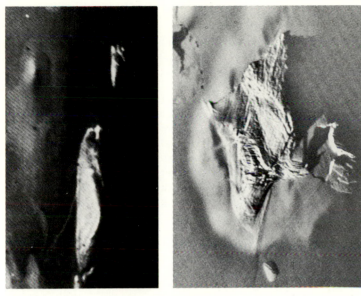

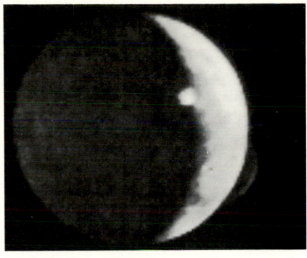

Fig. 21. These mountains are examples of the topography that can be glimpsed on Io under favorable lighting conditions (near the terminator or against the limb). Such peaks can be up to 10 km tall.

Fig. 22. The dome-shaped eruptive plume of Pele rises more than 300 km above the eastern limb of Io in the "discovery picture" that first alerted engineer Linda Morabito to the satellite's dynamic nature. Many other plumes were quickly identified on additional Voyager images. A second plume, overlooked at first, is the bright spot at center.

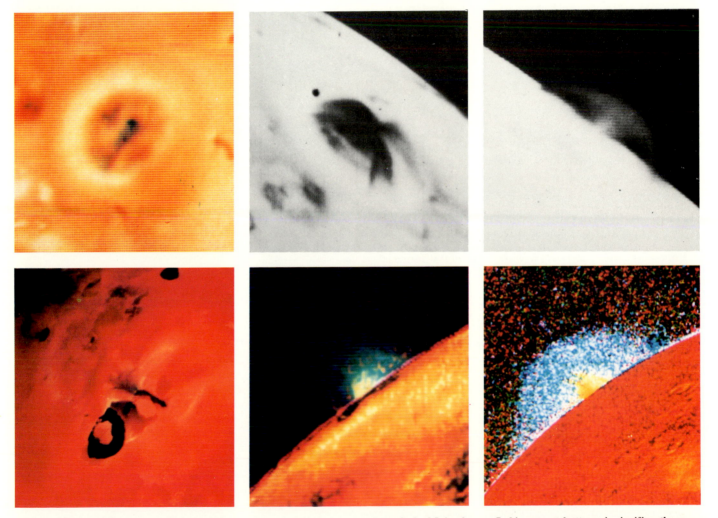

Fig. 23. The top three photographs are of Prometheus, one of the more symmetrical of Io's plumes; Loki, seen at bottom, is significantly asymmetrical due to the shape of its vent. In fact, the Loki plume likely consists of at least two components spewing forth from opposite ends of the large black fissure seen near the lava lake in Fig. 19. This plume also has an unusually large envelope, possibly made of sulfur dioxide gas, fine particles, or both, which appears prominently surrounding the more obvious core in ultraviolet-filtered images like the one at lower right.

changes had occurred on Io during the interim (Figs. 24 and 25). Of the eight volcanic plumes seen by Voyager 1, Voyager 2 reobserved seven of them and found six still erupting (Pele being the exception). Thus, some variation in volcanic activity may occur with time scales of months, but most of the eruptions remained remarkably constant over a four-month interval.

Most plume characteristics can be matched by ballistic ejection of volcanic material at vent velocities of 0.6-1.0 km per second, as illustrated in Figs. 26 and 27. Such high speeds imply that we are not dealing with terrestrial-style explosive eruptions, for on Earth vent velocities rarely exceed 0.1 km per second. The relatively constant, "sprinkler-head" appearance of the Ionian plumes also call to mind something more akin to a geyser than an explosive volcano. (Geologist Susan Kieffer notes that Yellowstone Park's "Old Faithful," if erupting under Io's low gravity into a vacuum, would send a plume of water and ice to an altitude of more than 35 km.)

Rapid phase changes in water (from liquid to steam, for instance) drive terrestrial geysers, but Io lacks water both on its surface and in its tenuous atmosphere. A different working fluid is required. The apparently ubiquitous nature of

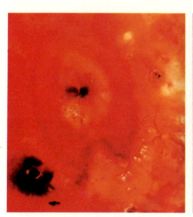
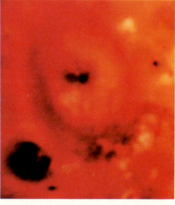

Fig. 24. The ring of ejected material surrounding Pele is more circular in the Voyager 2 image at right than in the Voyager 1 frame at left, made four months earlier. Other images of this feature showed that the eruptions that were so apparent to the first spacecraft had ceased by the second encounter.

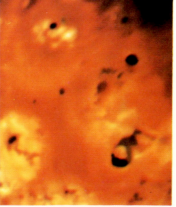

Fig. 25. The volcanic activity around Loki also changed between encounters, becoming considerably more extensive in the interim of four months. Note the blanketing of surrounding regions by light-colored deposits (*right*) that appeared by the time Voyager 2 arrived; another addition is the partial circle of bright material not seen at left.

sulfur and its derivatives on Io's surface and the discovery of sulfur dioxide gas near Loki lead to the suggestion that sulfur compounds are pivotal to the eruptions on Io. If estimates of the internal energy necessary for all this activity are correct, both sulfur and SO_2 will be molten at depths of no more than a few kilometers, and one model developed by Bradford Smith, Shoemaker, and others uses sulfur dioxide as the principal propulsive ingredient.

Models of this type can explain the major characteristics of the eruptive plumes, high exit velocities, constant velocity, fine particle "halos" and nearly continuous eruption. Not all of the eruptions matched the cases illustrated in Fig. 27 exactly; for some eruptions like Pele may require even higher velocities. In such cases, either higher initial temperatures for SO_2 or the use of sulfur as the driving gas may be required.

More subtle geyserlike activity may explain another feature on Io's surface. Fig. 28 shows a series of features which resemble erosion scarps. With no evidence for wind or fluid erosion on Io, these scarps were initially puzzling. One model suggests that erosion along cliff faces proceeds by the "sapping" or leakage of fluid SO_2 at the bases of the scarps. Since this substance may be fluid at very shallow depths, scarps of only 100 m or so may be needed. The result would be a small geyser of liquid SO_2 vaporizing in the near vacuum of Io's surface. Numerous white deposits seen spreading away from scarp faces may be evidence of this process in operation.

With all the various forms of volcanic activity described above, it is obvious that new surfaces are being created on Io far more rapidly than are impact craters. Estimates of the rate at which material is being brought to the surface in eruptive plumes alone suggest that a new layer at least 10 microns thick is being produced every year. If cratering rates at Io are similar to the Moon's, then an annual deposition of 1 mm or more may be required to erase the impact craters fast enough. This level of activity implies that Io's crust and upper mantle have been recycled by volcanic activity many times over during geologic history — a process that undoubtedly has affected Io's chemistry profoundly. The absence of water, carbon dioxide, and other volatiles on or around Io probably means that these compounds have emerged from the interior and escaped to space, particularly hydrogen from the dissociation of water. Only the heavier volatile species, such as sulfur and SO_2, remain, apparently concentrated in the crust and atmosphere.

The level of volcanic activity which Io exhibits is extraordinary by any standard, but it is particularly surprising for such a small planetary body. Io has essentially the same size and density as the Moon. Models of the thermal history for small planets and satellites show that the early heat from accretion and perhaps short-lived radionuclides such as aluminum-26 can melt all or part of the body (see Chapters 19 and 20). But this heat is lost relatively quickly, and other familiar sources of internal heat (such as long-lived radionuclides) lose much of their effectiveness early in a planet's history (about three billion years ago for the Moon).

Fortunately, just prior to Voyager 1's encounter, a new energy source was identified for Io which revised our perceptions of its thermal history. Stanton Peale and his colleagues proposed that tidal heating of Io is supplying as much or more energy than radioactive decay. This potenti-

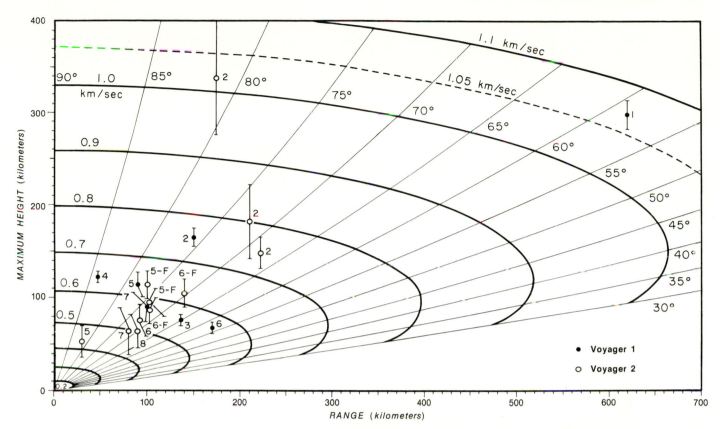

Fig. 26. On this graph by Robert Strom are plotted the ejection angles and velocities required to obtain the maximum plume heights and ranges observed by the Voyager spacecraft. Each point is labeled with the plume's number; vertical bars are uncertainties in the height measurements. In reality a range of ejection angles probably exists for each plume. Furthermore, the velocities generally fall between 0.5 and 0.6 km per second, although plume 1 (Pele) tops 1 km per second. Note the change in plume 2's (Loki's) height during the Voyager 2 observations.

ally large source of energy had never been seriously considered before because Io's mean orbit is almost exactly circular, thus providing no variation in the tides (raised by Jupiter) in Io's crust that would pump energy into the satellite's interior. However, Peale and his co-workers noted that orbital resonances among the inner Galilean satellites force Io into a significantly more eccentric orbit. Perturbed from its strictly synchronous (one rotation per revolution) relation with Jupiter, tides cause the satellite hemisphere facing the planet to flex methodically. Fig. 29 illustrates this situation, which results in heating of the satellite's interior and in torques acting on Io from different directions as it moves about Jupiter. Peale and his colleagues calculated that the tidal-heating process could melt most of Io and suggested that Voyager 1 might see evidence of volcanic processes. Rarely has such a prediction been so swiftly and overwhelmingly confirmed!

Another potential source of energy for Io may be its unique location in Jupiter's magnetosphere. Since Jupiter's rotation period is only 10 hours compared to Io's orbital period of 1.77 days, the magnetic field of Jupiter is continually sweeping past Io at a relative speed of about 57 km per second. If Io or its sparse atmosphere has a reasonable conductivity, the moving magnetic field generates a potential drop of some 600 kilovolts across Io's diameter. A number of magnetospheric theories suggest that currents driven by this potential difference flow along the magnetic field lines connecting Io to Jupiter. Such currents could carry close to one million amps; evidence for currents of this

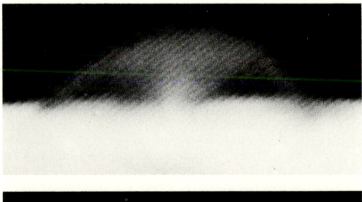

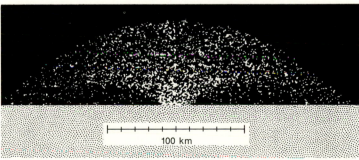

Fig. 27. Compare the Voyager 1 image of Prometheus' plume along the limb of Io with the computer simulation below it prepared by Schneider and Robert Strom. The latter assumes that material is ejected along ballistic trajectories at a velocity of 0.5 km per second and at angles of at least 55° from horizontal. The plume's actual base in both cases is 7 km below the "surface" because the source vent is turned 5° toward us from the limb.

Table 2. HEAT FLOW DATA

	Watts/m²	Remarks
Io	2 ±1	Global average
Earth		
average	0.06	Global heat flow through crust
geothermal area	1.7	Wairakei, New Zealand
Moon	0.02	Average of Apollo 15 and 17 landing sites

Table 3. IO-ASSOCIATED ATOMIC AND
IONIC SPECIES IN JUPITER'S MAGNETOSPHERE

Na	Exists in banana-shaped cloud leading Io along its orbit; first observed from Earth in 1972
Na⁺ ions	Detected in small amounts by Voyager instruments
K	Discovered soon after neutral sodium; probably associated with it in cloud
K⁺ ions	Probably detected by Voyager's charged-particle sensors
S⁺, S²⁺, S³⁺, O⁺, O²⁺ ions	Sulfur and oxygen ions exist throughout the magnetosphere, with low-energy ones concentrated in a torus centered roughly on Io's orbit; S⁺ and O⁺ were discovered from Earth
SO₂⁺, SO⁺, and related ions	Probably detected by Voyager's plasma experiments

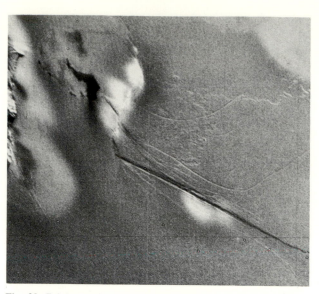

Fig. 28. Bright patches accompany what may be erosion scarps on the surface of Io. In locations such as these, sulfur dioxide may be escaping to the surface from below as blocks of crust slump or break away from the scarp's bases.

magnitude was provided by the magnetic field measurements made by Voyager 1 when it flew directly under Io's south pole. These currents have been associated with the control of Jupiter's radio emissions by Io and could conceivably heat the satellite if all the current's power (about one trillion watts) were dissipated in the interior.

When Voyager 1 flew past Io, it detected a number of anomalously warm regions associated with volcanic areas. For example, the infrared flux emanating from a dark region south of Loki was far in excess of our expectations; there the spacecraft recorded a temperature of about 300° K, compared with typical noontime "highs" along Io's equator of perhaps 120° K. Could this feature actually be a lava lake some 250 km across, possibly liquid sulfur with a somewhat cooler skin or crust? Perhaps. Lava lakes are a common feature of basaltic eruption on the Earth, such as those in the Kilauea caldera on the island of Hawaii, but on much smaller scales.

The infrared experiment also detected numerous smaller hot spots, including one 500° K reading over the very center of Pele. If such hot spots cover a significant fraction of the satellite's surface, their combined effect should be evident even in whole-globe infrared measurements made from Earth. And, indeed, observations of Io made during the early 1970's show a curiously shaped curve in its infrared spectrum and puzzling cooling trends during eclipses by Jupiter. These could be explained by emissions from hot spots covering about 1 percent of Io's surface; a similar result has been derived from Voyager data. Another Earth-based observation in early 1979 recorded outbursts of 5-micron energy far above expectations that lasted for only hours or days. The only reasonable source appears to be small regions of hot (500-600° K) surface material.

All of these infrared measurements indicate that Io emits an impressive 10^{14} watts of energy in this wavelength region — an average of about 2 watts per square meter over its entire surface. Table 2 relates this heat flow to those of the Earth, the Moon, and an active terrestrial geothermal area.

Note that Io's *average* emission is comparable to the *most* active areas on the Earth, making it truly the most volcanic planetary object (including the Earth) yet studied.

Io is so warm that magnetospheric currents cannot be a significant heating mechanism. In fact, even potent tidal energy must be converted very efficiently within the interior to match the observed thermal output. A partially molten interior is capable of dissipating enough tidal energy. However, the rate of dissipation is linked to properties of the Jovian interior and to the rate at which Io's orbit changes with time in response to tidal forces (just as the Moon's orbit is changing). Charles Yoder and Stanton Peale have suggested that the observed level of energy emitted by Io implies too rapid an orbital evolution. They propose that the level of volcanism (and therefore tidal energy dissipation) on Io may vary with time and that what Voyager observed is not typical of the last 4½ billion years. Io's volcanoes continue to puzzle planetary scientists.

AN ATMOSPHERE, TOO

Our current knowledge of Io's atmosphere suggests that it is "lumpy," with more gas over warm areas and active vents. Sulfur dioxide is the major known constituent, but other as yet unidentified gases could modify this picture. The first firm indication of an atmosphere surrounding Io came from Pioneer 10's discovery that the satellite possesses a reasonably dense ionosphere, with electron densities of 10-100 thousand per cubic centimeter. The later confirmation of sulfur dioxide on and above the surface identified at least one important atmospheric component, but spectroscopic studies have ruled out a number of other potential gases like water, hydrogen sulfide, and carbon dioxide. The virtual absence of heavy ions other than sulfur and oxygen in the surrounding magnetosphere also limits the contributions from nitrogen and inert gases like argon.

Sulfur dioxide on Io has a number of interesting properties. First, new gas can be abundantly supplied by volcanic vents. Second, SO_2 can condense anywhere on Io's surface;

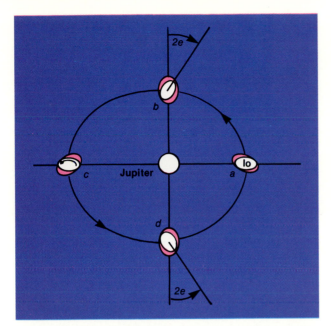

Fig. 29. Io's dynamic activity stems from an orbital resonance with nearby Europa that forces Io into a slightly eccentric orbit. Ordinarily, Jupiter's strong gravity would keep one hemisphere of the satellite facing the planet at all times. But the forced eccentricity makes Io move at different velocities along its orbit, and the side facing Jupiter nods back and forth slightly as seen from the planet. Tidal forces develop inside the satellite which generate heat through friction, and much of the interior remains molten as a result. This drawing is from a paper by Charles Yoder in *Nature, 279,* 1979, 267-70.

in the polar regions and at nighttime, temperatures are low enough that virtually all SO_2 will be frozen out of the atmosphere. This suggests that, to some degree, cold regions regulate the supply of gas, similar to the situation with condensed CO_2 in the polar caps of Mars. Under these conditions, atmospheric pressure should vary locally according to the surface temperature, since that controls the amount of SO_2 that can sublimate from solid to gas. The atmospheric pressure detected by Voyager near Loki is, in fact, close to the value expected from the locality's calculated temperature of 120° K. This simple equilibrium model is probably not correct in detail, however, because a number of mechanisms can add or subtract gas on a local or regional level. There is also the possibility of major atmospheric variations driven by changes in volcanic activity, the amount or composition of released gases, or the rates at which atmospheric species escape into space.

MAGNETOSPHERIC INTERACTIONS

All of the Galilean satellites are deeply embedded within the Jovian magnetosphere; Io, orbiting closest to Jupiter, lies deepest in this medium and plays host to a complex array of electromagnetic interactions. One of the earliest "connections" found between Io and the magnetosphere concerned radio bursts from Jupiter with dekameter (tens of meters) wavelengths. The bursts occur far more frequently when Io, Jupiter, and Earth align in a certain way. Still not completely understood, this phenomenon may be related to the electrical currents generated by Io's relative motion through the magnetosphere (discussed earlier in connection with the satellite's heating).

An even more striking indication of Io's magnetospheric involvement came in 1974 when Robert Brown announced the discovery of sodium D-line emission from the vicinity of Io. This emission was found to be coming from a "cloud" of neutral sodium surrounding Io and extending for tens of thousands of kilometers along its orbit.

Why is sodium, a relatively minor element, so obvious in Io's spectrum? The answer is that sodium is one of the easiest elements to observe under these conditions. Its atoms vibrate in a strong resonance with light at the wavelengths of 5890 and 5896 angstroms, and thus respond very efficiently to sunlight. Io's sodium, glowing brightly by reflected sunlight, dims and brightens periodically as the satellite's orbital motion induces Doppler shifts in sodium's two "rest" wavelengths. (The change in pitch from a passing locomotive's horn is another example of Doppler shifting.)

Other atoms have similar, but generally weaker, emissions (Fig. 31); other emissions can also be stimulated by electron impact in a plasma. Shortly after sodium was discovered, potassium was detected and then singly ionized sulfur. With the Voyager 1 encounter, the full extent of Io-related atomic and ionic species in Jupiter's magnetosphere became evident. Measurements by the various plasma, charged-particle, and ultraviolet detectors showed that heavy ions, particularly those of sulfur and oxygen, were important throughout the magnetosphere from the low-energy plasma to high-energy cosmic rays. Table 3 lists the various species, atomic and ionic, now known to be abundant in the Jovian environment. All of these magnetospheric species appear to originate in the dense plasma torus at Io's orbit, and ultimately from Io itself.

A major question lingering since the discovery of the sodium cloud concerns the process or processes that eject material from Io into the magnetosphere: Io has about the same surface gravity as the Moon, and velocities of at least 2½ km per second are needed to escape from Io's influence and go into independent orbit about Jupiter. Some sodium atoms have been observed leaving Io at well over 10 km per second. Dennis Matson and his colleagues suggested in 1974 that sputtering (ejection by the impact of charged particles on a surface) using magnetospheric ions might account for the escape of sodium, a theory reinforced by subsequent studies. The abundance of heavy S and O ions in the magnetospheric plasma provides a ready source of sputtering ions, and this idea remains one of the best candidates for removing material from Io. In fact, recent laboratory studies indicate that sputtering may also erode the icy surfaces of the other Galilean satellites.

Other processes for escape of material have also been suggested, especially for species more volatile than sodium. The escape of SO_2 or its components solely by virtue of their high temperature (and hence collisional energy) is possible but may not be efficient enough to supply the amount of material required by magnetospheric calculations. Ionization and sweeping by the magnetosphere is another possibility, although the observation of neutral oxygen in the magnetosphere indicates that this species as well as sodium and potassium may escape as neutral atoms. Sputtering directly from the atmosphere has also been considered.

Pictures of large volcanic plumes suggest that direct "blowing off" of material through volcanic processes might be possible, but thermodynamic considerations suggest that

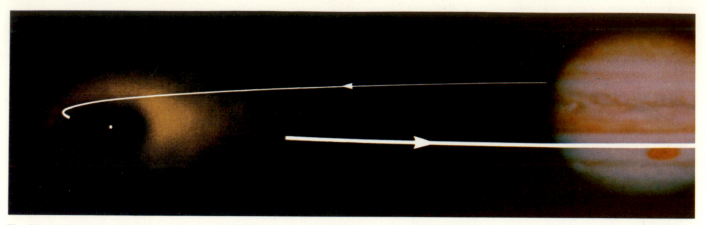

Fig. 30. A three-hour exposure reveals the presence of a yellowish cloud of sodium along the orbit of Io, seen here combined with an image of Jupiter. The dark area within the cloud is the occulting disk used to mask glare from Io, whose size is indicated by the white dot. The cloud was detected in 1977 by Bruce Goldberg and his co-workers.

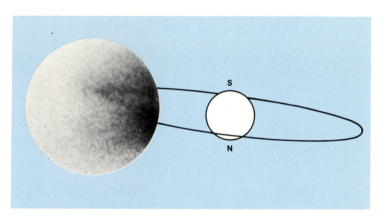

Fig. 31. The circular photograph contains a portion of the tenuous ring of ionized sulfur that surrounds Jupiter (indicated schematically). Sulfur around Jupiter was recognized spectroscopically by Israeli scientists in 1975, but this image made at a wavelength of 6731 angstroms by Carl Pilcher in 1979 clearly shows the ring's stucture. The ring lies in the plane of Jupiter's magnetic equator at a distance of 5.3 Jovian radii from the planet's center, farther out than the ring of particles discovered by Voyager 1.

it is difficult or impossible to drive the volcanic gases to the required escape velocity. It may be possible to remove some of the plume material by electrically charging and sweeping it away by electromagnetic interaction with the magnetosphere, but, again, it does not appear that this process can supply enough material.

All of these ejection theories have their problems, and it seems likely that more than one process may be operating. What is clear, however, is that Io is the ultimate source of these materials and that volcanic activity must play a major role in supplying fresh gas and volatile material to the surface and atmosphere. The total injection of ions into the magnetosphere has been estimated a number of ways and seems to be somewhere between 10^{27} and 10^{29} ions per second. If this rate were constant over geologic time, it would cause only a few hundred meters to perhaps a kilometer of total surface erosion, far less than the volcanic eruption rates seen today (millimeters of new material per year) can supply. Io seems to be in no danger of eroding away before our eyes.

FOUR NEW WORLDS

Despite over three centuries of telescopic observations, the new data about the Galilean satellites acquired by Voyager spacecraft represent an advance in knowledge almost as profound as the satellites' original discovery. During those hours of close observations in March and July of 1979, these four worlds, with a total land area equivalent to the Earth's, went from dots of light on our mental horizons to geologically and geographically known places — *terra cognita*. Totally different from previously studied planets, they present us with a stunning diversity of surface features, colors, compositions, geophysical, and geochemical processes. From frozen Callisto to bubbling Io, the satellites are a veritable laboratory for the study of planetary evolution. And, as always, we are left with a wide range of new theories, questions, problems, and controversies to fuel further exploration, such as that planned by Project Galileo in the late 1980's.

Titan

James B. Pollack

Saturn's largest satellite, Titan, is a unique and intriguing planet-size body. It is the only satellite in the solar system known to possess a substantial atmosphere (Fig. 1). In fact, the mass of its atmosphere above a given area of its surface exceeds that for Mars by a factor of about 500 and for Earth by a factor of 10! Titan is the only other object in the solar system besides the Earth to have molecular nitrogen as the principal constituent of its atmosphere. However, unlike the Earth, the other gaseous components in its atmosphere are highly reducing, that is, they consist of hydrogen-rich molecules like methane and higher-order hydrocarbons. A rich chemistry there leads to the production of such exotic molecules as hydrogen cyanide (HCN) and complex organic polymers; these form a deep smog layer of tiny particles that shroud the surface from view. This chemistry may ultimately shed some light on the chemical steps that led to the origin of life on the Earth. Finally, Titan's temperatures may permit the formation of methane ice clouds in its lower atmosphere and liquid methane oceans on its surface.

Much of the information presented in this chapter has been derived from the passage of Voyager 1 through the Saturn system in November, 1980. The spacecraft brushed to within a scant 5,000 km of Titan's surface, making detailed measurements that included transmitting radio signals through Titan's atmosphere to determine its temperature structure and surface pressure; obtaining infrared and ultraviolet spectra to determine the gaseous constituents of its atmosphere; taking closeup images to study the structure

Fig. 1. Opaque layers of particles in Titan's atmosphere (*left*) prevented Voyager 1 from seeing the satellite's surface during a 1980 flyby. Note the lighter color of clouds over the southern hemisphere and the dark hood at the north pole. Voyager 2 looked back at a crescent Titan (*right*); the extension of blue light around the moon's night side is due to scattering by smog particles in the sunlit portion.

of its smog layer; and conducting various measurements of magnetic fields and high-energy particles near Titan to investigate the interaction between its atmosphere and near-space environment. Of course, ground-based observations have also contributed to our knowledge. For example, in the 1940's, Gerard Kuiper detected the presence of methane gas on Titan and thus showed it had an atmosphere.

This chapter reviews our current knowledge about Titan's atmosphere, surface, interior, and nearby environment. It attempts to place the satellite's history within the context of the formation of the Saturn system and to describe possible evolutionary paths for its atmosphere and interior.

EARLY HISTORY

As shown in the table, Titan's orbit is almost exactly circular; it lies almost precisely in the equatorial plane of Saturn; and the satellite moves in the same direction that Saturn rotates. These characteristics suggest that Titan, as well as most of the other satellites of Saturn, formed within a flattened disk of gas and dust shed by the planet as it contracted during the earliest epochs of the solar system; that is, Titan formed within the Saturn system, rather than forming elsewhere and being subsequently captured. Let us now turn to consideration of the environmental conditions

PROPERTIES OF TITAN AND EARTH

	Titan	Earth
Mass (g)	1.35×10^{26}	5.98×10^{27}
Equatorial radius (km)	2,570	6,370
Surface gravity (cm/sec²)	144	978
Density (g/cm³)	1.89	5.52
Average surface temperature (Kelvin)	93°	288°
Mean distance from Saturn (km)	1.22×10^6	—
Orbital eccentricity	0.029	0.017
Orbital inclination	0.33°	—
Rotational period (days)	15.9	1.0

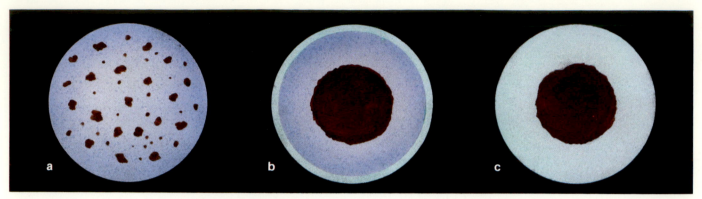

Fig. 2. A possible sequence in the evolution of Titan's interior. Soon after condensing from the nebula surrounding Saturn, the satellite's heterogeneous mixture of ices and silicates (*a*) begins to segregate, as heat created during formation mobilizes the interior. Rocky material sinks to the center, and water, ammonia, and methane rise to the top; at first, residual heat keeps all but the outermost portion liquid (*b*). Heat loss from the mantle through the ice crust soon freezes the entire mixture (*c*), probably within the first billion years of Titan's existence.

within "Saturn's nebula" that determined the chemical makeup of Titan's interior and, ultimately, its atmosphere.

Jupiter and Saturn are composed chiefly of gases, in approximately the same elemental proportions as exist in the Sun today (and presumably in the primordial solar nebula from which they formed). Currently, both planets are radiating to space about twice as much energy as they absorb from the Sun. Theoretical simulations of the evolutionary history of these planets by Allen Grossman, Harold Graboske, and the author indicate that this excess energy arises largely from their slowly cooling interiors, which were heated to high temperatures soon after formation by a fairly rapid contraction and attendant release of gravitational energy. According to these calculations, the excess luminosity of both objects was many orders of magnitude larger during their earliest history than now, at the same period when their satellites were forming. If so, the heat from these planets, rather than that coming from the early Sun, dominated the temperature conditions within their nebulae and hence profoundly influenced the composition of the satellites that formed within them.

The satellites of the outer solar system are believed to be composed of a mixture of "rock" (minerals rich in iron and silicates that constitute the inner terrestrial planets) and "ice" (chiefly water ice). We base this inference on the observation that a number of satellites, including Titan, have densities substantially less than 3 g per cubic centimeter, which is the lowest value expected for a body made only of rocky material. Presumably, the "ice" component of satellites like Titan originated from the condensation of water vapor, when the temperature of the nebula at the satellite's distance from its parent planet became cool enough. The timing of this condensation phase was dictated in turn by the evolution of the planet's excess luminosity, which gradually decreased after the planet formed.

Water ice has the highest condensation temperature of any cosmically abundant ice and hence was the first ice to form within these early planetary nebulae, appearing at progressively earlier times for more distant and therefore cooler regions of the nebula. As the temperature continued to decrease following the onset of satellite formation, ices containing first ammonia and then methane and perhaps some molecular nitrogen could have also formed and so been available for incorporation into the growing satellites.

These hydrogen- and helium-rich planetary nebulae are not present now and probably dispersed at some very early time. They may have been either blown out of the solar system by an intense solar wind, or incorporated into the planet.

Telltale signs of this scenario — early high planetary luminosity and a short-lived nebula lasting about one million years — may be reflected in the systematic decrease of the mean density of the Galilean satellites of Jupiter: the inner pair of moons are dense enough to consist almost entirely of rock, while the much lower densities of the outer two satellites imply a large ice component. Before the Jovian nebula dissipated, its temperatures may have been low enough to permit water-ice condensation near the orbits of the outer two Galilean satellites, but little or no condensation near the orbits of the inner two. Titan is located at a distance from Saturn comparable to the outer Galilean satellites' distances from Jupiter. Since Saturn's excess heat, at any given time, was about an order of magnitude less than Jupiter's, it is feasible that considerable water ice was available for incorporation into Titan before Saturn's nebula dispersed. In fact, ices containing ammonia, methane, and nitrogen may have also condensed and been drawn into Titan, whereas they may be absent, in any appreciable amounts, from the outer Galilean satellites.

TITAN'S INTERIOR

Estimates of the relative proportions of rock and ice in Titan's interior can be obtained by constructing mathematical models of the pressure, density, and temperature conditions in its interior that are constrained to match Titan's observed mean density, radius, and mass. Assuming that the density of the rock component of Titan matches that of Io, the innermost Galilean satellite, and that this component is segregated towards the center of the satellite, theorist Ray Reynolds estimates that rocky material constitutes some 55 percent of Titan's mass, with the remainder ice. A bit of the current ice component may have arisen from the dehydration of minerals that initially contained water bound into their crystal structures, so the initial mass fraction of ice may have been somewhat smaller than 45 percent.

Because Titan is a planet-size body (slightly larger than Mercury), it may have achieved internal temperatures during its lifetime high enough to initiate substantial internal differentiation, with denser rock sinking toward the center

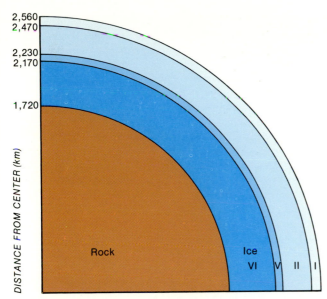

Fig. 3. A possible present-day configuration for Titan's interior has been proposed by Claudia Alexander and Ray Reynolds. Outside the rocky core, variations in pressure and temperature have produced ice layers with differing crystal structures. The rock, ice VI, V, II, and I sections have densities (in g per cubic centimeter) of 3.52, 1.43, 1.28, 1.18, and 0.94, respectively.

and lighter ice moving towards the outer layers. Three potential mechanisms could have produced such differentiation: heat of accretion, heat generated from the decay of long-lived radioactive elements in the rock fraction (certain isotopes of uranium, thorium, and potassium), and heat produced by solid-body tides raised by Saturn. Tidally derived heat is believed to be the principal agent responsible for active volcanism on Io. However, because Titan is much farther from Saturn than Io is from massive Jupiter, the satellite's tidal flexing probably provides only a minor source of interior heat, an amount probably insufficient to cause extensive differentiation. Radioactive heating is unlikely to have caused much differentiation either.

Most of the heat for differentiating Titan's interior probably arose during accretion, from the accumulation of many tiny objects into a single large one. As the satellite grew from continual low-velocity impacts with small bodies, some of their kinetic energy converted to heat during the collisions. If more than 10 percent of the impact energy made this transformation, some parts of the satellite's interior then became hot enough to melt the ice component and the settling-out process began. If the heat conversion occurred at efficiencies of several tens of percent or more, which seems plausible, Titan would have undergone almost complete differentiation within the first billion years of its 4.6-billion-year history.

Soon thereafter, the satellite would have developed a three-layered structure consisting of a rocky core, a mantle of liquid water (with some dissolved ammonia and methane), and a thin icy crust. The combination of accretional heat and radioactive decay may have raised temperatures in the core enough to drive water from hydrated minerals residing there; the released water would then percolate upward to the mantle. If the mantle lost heat only by conduction through the icy crust (a rather slow process), a substantial liquid zone might remain beneath Titan's surface

even today. However, it is much more likely that solid-state convection — a slow, plastic deformation of all but the frigid, outermost portion of the icy crust — carried heat away from the liquid mantle in a much more efficient manner. If true, the mantle would have become frozen throughout long before now.

Fig. 2 illustrates this evolutionary scenario, beginning with an initially hot interior having a more-or-less uniform distribution of ice and rock, proceeding to a differentiated object, with an extensive liquid water mantle, and ending with the frozen mantle of today. Fig. 3 shows a more detailed profile of Titan's interior structure at the present time, displaying the possible locations of the rock/ice interface and the various phases of water ice, whose occurrence is dictated by the distribution of pressure and temperature.

ATMOSPHERIC CHEMISTRY

Prior to 1980, most planetologists concerned with Titan thought its atmosphere consisted largely of methane; they likened it to a primordial Earth, a terrestrial planet in a "deep freeze." However, to the surprise of many, Voyager found that nitrogen — not methane — constitutes 80-95 percent of all molecules in the atmosphere, with methane and perhaps argon comprising nearly all of what's left over. Other gases either detected or inferred there are produced from methane and nitrogen through a series of chemical reactions triggered by ultraviolet sunlight. A UV photon can dissociate a polyatomic molecule into smaller fragments or free radicals; it can also create an ion by stripping the molecule of an outer electron. For example, sunlight at wavelengths shorter than about 1600 angstroms dissociates methane (CH_4) into a methyl radical (CH_3) and a single hydrogen atom. Nitrogen dissociation requires even shorter wavelengths and occurs at higher, more rarified altitudes; consequently, its photolysis rate is much slower than methane's.

Methane and its dissociation products dominate the photochemistry of Titan's middle atmosphere. Derived gases like ethane (C_2H_6), acetylene (C_2H_2), and ethylene (C_2H_4) attain abundances of several parts per million, through reactions that may be accelerated by the catalytic action of acetylene and polyacetylene ($C_{2n}H_2$).

Higher in Titan's atmosphere, radiation and magnetospheric particles break down molecular nitrogen primarily into N_2^+ ions and electrons. Subsequent reactions between these by-products and the methane family generate hydrogen cyanide, present at about one part per million. Laboratory simulations of the chemical steps that led to the origin of life on Earth have used HCN as a starting material. However, while HCN undoubtedly spawns more complicated compounds in the Titanian atmosphere, we do not expect it to produce those capable of giving rise to life. There is a missing chemical link — oxygen — which is scarce because its parent molecule — water — lies inescapably frozen to Titan's frigid surface, largely unable to evaporate and interact with other compounds. Nevertheless, the molecules produced in Titan's atmosphere may provide some insight into the chemistry that gave rise to life on Earth.

The dissociation of methane can also lead to more complex hydrocarbons in Titan's atmosphere. HCN may enter into some of these reactions. The net result of this chemistry is a set of complicated molecules that have such low vapor pressures at the ambient temperatures that they condense into very small particles. Through coagulation — sticking together after low-velocity collisions with one another — somewhat larger particles result. These "smog" particles, averaging about 0.1-0.5 micron in radius, form a layer that totally shields the surface from view. The main smog layer enshrouds the entire globe of the satellite and stretches from the surface to an altitude of about 200 km. Thus, in marked contrast to the smoggy hazes that form within 1 km of the Earth's surface, the smog layer in Titan's atmosphere is generated at very high altitudes. We believe that the number of smog particles per unit volume of "air" is much lower in the case of Titan's atmosphere, so that the visibility at ground level might be many tens of kilometers rather than a few kilometers, as is typical near urban areas on Earth. But, because the smog layer is much deeper on Titan, its surface is obscured from our view; fortunately, the Earth has not advanced to that point yet.

Voyager pictures reveal subtle brightness variations in Titan's clouds. High resolution views show a faint, high layer of haze centered about 100 km above the main haze layer. In addition, the northern hemisphere of Titan is somewhat darker than the southern one, with the transition in brightness occurring gradually across the equator. The north polar region looks even darker than the rest of the hemisphere on Voyager 1 photographs, and the main smog layer extends to higher altitudes here, blending into the high altitude haze layer. These morphologic features are not well understood at present, although seasonal effects could be responsible for the hemispherical asymmetry, and interactions with the Saturnian magnetosphere might engender the dark north polar cap. At the time of the Voyager 1 encounter, the northern hemisphere was just beginning its spring season after a winter that lasted for 7½ Earth years. (A year on Titan is 30 of ours. Since we suspect that Titan's axis of rotation is aligned with Saturn's, it will be tilted by about 27° to the orbit of the Saturn system around the Sun.)

With less sunlight falling on the winter hemisphere than the summer one, the photolysis rate of methane and the production of aerosols (particles) would vary between hemispheres. Thus, aerosols' characteristics (like size and composition) may also change with the season, leading to the hemispherical brightness variation we observe. While ultraviolet sunlight is the chief energy source for generating aerosols, some additional production may result from the precipitation of high-energy particles from Saturn's magnetosphere onto Titan's north pole; this could also account for the dark polar hood.

Gravitational forces and atmospheric turbulence force

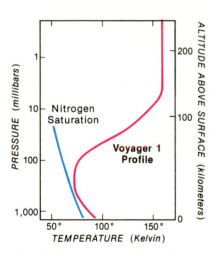

Fig. 4. The red curve, from Voyager data, indicates that a temperature minimum occurs about 40 km above Titan's surface. At that altitude, it is probably slightly warmer than the nitrogen saturation temperature (blue) at which clouds of nitrogen would form in equilibrium with the surrounding gas.

the aerosols produced at high altitudes in Titan's atmosphere to fall gradually through the air mass, reaching the surface in one to several years. There they remain, since the surface is too cold for them to evaporate back into the atmosphere and they probably would not decompose even in hypothetical oceans of liquid methane. Over the lifetime of the satellite, a thick deposit of these precipitating, complex organic molecules has either blanketed the surface of Titan to a depth of about 1 km (if the surface is solid) or has been mixed with the "oceans" of Titan (if the surface is liquid). Conceivably, the surface of Titan could possess the richest deposit of hydrocarbons in the solar system!

The continual removal of aerosols from the atmosphere has important consequences for several of its gaseous constituents. First, it implies a long-term drain on methane. In fact, aerosols produced over the lifetime of Titan would consume much more methane gas than currently exists there. It is possible, therefore, that the methane gas now in the atmosphere is being buffered by the evaporation of massive deposits of methane ice or liquid at the surface: as the gas is removed by aerosol formation, it is replaced by surface methane. Thus, aerosol formation may ultimately occur at the expense of this surface layer of methane.

The ratio of hydrogen to carbon atoms in the aerosols is probably smaller than for gaseous methane because excess hydrogen is left over as aerosols are generated. This leads to a buildup of molecular hydrogen in Titan's atmosphere. However, the low mass of H_2 (and its dissociation product H) and the low gravity of Titan (some one-seventh of Earth's) allow the hydrogen to escape rapidly into space. Eventually, a balance is established between the production of hydrogen from methane photolysis and its loss by escape to space. Calculations by Donald Hunten indicate this balance results in a fractional abundance of several tenths of a percent H_2 in Titan's atmosphere. Some direct evidence for the validity of this chemical cycle of hydrogen comes from Voyager 1's detection of a diffuse cloud of hydrogen atoms that spans a region from somewhat outside of Titan's orbit to well within it.

Ordinarily, nitrogen atoms and molecules move too slowly at the temperatures of Titan's upper atmosphere to escape into space at an appreciable rate. However, their velocities can increase to the point of escape when nitrogen participates with ions in chemical reactions that release energy. In fact, Darrell Strobel estimates that enough nitrogen to equal one-tenth that currently in Titan's atmosphere could have been lost in this manner over the satellite's lifetime. Additional losses may result from atmospheric interactions with Titan's near-space environment, discussed later in this chapter.

ATMOSPHERIC TEMPERATURE

Fig. 4 illustrates the variation of temperature with pressure and altitude in Titan's atmosphere found by the radio occultation, infrared spectrometer, and ultraviolet spectrometer experiments on the Voyager 1 spacecraft. (Prior to the Voyager 1 flyby, estimates for the surface pressure on Titan ranged from 20 millibars to 20 bars!) The Voyager radio observations demonstrated that the surface pressure is roughly 1.6 bars.

The atmospheric temperature achieves a minimum value of about 70° K at an altitude of about 40 km (the tropopause) and steadily increases at lower altitudes (the troposphere or lower atmosphere) to a value of about 93° K at the surface. The temperature also increases at higher altitudes (the stratosphere or middle atmosphere) towards about 170° K, a value that holds at still higher altitudes (the upper atmosphere). By comparison, the temperature at the surface of the Earth at mid-latitudes averages about 288° K.

As indicated in Fig. 4, temperatures throughout the troposphere are somewhat higher than the condensation temperatures of molecular nitrogen, so no condensation clouds of nitrogen are expected near the tropopause. Nor is it likely that nitrogen ice exists at the surface, a point that suggests the total atmospheric pressure is limited not by the cold temperature of Titan's surface but by the availability of nitrogen.

The temperatures may be cold enough for *methane* to condense both in the troposphere and at the surface (Fig. 5). We can't be sure, because of uncertainties in our knowledge of the fractional abundance of methane in the lower atmosphere. But what we find in the middle atmosphere (about 1 percent) is fairly close to the abundance required for methane clouds to exist near the tropopause. If methane

achieves about 7 percent at the surface, its condensation clouds should be present throughout the troposphere, with much more on the ground as ice, liquid, or both. The amount of methane in the atmosphere would then be controlled by the temperature of the surface. Such a situation seems likely in view of earlier discussion about the substantial endowment of Titan's interior with methane-containing ices, which became concentrated towards the surface by internal differentiation. The temperature at the surface of Titan is quite close to the triple point of methane — 90.7° K — at which all three phases of methane can exist simultaneously. If the temperature is in fact a bit higher (as now seems likely), liquid-methane oceans could cover much of the surface, with perhaps methane glaciers occurring in the slightly colder polar regions.

Titan absorbs about 80 percent of the sunlight incident upon it. The smog particles and methane gas provide nearly all of this absorption, although a small but significant fraction (some 5-10 percent) may reach the surface and be absorbed there. If a single temperature occurred throughout Titan's atmosphere and at its surface, and if Titan radiated to space equally well at all infrared wavelengths, this

Fig 5. Although scientists did not glimpse the surface of Titan during Voyager 1's brief flyby, they learned enough to speculate on the appearance (depicted here by artist Don Davis). Somewhere in the frozen desolation near Titan's north pole, the Sun's dull glow along the horizon marks the arrival of spring. Opaque clouds hang ceilinglike 40-50 km above the ground, trailing a constant drizzle of methane ice laced with brownish organic compounds. As they fall, the methane crystals may evaporate in the warm lower atmosphere, coagulating the organics into drifts of cindery debris that rain onto the surface. Or they may reach the surface, perhaps changing into liquid droplets close to it. Muted crater forms bear witness to occasional meteoric collisions long mantled by the continual deposition from the clouds. One large pit, excavated recently, reveals fresh exposures of water ice. Here and there puddles of liquid methane make a gradual, steamy return to the atmosphere.

temperature would be about 85° K. The much higher stratospheric temperatures result because the small smog particles are much less effective emitters of long-wavelength infrared radiation than they are absorbers of incoming shorter-wavelength radiation. Consequently, the stratosphere heats up until it strikes a balance between warming by the absorption of sunlight and cooling by the emission of thermal radiation. Since much of the sunlight is absorbed in the middle and upper reaches of the smog layer, well above the tropopause, the lower stratosphere and troposphere tend to be cooler. But the increase in temperature from the tropopause to the surface is due to a modest "greenhouse" effect created by broad absorption bands of nitrogen, hydrogen, and methane induced by the high-pressure conditions. These infrared bands prevent some of the heat radiated by the surface and lower atmosphere from escaping to space and thus elevate the temperatures there.

Except at high altitudes, heat transport by winds from sunnier equatorial regions to the poles is expected to be very efficient, according to rough calculations made by Conway Leovy and the author. Consequently, we suspect that the surface and lower portions of the atmosphere experience little change in temperature with varying latitude or time of day. This near-isothermal condition is due to the relatively cold temperatures and weak sunshine on Titan; it takes much longer for radiation to cool or warm the atmosphere. Thus, very modest wind speeds can largely erase substantial temperature variations with latitude, except at very high altitudes where the air is too rarefied to store heat for very long. Moreover, little cooling occurs during the night.

The dynamics of the solar wind vary considerably on time scales of days to years in response to changes in solar activity (see Chapters 2 and 3). As a result, the magnetopause of Saturn, as well as those of other planets, fluctuates in position; it is pushed closer to Saturn when the solar wind blows hard. Depending on its orbital position and the condition of the solar wind, Titan resides mostly within Saturn's magnetosphere, but sometimes lies in the magnetosheath region of the solar wind or even "outside" in the undisturbed solar wind. In all three cases, Titan's lack of an appreciable magnetic field allows nearby ions and magnetic fields to interact directly with the satellite's upper atmosphere. Neutral gases in Titan's upper atmosphere become ionized through impacts with high-velocity ions and electrons. Once that happens, the magnetic field picks up the newly ionized gases and sweeps off with them. Passing through the "wake" region on the downstream side of magnetospheric flow past Titan, Voyager 1 may have detected the presence of ionized gases recently removed from the satellite's atmosphere.

Conversely, Titan affects its surroundings by slowing down the motion of the gases and magnetic field near it. In addition, gases escaping from Titan may provide much of the material populating Saturn's magnetosphere.

ATMOSPHERIC EVOLUTION

The primary gases in Titan's atmosphere include nitrogen and methane, with the amount of methane probably controlled by the vapor pressure of solid or liquid methane at the temperature of the surface. While molecular nitrogen is also found in the atmospheres of Venus, Earth, and Mars, its abundance in Titan's atmosphere appears surprising at first glance. We expected nitrogen to reside mostly *inside* Titan, not as gas but principally as ammonia ices. Sushil Atreya, Thomas Donahue, and William Kuhn have proposed this scheme to generate molecular nitrogen from ammonia: Ammonia gas readily dissociates in relatively long-wavelength ultraviolet sunlight, with subsequent chemical reactions leading to the production of molecular nitrogen and molecular hydrogen. Once formed, molecular nitrogen cannot be converted back to ammonia under the pressure and temperature conditions of Titan's atmosphere.

The abundance of ammonia in Titan's atmosphere is severely limited by the low vapor pressure of ammonia ice at the temperature of the surface. However, like the situation for methane, conversion of atmospheric ammonia to nitrogen ultimately takes place at the expense of the surface ammonia ice, with a steady supply of vapor being maintained as long as ammonia ice is in contact with the atmosphere. Thus, atmospheric nitrogen may have gradually accumulated over the lifetime of the satellite as sunlight transformed ammonia vapor irreversibly into nitrogen and hydrogen. However, we need additional calculations to determine whether the observed 1½ bars of nitrogen can be generated at the current atmospheric temperatures, and to assess the impact of the possible loss of nitrogen to the near-space environment. Small quantities of nitrogen-containing ices incorporated into the satellite represent an additional source of the atmospheric nitrogen. Early heating of the interior could have released much of this constituent into the atmosphere.

Several factors enabled Titan to be the only satellite with a large atmosphere. First, the temperature at its surface is high enough to maintain significant amounts of methane and ammonia gas in equilibrium with their corresponding surface ices. By contrast, Triton, the largest satellite of Neptune, has a mass comparable to Titan's and may have a substantial endowment of methane and ammonia ices. But its surface temperature is much lower because Triton is located farther away from the Sun. Second, Titan formed from material that could generate an atmosphere at the temperature of its surface — ices containing methane and ammonia. But the Galilean satellites, formed in the hotter nebula of Jupiter, probably lack significant quantities of methane or ammonia ices. Even though the surface temperatures of the Galilean satellites are somewhat higher than Titan's, water vapor produced from exposed deposits of water ice has a vapor pressure too low to generate a substantial atmosphere.

Third, Titan's atmosphere is not subjected to extremely large losses to its near-space environment, as compared to Io, which is located deep inside Jupiter's more vigorous magnetosphere. And finally, Titan is massive enough to have undergone some internal differentiation, which concentrated ices towards its surface; its gravity is great enough to retain all but the lightest gases.

It is interesting to note that many of the satellites of the outer solar system have been endowed with a much larger quantity of atmospheric source materials than planets of the inner solar system, including Earth — tens of percent of the object's mass versus a few hundredths of a percent. Once again, we see the importance of sunlight, which warms the terrestrial planets and allows them the luxury (in our case, the necessity) of a substantial atmosphere.

The Outer Solar System

David Morrison and Dale P. Cruikshank

On old maps of the Earth large regions were designated *terra incognita* — unknown lands, beyond the frontiers of exploration. Until very recently, all of space beyond our planet was *terra incognita*. During the past two decades of solar system exploration the frontier has been pushed out to a distance of about 10 astronomical units (AU) from the Sun, beyond which no planets have been visited by spacecraft and even the largest telescopes can see only dimly. There, in a region of perpetual dusk, where the Sun is so small it would appear as a starlike point to the naked eye, lies still a *terra incognita*. These regions, from the orbit of Saturn outward to the edge of the solar system, are the subject of this chapter.

Beyond Saturn lie two large planets, Uranus and Neptune, and a variety of smaller objects that appear to be composed primarily of ices of different chemical compositions. These smaller icy worlds include Pluto and its satellite Charon, seven known satellites of Uranus and Neptune, and a few asteroids that may be only the largest members of undiscovered multitudes of planetary debris. In the following sections of this chapter we will discuss the discoveries of these objects, the nature of Uranus and Neptune, and finally the smaller objects of the outer solar system, concluding with the recent Voyager exploration of the icy satellites of Saturn.

THE ERA OF DISCOVERY

To ancient peoples, the planetary system ended with Saturn. All of the objects in trans-Saturnian orbits have been discovered since the invention of the telescope, some by luck and some as a result of deliberate and patient searches. The first major expansion in the solar system occurred on March 13, 1781, the product of a systematic visual survey of the heavens undertaken by the then little-known English musician and astronomer William Herschel. Observing with a 16-cm reflector, he found a nonstellar object of 6th magnitude, which he first supposed to be a comet. By the end of 1781, however, it was clear that the orbit was more nearly planetary than cometary, and that Herschel had discovered a trans-Saturnian planet, approximately doubling the known dimensions of the solar system. Herschel was appointed court astronomer to King George III, ensuring him the oppor-

tunity to pursue a scientific career. For about 60 years the new planet bore three names ("Georgian Sidus," suggested by Herschel; "Herschel," suggested by Joseph G. L. de Lalande, and "Uranus," suggested by Johannes Bode), until finally the latter, mythological term became generally accepted by astronomers.

Uranus orbits the Sun at a distance of 19.2 AU, almost exactly the position expected from the Titius-Bode rule based on the spacing of planetary orbits. This discovery aroused interest in the possibility of still more distant planets, but it was not until problems with the orbit of Uranus pointed the way that Neptune was discovered.

As early as 1790, positional astronomers began to have difficulty reconciling the calculated positions of Uranus with the observations. After 1830 there was increasing speculation that the problem might result from the gravitational perturbations of an unknown planet. The problem, then, was to calculate the mass and position of such a planet with sufficient precision to permit its telescopic discovery. The common assumption at the time was that the perturbing planet would follow "Bode's law" and be at a distance of 39 AU, and that its orbit would lie in the ecliptic plane, but the other parameters would have to be calculated from the observations of Uranus.

By 1843 John Couch Adams, in England, had obtained a preliminary solution to this problem, which he revised in 1845. He communicated his prediction to the British Astronomer Royal, George Airy, but Airy reacted negatively and no attempt was made to look for the object — an omission that seems strange to modern astronomers, given that the observational test was so simple and definitive. Meanwhile, the Frenchman Urbain Leverrier carried out similar calculations and presented his results to the French Academy on June 1, 1846, with an update on August 31. But Leverrier had no better luck with the French observers than had Adams with the English, and it remained for Johann Gottfried Galle, at Berlin, to locate the new planet on September 23, 1846, using Leverrier's tables of predicted positions.

Adams and Leverrier are both given credit for the discovery of the eighth planet, Neptune. Their predictions were based on sound analytic techniques and agreed with each other and with the location where it was found to within a degree, even though Neptune does not follow the Titius-Bode rule and is at 30.1 AU, rather than 39 AU.

An interesting footnote to this history is provided in a study recently published by astronomer Charles Kowal and Galileo scholar Stillman Drake. In an effort to locate prediscovery observations of Neptune, they noted from calculations by Steven Albers that in January of 1613 Jupiter passed extremely close to Neptune in the sky. At this time there was only one telescopic observer in the world, but he was an extraordinarily gifted one — Galileo Galilei. A search of his notebooks revealed two drawings in that month of "stars" in the same telescopic field with Jupiter that do not appear in modern catalogs (Fig. 1).

The positions of both are drawn with respect to the Jovian satellites, and both agree to within one arc minute with the position we calculate for Neptune on those dates. Remarkably, Galileo even notes in his observing log that on one night the "star" seemed to have moved, relative to other stars, from the position of the previous night. Thus it now appears that Galileo himself saw Neptune fully 233 years before it was identified at Berlin by Galle.

Emboldened by the successful prediction of Neptune, astronomers began to search for evidence of a still more remote planet. The data with which to work were sparse, however. Neptune had been observed over too short an arc of its orbit for perturbations to be measurable, so it was only those residuals in Uranus' position (remaining after correction for the perturbations of Neptune) that might point the way toward a new planet. Several suggestions of new planets were made in the late 19th century, but it was primarily the efforts of two proper Bostonians — William Pickering and Percival Lowell — that gave respectability to this effort in the first two decades of the 20th century. Using independent methods derived from those of Adams and Leverrier, they both deduced the presence of trans-Neptunian planets in approximately the same region of the sky. Lowell, who called his object "Planet X," published his final prediction in 1915. Pickering gave the first position for his "Planet O" in 1909, with an update in 1919, and later expanded his calculations to include a second "Planet S." Meanwhile, unsuccessful searches were carried out starting in 1905 at both Lowell and Mt. Wilson observatories.

Before his death in 1916, Lowell urged that a special wide-field camera be constructed to carry out a systematic search for Planet X, but it was not until 1929 that the survey with that camera was begun at Lowell Observatory. The specific observing and plate examination techniques were developed by Clyde W. Tombaugh, who had joined the observatory staff to carry out this project. On February 18, 1930, he first identified the images of the new planet, disappointingly faint at magnitude 15 and showing no measurable disk. The discovery was announced on March 13th, the 149th anniversary of the discovery of Uranus, and the new planet was soon given the mythological name Pluto, a word with Percival Lowell's initials

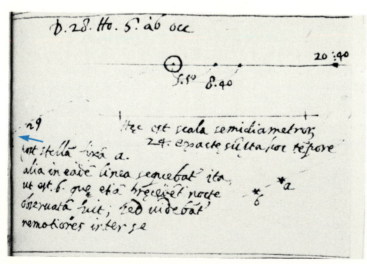

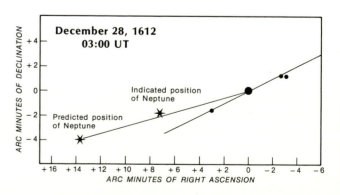

Fig. 1a (left). Galileo's observations of Neptune, 2334 years before astronomers recognized it as a planet. On December 28, 1612, and January 28, 1613, while observing Jupiter and its satellites, he recorded the position of Neptune (arrowed) which he thought to be a fixed star ("fixa," in Latin). Using a calendar in which the days began at sunset, he labeled this entry December 27th instead of the 28th. One month later, Neptune was so far from Jupiter that Galileo could not show the two planets in their relative positions; instead, he made a separate drawing for Neptune and a nearby star. Galileo noted the pair's relative motion, but did not realize he had observed a new planet. Fig. 1b (right). This diagram indicates the positions for Jupiter, three satellites (from left to right: Ganymede, Europa, and Callisto), and Neptune for Galileo's first sighting.

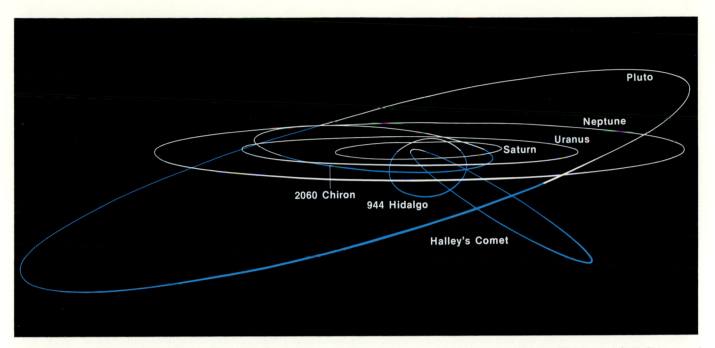

Fig. 2. This computer-generated diagram shows the orbits of Saturn, Uranus, Neptune, and Pluto; the asteroids 944 Hidalgo and 2060 Chiron; and Halley's Comet. The view is from 5° above the ecliptic, 6.7 billion km from the Sun. Saturn, Uranus, and Neptune travel nearly within the plane of the ecliptic, but Pluto, Hidalgo, and Chiron have orbits that are both noticeably inclined and eccentric. The extremely elongated path of Halley's Comet brings it above the ecliptic only near perihelion.

as its first two letters. The new planet orbited the Sun in 240 years, at an average distance of 39.4 AU (Fig. 2).

From its faintness and the absence of a visible disk, Pluto was immediately perceived to be different from the giant planets. Indeed, suspicions were raised almost at once that it was too small to have caused the perturbations in Uranus' orbit used by Lowell and Pickering. Tombaugh continued his systematic search for other trans-Neptunian planets, demonstrating over the next 13 years that no planet could exist brighter than magnitude 18. For a time many astronomers believed that Pluto might be extremely dense, as this seemed the only way to reconcile its small diameter with the need for a massive object. When its mass was finally measured and found to be about 100 times smaller than had been indicated from the prediscovery analyses, astronomers were forced to accept the fact that the original calculations were spurious and that the discovery of Pluto resulted from the careful survey carried out by Tombaugh, rather than representing a triumph of predictive science.

The other, smaller objects discussed in this chapter — the satellites and asteroids of the outer solar system — were discovered by observers over a span of many decades. The most recent additions have been Pluto's satellite Charon (technically, satellite 1978 P1), found by James Christy and Robert Harrington in 1978, and the similarly named asteroid 2060 Chiron, found by Charles Kowal in 1977. We will return to these objects after discussing the physical properties of Uranus and Neptune.

URANUS AND NEPTUNE

Although similar to each other in general physical characteristics, Uranus and Neptune are quite different from any of the objects encountered nearer the Sun. They have nearly identical diameters of 50,000 km and atmos-

pheres of hydrogen, helium, and methane. They are often grouped with Jupiter and Saturn to make up the four giant or Jovian planets. Superficially, this classification has the value of distinguishing between those bodies with reducing (hydrogen-rich) atmospheres as compared with oxidizing (oxygen-rich) atmospheres, those planets of large size as compared with the smaller, Earthlike bodies, those with many satellites instead of few, and those which were formed in the colder regions of the solar nebula.

When we look at details of Uranus and Neptune in comparison with Jupiter and Saturn, however, we begin to see distinct differences. The two outer planets are both much smaller (with masses only five and six percent that of Jupiter), and their greater densities suggest a difference in composition. Whereas Jupiter and Saturn are composed almost entirely of hydrogen and helium in approximately cosmic proportions, theoretical arguments show that Uranus and Neptune must contain primarily heavier materials. Oxygen, nitrogen, carbon, silicon, and iron are among the most abundant elements, and probably constitute most of the pair's bulk, according to the pioneering work of Rupert Wildt at Yale University in the 1930's.

The nature of planetary interiors cannot be determined directly, but must be inferred from computed models that attempt, in effect, the "experimental" synthesis of a planet from different initial conditions. Each model produces a prediction about the present state of the planet, which can then be compared with observations. The primary uncertainty in such models arises from our limited knowledge of the behavior of matter at the immense pressures found in planetary interiors, but advances in high-pressure physics give us new confidence that we can calculate meaningful representations.

Detailed models for Uranus and Neptune have recently been developed by William Hubbard and J. MacFarlane.

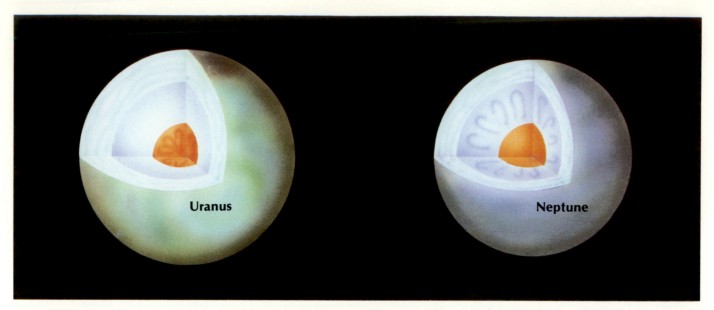

Fig. 3. **William Hubbard and J. J. MacFarlane propose these models for the interiors of Uranus and Neptune. In their calculations, they combined what we know of the planets' bulk properties with the expected response of hydrogen, helium, and rock to high pressure and temperature. Here the rocky cores contain metals and silicates, with methane, ammonia, and water concentrated in the "ice" mantles. Hydrogen and helium, tremendously compressed due to gravity, form thick "crusts" that grade into atmospheres at their outer boundaries. Studies of Uranus' evolution by Michael Torbett and Roman Smoluchowski suggest that its core may be fluid and in convective motion, creating a magnetic field; such a field has not yet been detected conclusively. Neptune's greater density may create enough pressure in the core to solidify most of it. Yet excess heat emanating from the planet indicates some kind of convective activity, and this may occur largely in its icy mantle; water, in motion and compressed into an ionized state, may conduct electricity well enough to initiate a mantle dynamo. In fact, Neptune is the only planet suspected of having a magnetic field generated outside its core.**

They show that these planets probably have similar, three-layer structures, shown in Fig. 3, with each region having a significantly different chemical composition. The core of each planet is presumed to consist of an amalgam of heavy elements, chiefly silicon and iron, termed "rock." The rocky core is surrounded by a liquid mantle of water, methane, and ammonia, and this is in turn enveloped by a low-density layer of mostly hydrogen and helium gas, as is observed from Earth. The exact proportions of the three component layers are difficult to ascertain, however, and depend in part on the conditions of temperature and pressure in the solar nebula at the time and location of the condensation of the protoplanets that eventually became the giant outer planets.

According to these models, Uranus and Neptune should be very similar, with central pressures of about 20 megabars (20 million times the atmospheric pressure on Earth) and temperatures of roughly 7,000° K. The difference in density (1.66 g per cubic centimeter for Neptune and 1.19 for Uranus) indicates a larger core and higher central density and pressure for Neptune, but the difference is not great. Surprisingly, then, the one additional observational test we have of interior conditions — the heat flow — shows a dramatic difference between the two planets.

For several years we have known of major internal heat sources in Jupiter and Saturn, as discussed in Chapter 12. It is believed that the source of this energy is heat stored from the initial gravitational contraction of the planets, and heat generated by the gravitational "unmixing" or separation of heavy elements from the lighter ones. Both possibilities predict that all four giant planets existed at high temperatures, hence high luminosities, in their early histories, with subsequent cooling regulated by the pro-

cesses of conduction and convection in their interiors. While the smaller sizes of Uranus and Neptune would suggest smaller heat sources, there is no reason to expect one to have been heated much more than the other. Yet observations show that Neptune is emitting excess heat at a rate of 0.03 microwatts per ton of mass, while Uranus is not. As a result, Neptune has the same effective temperature as Uranus (57° K) in spite of its much greater distance from the Sun. Why this difference?

Hubbard has sought to explain the presence and absence of observable excess heat fluxes on the basis of the atmospheres of these planets and on the amount of incident sunlight that each receives. He finds that the dense atmospheres operate as effective valves on the flow of interior heat convected up through the liquid mantles, and that the effect of the incident sunlight is to control the thermal evolution of a planet whose interior is in convective motion. The greater amount of sunlight received by Uranus (some 2½ times more than Neptune) prevents the internal heat from escaping in a form observable from Earth. There are other possible causes for the difference between the two planets, and as observations from spacecraft, at the telescope, and in the laboratory improve, better understanding of the interior structures, heat flows, and compositions will result.

The interaction of a planet with its space environment, especially the stream of particles flowing outward from the Sun, is governed by the presence of a magnetic field and its strength. Studies of the internal structures of the outer planets bear heavily on the possibilities of magnetic fields associated with these bodies. Jupiter and Saturn have large fields, presumed to be generated from electric currents circulating in their conductive, rapidly spinning

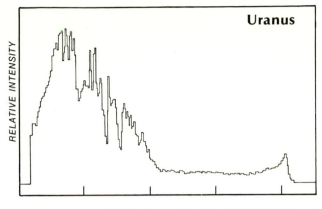

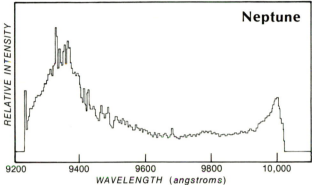

Fig. 4. These near-infrared spectra of Uranus and Neptune show an absorption due to methane from 9600 to 10,000 angstroms. Harold Reitsema, Bradford Smith, and Stephen M. Larson obtained this data with the 4-m telescope at Kitt Peak National Observatory.

cores. It seems reasonable to expect similar processes to act on Uranus and Neptune, but we have not yet detected magnetic fields on either planet. Evidence for such fields might be obtained from searches with high sensitivity for long-wave radio emission from charged particles in their magnetospheres, or we may have to wait for direct sensing by flyby spacecraft.

The atmospheres of Uranus and Neptune, unlike their interiors, can be studied directly by astronomical tech-

niques. The spectra of these planets are crossed by many strong absorption bands like those in Fig. 4, which were linked to atmospheric methane by Rupert Wildt in the 1930's. More recent spectroscopic studies have shown hydrogen in the dense atmospheres of these planets, and helium has also been deduced. These atmospheres, rich in hydrogen and hydrogen compounds, are called chemically reducing as opposed to the oxidizing atmospheres of the terrestrial planets. While the compositions and temperatures of the atmospheres of Uranus and Neptune are generally similar, their atmospheric structures, that is, the variation of temperature with height and the presence of clouds, are apparently quite different. We know, for example, that the atmosphere of Uranus is cold and very clear to great depths. There is rarely any haze and no observed clouds, and the strong absorption of certain wavelengths by the great quantities of methane in the atmosphere gives the planet a greenish color.

In contrast, Neptune's atmosphere contains a variable haze of aerosol particles or ice crystals of unknown composition (recent images appear in Fig. 5). The haze contributes to a warming of the upper atmosphere by the absorption of sunlight, giving rise to a temperature inversion of the kind also found in the atmosphere of Titan. Infrared spectroscopy and photometry show that at times nearly half of the planet is enshrouded in a thin atmospheric haze, and that the haze dissipates and reforms in a matter of weeks or perhaps days. In 1979, astronomers G. Wesley Lockwood and Don Thompson called attention to observed changes in the brightness of Neptune (and Titan) in synchronism with solar activity; increased solar activity during the cycle that peaked in 1980 resulted in an overall decrease in the observed brightness of these two bodies. The implication is that solar activity affects the formation and dissipation of hazes in the atmospheres of both Neptune and Titan, but the nature of the relationship is scarcely understood.

Recent photometric studies of both Uranus and Neptune have yielded better information on the rotation periods of these planets. The rotation periods have been poorly known, because the disks of these planets are so

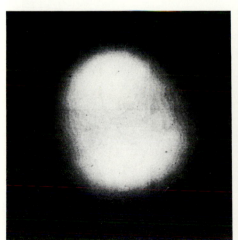
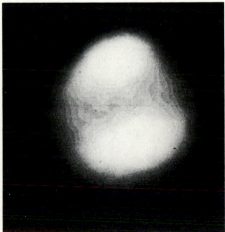
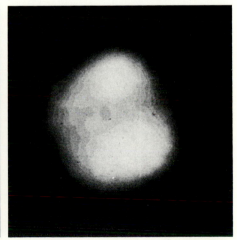

Fig. 5. Taken through a filter which transmits lights only at the strong, 8900-angstrom absorption band of methane, these images of Neptune show bright, high clouds of ice crystals in the northern and southern hemispheres. The dark equatorial band corresponds to deeper layers of methane. Harold Reitsema obtained these images with a charge-coupled device (CCD) attached to the 1.54-m telescope at Catalina Observatory in Arizona.

small (even in the largest telescopes), and because there are no substantial visible surface features. Even measurements of the Doppler shift of the spectral lines from the receding and approaching limbs of these planets give ambiguous results because of the difficulties inherent in the observations. Studies in the last few years at the University of Texas' McDonald Observatory show that the atmosphere of Uranus has a rotation period of 23.9 hours. Similar studies of Neptune carried out at Texas, Kitt Peak National Observatory, and Mauna Kea Observatory show that Neptune's upper atmosphere (where the haze forms) rotates once in 17.7 hours, but that a bit lower in the atmosphere the rotation period is apparently 18.4 hours. These results suggest that there are strong wind shears in Neptune's atmosphere with important implications for the planet's atmospheric dynamics.

For reasons that are not at all understood, the pole of rotation of Uranus is tipped some 98° from the vertical (vertical means perpendicular to the orbital plane, which, for Uranus, is almost exactly in the plane of the ecliptic). Thus as Uranus revolves about the Sun with its 84-year period, the north and south poles periodically point approximately in the direction of the Earth and Sun, causing regions in the polar latitudes to remain alternately in sunlight and darkness for periods up to 42 years. The importance of this extreme seasonal effect on atmospheric circulation, weather, and climate are topics of considerable interest, but so far there is little direct information and only some educated speculation.

ICY BODIES

In addition to the two large, fluid planets, there are numerous smaller, solid bodies in the outer solar system. All appear to be composed in large part of ices of water and other compounds, and all are presumed to be aggregates of the solids that condensed from the solar nebula far from the heat of the proto-Sun.

We begin our discussion with the best known of these

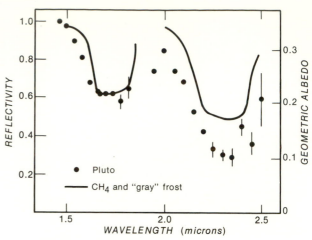

Fig. 6. The surface brightness of Pluto from 1.5 to 2.5 microns, as measured by B. T. Soifer, G. Neugebauer, and K. Matthews in March, 1974. Geometric albedo is the ratio of the observed light to that expected from Lambert's law (a means of relating photometric observations to a hypothetical, "standard" reflecting surface). Adapted from *Astronomical Journal 85*, 1980, pages 166-167.

icy worlds, Pluto. Although Pluto is classed as one of the nine major planets, it is by far the smallest, with a diameter of only about 2,500 km and a mass only 0.0017 that of the Earth. It also has an unusual orbit, with both greater eccentricity (0.25) and higher inclination (17°) than any other planet, and it is "planet crossing," coming inside the orbit of Neptune near perihelion. For all of these reasons, it is clear that Pluto has little in common with the other major planets. Astronomer Brian Marsden even went so far to suggest, at a meeting celebrating the 50th anniversary of Tombaugh's discovery of Pluto, that it should properly be "demoted" to the status of a minor planet or asteroid.

The diameter of Pluto has not been measured directly, but has been inferred from several indirect lines of

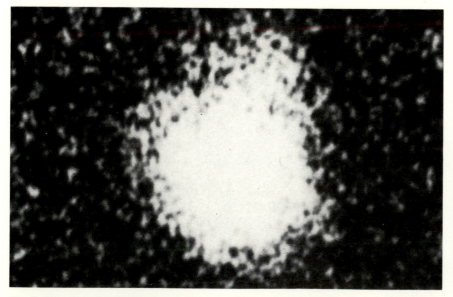

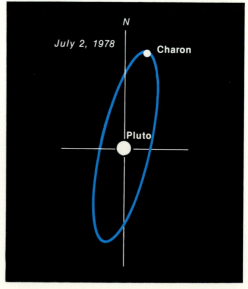

Fig. 7. The discovery photograph of Pluto's satellite Charon, taken on July 2, 1978, with the 1.54-m reflector of the U. S. Naval Observatory. Charon appears as a bump on the upper-right edge of Pluto. The accompanying diagram shows Charon's position in its orbit at the time of the photograph, as well as the true relative sizes of the two objects.

evidence. It is almost certainly between 2,700 km, a value that would give it a density of 1, and a lower limit near 2,000 km, corresponding to a highly reflective surface. The first information on its composition was obtained in 1976, when we with our colleague Carl Pilcher used the Kitt Peak 4-meter telescope to detect frozen methane on the surface of Pluto, a result later confirmed by other observers (Fig. 6). Methane can freeze only where temperatures are less than about 60° K, and the presence of this substance on Pluto implies that the planet has a tenuous atmosphere of methane and some other heavy gas, such as neon. The other gas is required to "bind" methane gas to the small planet, since even a solid ball of methane ice the size of Pluto would evaporate in the lifetime of the solar system. In 1980, Uwe Fink and his colleagues detected methane spectroscopically and confirmed the expectation that the partial pressure of this gas at the surface of Pluto is about 0.1 millibar, or about 1/60 the atmospheric pressure on the surface of Mars.

In 1978 the discovery of a satellite of Pluto greatly increased our interest in this small, cold world. The satellite, tentatively named Charon and seen in Fig. 7, has a maximum elongation from the planet of only 0.8 arc second, and it can only be viewed under excellent seeing conditions as an elongation of the image of Pluto. It orbits at a distance of about 17,000 km from the planet in a period of 6.39 days, locked into the same period as the rotation of Pluto deduced earlier from its periodic light variations. Judging from its brightness, Charon has a diameter about a third that of Pluto, perhaps 800 km. This makes it the largest satellite, relative to its primary, in the solar system. This is also the only planet-satellite pair that both rotate and revolve synchronously — truly a double planet, or should we say a double asteroid?

Uranus and Neptune have, between them, seven known satellites. The two brightest satellites of Uranus, Titania and Oberon, were discovered by William Herschel in 1787, and in 1851 William Lassell added two more, Ariel and Umbriel. A century later, in 1948, Gerard P. Kuiper discovered the much fainter Miranda with the 2.1-meter McDonald reflector. Neptune's largest satellite, Triton, was found by Lassell on October 10, 1846, less than a month after the first sighting of the planet itself. Nereid was found by Kuiper in 1949 in a systematic search for undiscovered satellites of the outer planets.

The satellite systems of Uranus and Neptune set these planets quite apart from one another. Uranus, with its five known satellites in circular orbits nearly all in the same plane, resembles the Saturn system (Fig. 8). The similarity is especially striking in view of the fact that both planets have rings, although the nature of the rings is quite different, as discussed in Chapter 13. Seen in Fig. 9, Neptune's two known satellites are irregular, with Triton, in a retrograde orbit of high inclination, and tiny Nereid in an orbit both inclined and highly eccentric (Fig. 10).

The three brightest satellites of Uranus, Ariel, Titania, and Oberon, all show strong ice absorptions in their spectra, and their surfaces must be almost completely covered with ice or frost, giving a high albedo or reflectivity. The surface of Umbriel, however, may have bare rock or soil exposed on its surface, because the ice absorption bands are substantially weaker than on the other satellites. The appearance of a smaller fractional cover of ice or frost on its surface may in

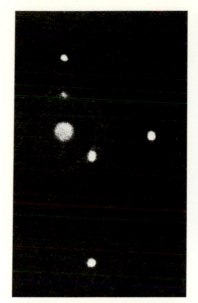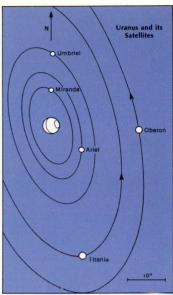

Fig. 8. At left are Uranus and its satellites, as photographed at infrared wavelengths by William Sinton from Mauna Kea Observatory in Hawaii. At right, the five satellites' orbits are shown at the same scale.

part account for the fact that Umbriel is fainter than its three large companions. Miranda has not been studied in the infrared where the water-ice signature lies. Except for comets, the discovery of water ice on the Uranian satellite represents the greatest distance from the Sun that water has yet been found in the solar system.

While it is not possible to determine directly the individual diameters and masses of the satellites of Uranus, progress has been made in the determination of the products of the masses of various satellite pairs on the basis of dynamical arguments related to the motions of the satellites in their orbits. Combining this new information with plausible values of the mean density (between 1 and 2) and diameters based on an assumed albedo compatible with the water frost observed there (0.3-0.5), we find that the Uranian satellites fall within the range of objects too small to have melted since their formation. That is, for an expected composition of silicate minerals comparable to those found in chondritic meteorites and a large fraction of water and other ices, the central pressures and temperatures have never been high enough to allow melting.

Neptune's Triton is one of several unique bodies in the solar system. It is the largest satellite of the most distant large planet and moves in a very peculiar orbit; dynamical studies show that this orbit is rapidly decaying with time such that within 10-100 million years Triton will be tidally ruptured as it approaches Neptune. While most astronomers agree with the studies that lead to this conclusion, it is regarded as highly coincidental and curious that this cataclysmic event is happening in the present epoch of solar system history.

Although Triton has long been known to be a planetary body of sufficient size to possess an atmosphere of heavy gases, and to have an interior that is melted and differentiated, very little direct information exists on the composition of its surface, the possibility of an atmosphere, or the state of its interior. Recent spectroscopic studies have given evidence that the surface is one characterized by rocks and

Fig. 9. Neptune and its two satellites, photographed by G. P. Kuiper on May 29, 1949. In the long exposure used to record Nereid (arrow), Neptune appears greatly overexposed; the spikes result from diffraction of light within the telescope. Triton is visible at Neptune's top edge, between the upper pair of diffraction spikes.

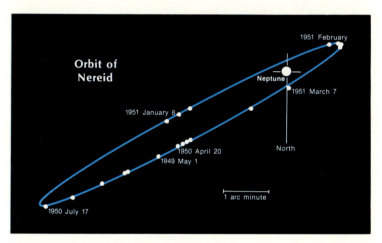

Fig. 10. The orbit of Nereid, from telescopic observations made earlier this century. The satellite moves in an inclined, highly eccentric orbit with a period of about 360 days. At its closest and farthest points from Neptune, Nereid is 1,390,500 and 9,733,500 km away, respectively.

soil, as opposed to ice, and that Triton has a tenuous atmosphere of gaseous methane with a surface pressure of about 0.1 millibar, similar to that of Pluto. In fact, Triton and Pluto appear to resemble each other closely in many ways, and it seems likely that Triton will also possess methane frost on its surface.

In Chapter 10, the asteroids or minor planets are described as rocky objects, found primarily between the orbits of Jupiter and Mars. But the term has been extended to any small object orbiting the Sun that is not a comet, and there is no reason to expect that such debris will be absent in the outer solar system. Nevertheless, only two asteroidal objects are known that penetrate as far as the orbit of Saturn. The first, 944 Hidalgo, was discovered by Walter Baade in 1920. Its orbit carries it from a perihelion inside Mars (2.0 AU) to an aphelion near Saturn (9.7 AU), similar to that of a number of short-period comets. Its spectral reflectance is unusual and we have only a rough estimate of its diameter —

Fig. 11. On this enlarged portion of Charles Kowal's discovery plate, taken in October, 1977, the distant asteroid 2060 Chiron appears as the arrowed trail. Note how much shorter it is than the track of a nearer, more typical asteroid at upper right.

about 40 km. Perhaps it is an extinct comet, although one of unusually large size.

Asteroid 2060 Chiron, discovered in 1977, is more unusual. A true inhabitant of the outer solar system (Fig. 11), Chiron has a perihelion just inside Saturn at 8.5 AU and an aphelion near the orbit of Uranus at 18.9 AU. Dynamicists consider this to be an unstable or "chaotic" orbit; Chiron's ultimate fate will be either collision with a planet or ejection from the solar system. There is little certain information about the physical properties of this object. If it is icy, its diameter must be about 100 km, but if rocky, it could be as large as 300-400 km. If Chiron is a dormant comet, it is a remarkably large one. What a spectacular show it would produce if it ventured near the Sun! Chiron might also be an escaped satellite of Uranus or Saturn; at this time, no one can say with certainty.

A FIRST CLOSE LOOK AT ICY WORLDS

The cloak of mystery concerning the icy bodies of the outer solar system began to part during the second week in November, 1980, when the Voyager 1 spacecraft sailed through the system of satellites of Saturn, providing close encounters with Mimas, Tethys, Dione, and Rhea. Nine months later Voyager 2 traced a similar path, concentrating on Enceladus, Iapetus, and Hyperion (see Chapter 11). Each of these satellites was already known to have water ice on its surface, and circumstantial evidence argued for ice as the primary bulk constituent as well. In size, these satellites of Saturn resemble Charon, Chiron, and the Uranian satellites, with diameters of up to 1,500 km (in the cases of Rhea and Iapetus).

What did Voyager find as it approached these previously mysterious worlds? Perhaps the most important revelation is that they really *are* icy, inside and out. The densities of Mimas, Tethys, and Rhea, as computed from newly determined masses and diameters, range between 1.1 and 1.4, values suggesting that 50 percent or more of their interiors must be composed of ice.

The surfaces of the Saturnian satellites are revealed in Voyager photographs to be heavily cratered. At the temperatures below 100° K that prevail at this distance from the Sun, ice is nearly as strong as rock and behaves similarly when struck by a meteorite or comet. Thus the high-reso-

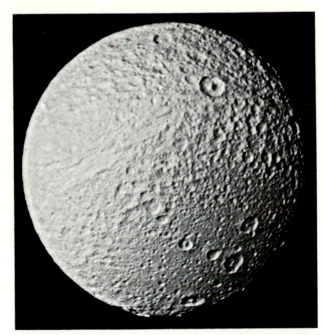

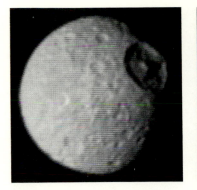

Fig. 13. Two Voyager 1 views of crater-pocked Mimas reveal different hemispheres of this inner, 400-km-diameter satellite of Saturn. The high density of craters indicates that the surface we see is an old one. One mammoth pit, roughly 135 km across, is the remaining scar from a cataclysmic impact that probably came close to shattering the entire satellite.

Fig. 12. Tethys, as seen by Voyager 2, appears more heavily cratered at upper left than at lower right, an indication that part of its surface has been modified by internal activity. A large, relatively fresh crater lies near part of an immense trench that here stretches from above center to extreme left.

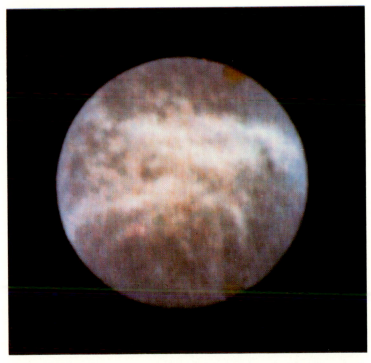

Fig. 14. Wispy white streaks, perhaps deposits of snow exuded from fractures in its crust, crisscross the surface of Rhea in the pattern seen here. The arc probably not crater rays, which would appear in radial patterns. Like Mimas, Rhea has a heavily cratered surface (not evident in this view) but has a diameter of about 1,530 km. A closeup of Rhea appears on page 204.

lution pictures of Mimas, Tethys, and Rhea (Figs. 12-14) look remarkably similar to spacecraft images of Mercury or the lunar highlands, although the surfaces are brilliant white ice rather than dark brownish rock. The story is more complicated than one of simple impact cratering, however. Dione (Fig. 16) and Tethys both have long, branching valleys tens of kilometers wide and hundreds of kilometers long, as well as regions of lower crater density indicative of resurfacing by internal processes early in their history. These have not always been dead worlds; they have experienced geological activity and evolution on a global scale at some period during their existence.

Enceladus is even more remarkable, displaying a surface dominated by what appears to be water volcanism. As seen by Voyager 2 (Fig. 15), parts of the satellite showed impact craters up to 35 km in diameter, but there are no larger craters, and even the most cratered regions are younger than the surfaces of the other Saturn satellites. Broad swaths of Enceladus have no visible craters at all, indicating major resurfacing events in the geologically recent past (since the age of the dinosaurs on Earth, for example). The variety of surface ages clearly suggests continuing internal processes, leading to the conclusion that the interior of Enceladus remains liquid even today. In addition, the satellite's fresh, uncontaminated ice surface is more reflective than that of any other known planetary body. Unexpected though they may be, the Voyager data lead scientists to conclude that some presently unknown process is supplying heat to Enceladus and maintaining its high level of geologic activity.

Iapetus, which orbits Saturn far beyond Titan and the icy inner satellites, is another bizarre and unique body. Since its discovery in 1671 by Cassini, Iapetus has been an enigma, for the trailing hemisphere (that which always faces where

the satellite has just been in its orbit) has a surface of water ice not so different from the inner satellites, but the leading face has been dramatically darkened (Fig. 17). Voyager has shown that the density of Iapetus is low, indicating a bulk composition primarily of ice; by implication, then, the leading hemisphere has but a thin veneer of dark material, perhaps originating from elsewhere. Spacecraft pictures only accentuate the strangeness of this 1,450-km-diameter world.

In the bright leading hemisphere and polar regions, the cratered surface resembles Rhea, but the dark hemisphere is almost featureless, with no evidence of bright crater ejecta as would be expected if the black veneer were thin. With an albedo of only about 5 percent, the dark material is ten

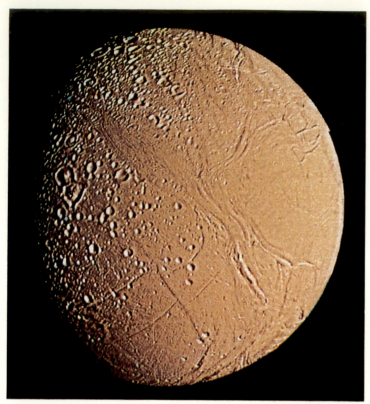

Fig. 15. At least five different terrain types have been identified on icy Enceladus based on this mosaic of Voyager 2 false-color images. Features as small as 2 km across are visible. Crater counts give widely varying ages for different surface regions; geologists believe Enceladus has undergone several episodes of resurfacing due to internal activity.

Fig. 16. Thousands of craters pepper the surface of Dione, about 1,150 km in diameter. Bright streaks similar to Rhea's appear on the other hemisphere, and a fracture can be seen here near the terminator. This evidence suggests that the satellite has had an active evolution since its formation.

times darker than the icy hemisphere. Furthermore, the dark hemisphere is oriented almost exactly toward the apex of motion — a fact that strongly suggests an external origin for the veneer. Yet, details in the Voyager images, such as dark crater floors within the bright hemisphere, appear to indicate that dark material may have come from inside Iapetus. Surely no more baffling place has been revealed by spacecraft exploration of the planets.

One of the interesting aspects of the Voyager results is that they show no obvious pattern in the differences among Saturn's satellites. Thus Rhea and Iapetus have almost identical sizes and masses but present entirely different appearances, perhaps related to their very different orbits. Another pair in terms of diameter are Mimas and Enceladus, but these two bodies have little in common in terms of internal heat sources and resulting degrees of geological activity. Relationships among mass, composition, orbital position, and evolution must exist in the Saturnian system, but at present they remain obscure.

Above all, the Voyager glimpses of Saturn's satellites caution us against easy generalizations. These relatively small icy worlds have their own histories, and each differs from the others. They were not all born alike, and internal forces as well as impact cratering have molded their surfaces. If the outer solar system seems simple to us, it is only an illusion based on our ignorance. The *terra incognita* of the solar system undoubtedly conceals many wonders to be revealed to later generations of explorers.

Fig. 17. Voyager 2 viewed enigmatic Iapetus at 20-km resolution (the large crater on the terminator is near the north pole). The moon has a bright, heavily cratered hemisphere facing away from the direction of travel and a much darker, apparently smoother opposite hemisphere. Dark material is also visible within some craters on the bright side.

Comets

John C. Brandt

The appearance in the sky of a bright comet often triggers great interest from scientists and the public alike. A recent example was Comet West in 1976 (Fig. 1a), a spectacular sight in the morning sky. Astronomers, space physicists, geologists, and even biologists study the various physical phenomena involved in these dramatic sights. To the eye, the tails of comets are their most distinguishing feature; these extensions routinely stretch several tens of millions of kilometers — and occasionally an astronomical unit (about 150,000,000 km) or more in length.

Public interest in comets is based on a curious mixture of scientific curiosity and awe, and the spectacle of a bright comet often evokes strong reaction. During the three-week period of 1974 when Comet Kohoutek was brightest, the Hayden Planetarium of the American Museum of Natural History in New York received a total of some 20,000 letters and recorded an average of 1,000 phone calls daily.

Comets can affect the Earth in more direct ways, too. For example, cometary debris is responsible for most meteor showers. Larger chunks may even hit the Earth from time to time. Researchers point to Tunguska, Siberia, as the prime candidate for such an event. There, in 1908, a dazzlingly bright daytime fireball was seen, and over the site of its fall flames and a cloud of smoke appeared. Deafening explosions, heard up to 1,000 km away, swayed the recording pens of seismographs and barographs. Trees were blown to the ground and stripped of their foliage (Fig. 2), and many other effects occurred in this sparsely inhabited area. But no crater was formed, a fact that suggests the collision was not with a meteor, asteroid, or other similarly dense object. The most promising theory contends that a piece of Comet Encke about 100 m across exploded in the atmosphere.

Although this event was obviously quite destructive, the possibility of the Earth being struck by a large cometary fragment should not cause great concern. First, calculations show that events like Tunguska should occur only once in 2,000 years on the average. Second, only a tiny fraction of the Earth's surface would be affected. (Of course, this becomes more important if the affected locality is, say, Chicago instead of Siberia.)

Comets are, nevertheless, of great importance to us, for they represent key pieces in our efforts to understand the origin and early evolution of the solar system. The scientific

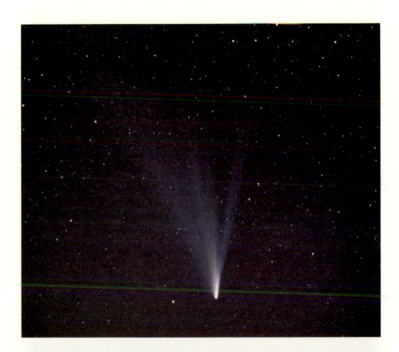

Fig. 1a (top). Comet West, which graced the predawn sky in the spring of 1976, was one of the most spectacular comets in recent years. This photograph, taken by Dennis di Cicco on March 8th, shows a straight, blue gas tail rising nearly vertically from the brilliant head; and a broad, pearly hued dust tail sweeping to the left (north). Fig. 1b (bottom). Comet Kohoutek was a disappointment to visual observers but a rewarding subject for astronomers. On January 13, 1974, Charles Kowal obtained this blue-light photograph, which reveals a wealth of fine structure in the 6½°-long gas tail.

Fig. 2. On the morning of June 30, 1908, an extremely violent explosion shook the Tunguska region of Siberia. This photograph shows a scene typical of the destructive results: trees, stripped of their foliage, have been knocked to the ground pointing directly away from the site of the blast. The most adequate explanation to date is that part of Comet Encke entered Earth's atmosphere and exploded here just above the ground. (Courtesy Sovfoto.)

Fig. 4. This photograph of Comet Bennett was taken on April 16, 1970, less than a month after the comet reached perihelion. The colors of the plasma tail (blue, lower) and dust tail (yellow, upper) are apparent.

Fig. 3. A schematic representation of the basic parts of a comet.

study of comets is increasing, after decades of relative obscurity. We are beginning a most significant decade, during which Halley's Comet will appear. This object will be discussed later, along with prospects for space missions that would return valuable *in situ* measurements.

OBSERVATIONS

All evidence available at present indicates that the central body and the ultimate source of all cometary phenomena is a ball of snow and dust about 1 km in diameter. The ability of this "dirty iceball" to interact with solar radiation and the solar wind to produce features of up to one astronomical unit in length is remarkable indeed.

The principal parts of comets have been determined by many different lines of evidence. These are (in decreasing size) the tail, hydrogen cloud, coma, and nucleus. All are illustrated schematically in Fig. 3 and discussed individually below.

Tails. Photographs of comets usually show two distinct kinds of tails, one containing dust and the other plasma. In Fig. 4, the dust tail appears yellow because the light reaching us from it is reflected sunlight. The plasma tail looks blue because radiation emitted by ionized carbon monoxide (CO^+), contained in the tail, peaks at about 4200 angstroms. Dust and plasma tails can be found alone or together in a given comet.

Usually observed as sweeping arcs, dust tails are homogenous and have lengths ranging from 1,000,000 km to perhaps 10 times that. Polarimetric studies show that the component particles are typically about one micron across and probably consist of silicate minerals. The so-called sunward-pointing "antitails" (seen, for example, in comets Arend-Roland in 1957 and Kohoutek in 1973) are not directed at the Sun at all, but merely result from our seeing a dust tail projected ahead of the Earth-comet line.

Plasma tails, usually straight, contain a great deal of fine structure and attain lengths roughly 10 times that of their dusty siblings — up to 100,000,000 km. The plasma's orientation is nearly antisolar, pointing away from the Sun. Consisting of electrons and molecular ions, the plasma tail seems to originate from a limited zone on the Sun-facing side of the nucleus. Fine structure can be characterized by local concentrations of plasma into thin bundles called rays or streamers. Such ubiquitous details provide convincing evidence for a magnetic field threading the tail's entire length.

ATOMS, IONS, AND MOLECULES IN COMETS

Coma	Plasma Tail
H, OH, O, S	CO^+, CO_2^+
C, C_2, C_3, CH, CN, CO, CS	H_2O^+, OH^+
NH, NH_2, HCN, CH_3CN	CH^+, CN^+, N_2^+
Na, Fe, K, Ca, V, Cr, Mn, Co, Ni, Cu	C^+, Ca^+

Hydrogen cloud. In 1970, observations above the atmosphere made at the Lyman-α wavelength of 1216 angstroms indicated that comets Tago-Sato-Kosaka and Bennett were surrounded by huge hydrogen clouds. Similar clouds have accompanied several other comets and span many million kilometers, substantially larger than the Sun. Comet Kohoutek's hydrogen envelope appears in Fig. 5, along with a visible-light photograph reproduced at the same scale.

Astronomers have estimated the production rate of hydrogen for several bright comets approaching the Sun. By the time they cross the Earth's orbit, these comets were found to produce more than 10^{29} hydrogen atoms per second! This escaping material cannot originate directly from the icy nucleus because the cloud's observed outflow speed is roughly 8 km per second, about 10 times faster than predicted for material sublimating from the nucleus' surface. Instead, most of this hydrogen probably comes from photodissociation (by sunlight) of the hydroxyl radical OH.

Coma. This spherical envelope of gas and dust surrounds the nucleus, extending from 100,000 to 1,000,000 km from it and flowing away at an average speed of 0.5 km per second. It is the outflow of coma gas that drags dust particles away from the nucleus. Comas usually don't appear until comets come to within about three astronomical units of the Sun.

The gaseous constituents within the coma are principally neutral molecules, some of which have been detected spectroscopically. For instance, Figs. 6 and 7 show spectra of comets Seargent and Bradfield from 1200 to 3400 angstroms. The visual spectrum (3000 to 5600 angstroms) of Comet Bradfield in Fig. 8 exhibits many emission lines frequently observed in comets from Earth.

Neutral molecules and atoms detected in comas up to the present are listed in the table, along with the ionized molecules found in plasma tails. Note the presence of relatively complex compounds like HCN

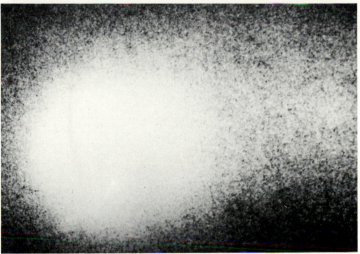

Fig. 5. Photographs of Comet Kohoutek obtained with a rocket-borne telescope are reproduced here at the same scale. The comet is seen at top in visual light, and below, three days later, in Lyman-a light (a strong ultraviolet emission of neutral hydrogen at 1216 angstroms), which reveals the comet's enormous hydrogen cloud.

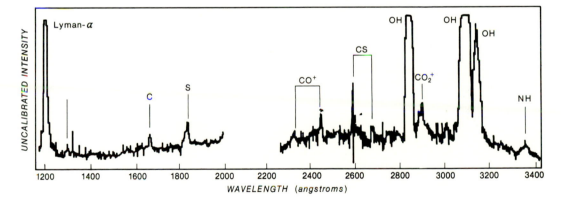
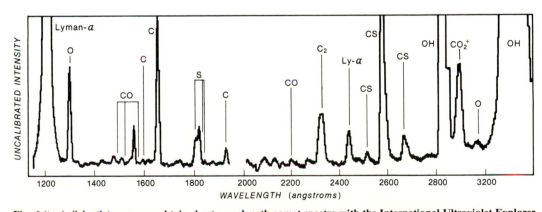

Fig. 6 (top). Scientists can now obtain short-wavelength comet spectra with the International Ultraviolet Explorer (IUE) spacecraft. The emissions seen here identify major constituents in the coma and tails of Comet Seargent. After Jackson *et al.*, *Astronomy and Astrophysics*, **73**, L7, 1979. Fig. 7 (bottom). This IUE emission spectrum of Comet Bradfield 1979 X has been adapted from Feldman *et al.*, *Nature*, **286**, 132-35, 1980.

and CH₃CN and the absence of water. (The detection of
H₂O has been reported but is the subject of controversy;
however, H_2O^+ exists in some plasma tails.) Heavy metals
begin to appear in the coma as a comet nears the Sun.

Nucleus. Although we have no photograph of a cometary
nucleus, circumstantial evidence strongly implies the existence of a central body, the source of all cometary gas and
dust. The nucleus is probably irregular in shape, ranging in
diameter from a few hundred meters to 10 km. Estimated
nuclear masses fall between 10^{14} and 10^{19} grams, and are
thought to be split roughly equally between ices and dust. If
this latter assumption is true, the average density would be
roughly 2 g per cubic centimeter (twice that of water).

A summary of the observed properties of comets is given
schematically in Figs. 9 and 10. An acceptable theory of
comets must explain these features and the ways in which
they change with heliocentric distance.

PHYSICAL MODELS

The purpose of modern cometary theory is to explain the
observations in a simple way and to point out critical observations or measurements to test the theory. A space mission
to comets appears to be necessary for substantial progress.
While we believe that our general understanding is in good
shape, surprises could easily occur.

The basis of current theory is Fred L. Whipple's "icy conglomerate" model of the nucleus as extended by Armand
Delsemme. A schematic illustration of their model nucleus is
given in Fig. 11. As a comet approaches the Sun in its orbit,
sunlight falls on the surface of the nucleus and heats it.
When the comet is far from the Sun, all the radiant energy
goes into heating the nucleus. But the physical situation
changes as the comet approaches the Sun, because eventually the temperature of the surface layers increases to a value

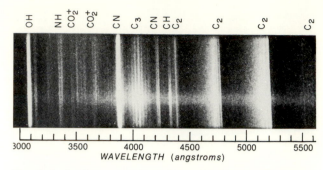

Fig. 8. A visual spectrum from Comet Bradfield 1979 X exhibits
most of the major emission lines seen in comets. The dark horizontal band corresponds to the comet's nucleus, where the emissions
are generally much stronger. The top edge of the spectrum marks a
point in the tail 43,500 km from the nucleus.

where sublimation of the ices occurs. Now almost all solar
radiation goes into sublimation of the ices. As the ices sublimate, a dusty crust forms that insulates the deeper layers
and regulates the sublimation process now occurring a few
centimeters below the surface.

The fact that comas appear when comets are near three
astronomical units is more consistent with water ice than
with other substances being the principal constituent of the
nucleus. Even though a definitive observation of neutral
water has not been made, the circumstantial evidence is
strong, particularly since many substances derivable from
water (for example, H_2O^+ in the plasma tail) have been
observed. Moreover, many emissions such as the bands of
CN and C₂ are observed in the spectra when the coma first
appears. It has been postulated that the ice occurs as a
clathrate hydrate, in which minor constituents can be
trapped in cavities within the water-ice crystal lattice. Thus,

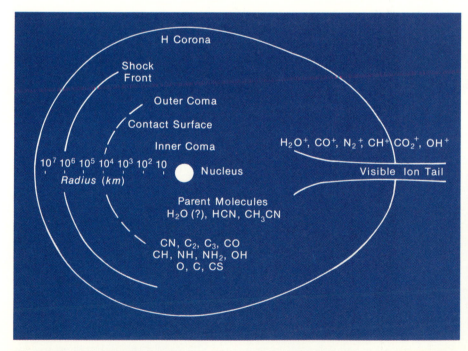

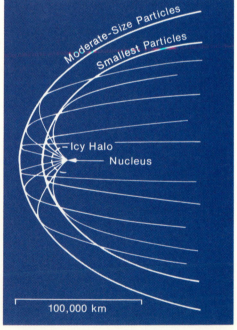

Fig. 9 (left). This schematic drawing shows the principal gaseous features of a typical comet on a logarithmic scale, so that both the nucleus (less
than 10 km across) and the huge hydrogen cloud (millions of times larger) are included. Frequently observed ions are shown as well. Fig. 10 (right).
The principal particulate features usually found in comets, here plotted linearly, result from dust grains streaming from the nucleus and becoming
pushed away by the solar wind. Their parabolic trajectories vary with particle size. A halo of sublimated nuclear ices is also indicated.

the sublimation of the ice also releases the minor constituents, and the thermodynamic properties of water ice control the escape of all substances from the nucleus.

If this view is correct, the temperature of the sublimating ice for a comet near the Earth's orbit will be about 215° K. There are in fact areas on our planet with ice fields at this temperature. An example is Plateau Station in Antarctica (latitude 78°S, longitude 41°E, altitude 3,624 m, ice thickness 3,165 m). These areas may provide the opportunity for important tests of instrumentation, designed to probe below the surface, on some future rendezvous mission to a comet.

The dominance of water ice in cometary nuclei may not apply to comets approaching the inner solar system for the first time. The storage capacity of the ice lattice is limited to approximately 17 percent of the number of molecules forming the lattice. If other substances (like carbon dioxide) exceed this value, their sublimation becomes controlled by their own thermodynamic properties rather than those of water. Since CO_2 and most other plausible minor constituents sublimate at lower temperatures than water does, their presence in fresh or new comets could account for abnormally bright comets observed at heliocentric distances greater than three astronomical units.

The neutral molecules continuously produced by the sublimation of the nuclear ices flow away from the nucleus in a manner physically analogous to the flow of the solar wind away from the Sun. They drag some of the dust particles with them, and this dusty gas forms the coma. The relatively simple molecules observed in cometary spectra may not be the ones initially released because the gas densities near the nucleus are high enough that gas-phase chemical reactions could occur.

The gas flowing away from the nucleus ultimately has an interaction with the solar wind; the importance of this interaction was demonstrated by Ludwig Biermann in the early 1950's. Actually, the existence of a solar wind was then unknown; it was inferred by Biermann from observations of plasma tails. The magnetic field carried along by the solar wind plays a vital role in the interaction, as discovered by Hannes Alfvén in 1957. Some ionization of the molecules in the tail is caused by solar radiation, and these ionized molecules are trapped in the magnetic field lines causing a deceleration that occurs only in the vicinity of the comet. The field lines with the trapped plasma wrap around the nucleus, like a folding umbrella, to form the plasma tail. In this scenario, the plasma tail is normally attached to the region near the nucleus by the magnetic field captured from the solar wind.

These phenomena can be photographed because trapped molecular ions such as CO^+ serve as tracers of the field lines. A sequence of photographs (Fig. 12) shows the folding of field lines around a comet. Photographs in Fig. 13 also show that the entire plasma tail occasionally disconnects from the comet. This is thought to occur when the polarity of the solar-wind magnetic field changes (at what is called a sector boundary), a situation that disrupts the connection to the near-nuclear region.

Solid particles in the nucleus have varying fates. Smaller particles or dust are blown in the antisolar direction by the Sun's radiation pressure to form the dust tail. Somewhat larger particles, not as strongly affected by radiation pressure, orbit the Sun and reflect sunlight that is seen by us as

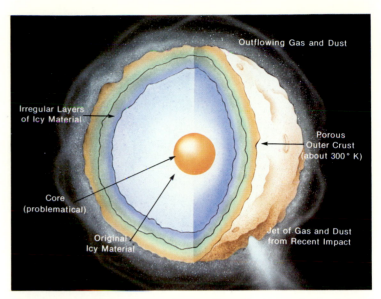

Fig. 11. Fred Whipple's "icy conglomerate" model, as extended by Armand Delsemme, is the basis for Charles Wheeler's rendering of a cometary nucleus. Comet nuclei may range from a few hundred meters to 10 km across.

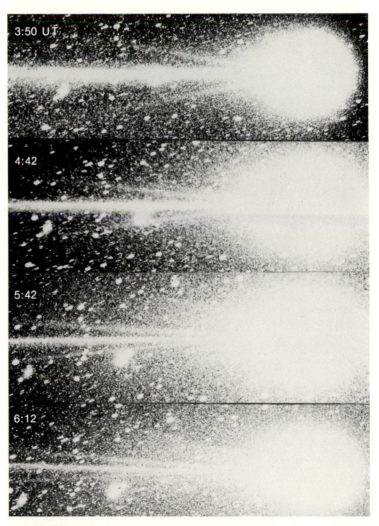

Fig. 12. Photographs of Comet Kobayashi-Berger-Milon taken on July 31, 1975, show the plasma tail's capture of magnetic field lines from the solar wind. The two most prominent streamers on either side of the tail's axis can be seen to lengthen and turn toward the axis in this sequence.

Fig. 13. Obtained during a three-hour period on September 30, 1908, these photographs record the plasma tail of Comet Morehouse detaching from the head of the comet.

a faint glow in the night sky called the zodiacal light.

The largest solid particles would probably be produced when a comet has spent a long time near the Sun. This occurs if a comet travels a short-period orbit (for example, Comet Encke has a period of 3.3 years) or even if it makes many perihelion passages in a long-period orbit. For the latter case, a comet loses roughly 1 percent of its mass on each round; for the largest comets, the figure may be 0.1 percent. Calculations show that for a typical comet with a diameter of 2 km, a passage results in the loss of the outer layers to a depth of 3 m. Sooner or later, all the ices sublimate and the comet becomes a mass of rocky material.

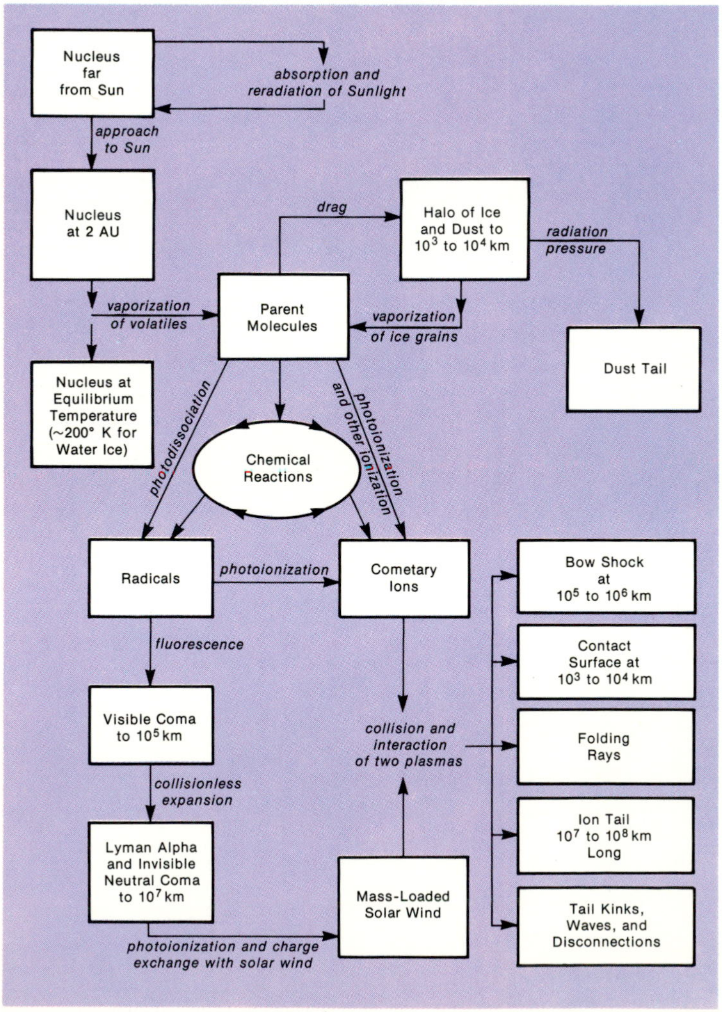

Gravitational perturbation disperses the rock along the comet's orbit, and some of these remnants enter the Earth's upper atmosphere, producing meteor showers. Most of the rocks responsible for meteors are light and perhaps even fluffy. Fragments liberated from the postulated core would be denser. Unfortunately, the form of the chunk from Comet Encke thought to be responsible for the Tunguska event is not clear.

A summary of the physical processes believed to be important in comets is given in Fig. 14.

THE ORIGIN OF COMETS

Here theory attempts to answer the question, "Where do these icy bodies originate?" Much of the information on this subject comes from the study of cometary orbits, and detailed orbital information is available for over 600 individual comets. Of the total, approximately 100 have periods of less than 200 years and are classified as short-period comets. These have mostly direct (prograde) orbits with inclinations to the plane of the ecliptic of 30° or less. Most short-period comets have aphelia (the points where they are farthest from the Sun) near the orbit of Jupiter and probably assumed their present orbits after numerous gravitational interactions with that planet. Short-period comets are probably spawned from among the population of long-period comets (those with periods greater than 200 years).

The 500 long-period comets are relatively unaffected by gravitational interactions with the planets, and records of these objects indicate that their orbital planes are oriented approximately at random with respect to the plane of the Earth's orbit. Thus, there are essentially as many retrograde comets as there are direct comets. In addition, very careful examination of the original orbits (that is, the comets' trajectories prior to entering the inner solar system) discloses none that are hyperbolic — there are no initially interstellar comets. These facts strongly imply that comets are gravita-

Fig. 14. Features and processes involved in a comet's interaction with sunlight and the solar wind are shown schematically.

Fig. 15a (top). Halley's Comet, as recorded on May 13, 1910, 24 days after perihelion. This Lowell Observatory photograph captured the comet's 45°-long tail, as well as the planet Venus (seen at lower left). Fig. 15b (bottom). A more detailed view of Comet Halley, taken on May 7, 1910, reveals delicate structure in the 11° of tail closest to the nucleus. (Courtesy Lick Observatory.)

Fig. 16. The apparition of Halley's Comet as depicted in the celebrated Bayeux Tapestry, commemorating the Norman Conquest in 1066. The comet made a close and impressive approach to Earth that year, about the time that William the Conquerer invaded England from Normandy (France). The appearance of Comet Halley was considered an evil omen for King Harold of England; in fact, Harold was killed later that year during the Battle of Hastings. The Latin inscription *Isti Mirant Stella* translates "They marvel at the star."

tionally bound to the Sun like other members of the solar system and that they likely formed early in its evolution.

In 1950, Jan Oort interpreted the statistics of comet orbits as indicating that they reside in an essentially spherical cloud around the Sun with a radius of perhaps 10,000 to 100,000 astronomical units. (By comparison, the nearest stars to the solar system, the Alpha Centauri system, are 275,000 astronomical units distant.) Gravitational disturbances produced by passing stars have several effects on the cloud. Most important for us, the disturbances regularly send comets from the cloud into the neighborhood of the Earth, where they produce their multifarious phenomena and where they can be observed. In addition, such stellar disturbances tidally limit the size of the cloud and tend to randomize the orbits. Oort's comet cloud contains an estimated 200 billion comets with a total mass of one-tenth the Earth's mass. Obviously, these figures are very uncertain.

Determining the origin of comets is an active area in cometary research. At present, the consensus view is that they condensed from the solar nebula at approximately the same time as the formation of the Sun and the planets. Of course, many details of the process are lacking, but comets seem to be a natural by-product of the physical processes responsible for the creation of the solar system. Some theories ascribe additional roles for comets in the evolution of the solar system. They might have been an important source of the atmospheres of the terrestrial planets (see Chapter 6), and, further, they could have supplied the original organic molecules necessary for the initiation of life

on Earth. Many of these ideas, though unproven, serve to illustrate the breadth of the scientific interest in comets.

HALLEY'S COMET

The decade of the 1980's is most promising for the study of comets because it features an apparition of the most famous of these objects, Halley's Comet. Observations of Halley's Comet go at least as far back as Chinese records of its appearance in 240 B.C., and it has been observed at each perihelion passage for over two millenia.

Until the time of Edmond Halley, comets were tacitly assumed to make sporadic passes through the inner solar system, and no serious attention was given to the possibility of their periodic return. Halley used Isaac Newton's then-new theories of gravitation and planetary orbits to compute the orbits of several comets. He noted that the orbits of comets observed in 1531, 1607, and 1682 were quite similar and assumed that the sightings probably referred to the same object at successive apparitions. On this basis, Halley made his famous prediction of its return in 1758-1759. The return occurred as predicted, and the comet was named in Halley's honor. The physical study of Halley's Comet was begun by F. W. Bessel at the 1835 apparition. In 1910, it was observed by E. E. Barnard and there was an effort to organize the observations of the comet around the world.

The orbit of Halley's Comet has an average revolution period of 76 years with a perihelion of 0.59 astronomical unit and an aphelion of 35 astronomical units; the nucleus has an estimated radius of 2½ km. It displays the full range of cometary phenomena which includes a long, spectacular tail, shown in Fig. 15.

We owe the comet's repeated naked-eye visibility over many apparitions to a number of circumstances. First, the orbit is favorably placed with the perihelion between the Sun and the Earth's orbit. Second, the comet is large, and the resultant activity is unusually spectacular. These two points have combined through the centuries to make Halley's Comet readily visible and memorable. Among the many historical records of this comet are the representation of the 684 apparition in the *Nuremberg Chronicle,* the 1066 apparition on the Bayeux Tapestry (Fig. 16), and the apparition in 1301 rendered naturalistically by Giotto in one of his Arena Chapel frescoes in Padua. Some responses to the appearance of Halley's Comet have been quite humorous, as shown in Fig. 17.

Thus, Halley's Comet is unique: it possesses an unquestioned place in human history and is, at the same time, a large, dependable comet with exciting scientific possibilities. It is the only comet exhibiting the entire range of cometary phenomena and a predictable orbit that will make an apparition during this century. Therefore, efforts to launch a spacecraft toward a comet inevitably focused on Halley's Comet, which last reached aphelion in 1948 and will next pass perihelion on February 9, 1986.

MISSIONS TO COMETS

Many scientists regard comets as physically fascinating objects for a wide variety of reasons. As mentioned earlier, they are possibly remnants of the solar system's formation. If that is the case, this ancient material has been little altered over geologic time. Moreover, comets display a large range of chemical and physical processes, some of which are described in this article. Clearly, the understanding of such bodies is a delightful challenge to the astronomer and space physicist. Finally, comets are most valuable tools for probing the properties of the solar wind, particularly at locations and times inaccessible to spacecraft. An example is shown in Fig. 18.

Yet we have so much to learn. The foundation of our present view is the icy conglomerate model of the nucleus,

but we have no photograph of even a single comet nucleus. Nor do we have *in situ* measurements of atmospheric density or magnetic-field strength. In a relative sense, comets have been unexplored in the space age, and the need for a remedy is compelling.

The history of space missions to comets has suffered severe ups and downs, particularly with respect to the NASA program. Missions to comets fall broadly into two classes: rendezvous missions and flybys. Until late 1979, it

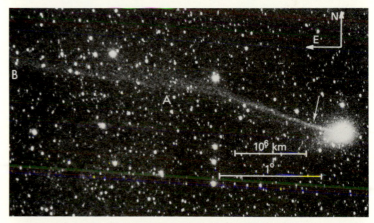

Fig. 17. "The Comet is coming!" exclaims a German postcard issued at the time of Comet Halley's appearance in 1910. Considerable public alarm was generated when it was learned that the Earth would pass through the comet's tail. Casual references by some astronomers to "cyanogen in the tail" produced nothing short of panic in some areas. Entrepeneurs took advantage of the hysteria, promoting gas masks and "comet pills."

Fig. 18. A remarkably rapid turning of the plasma tail of Comet Bradfield 1979 X on February 6, 1980. In a period of only 27.5 minutes, part of the tail (arrow) changed its orientation by an abrupt 10°. Scientists believe this "wind sock" action revealed the changing direction of the solar wind; the wind's velocity component perpendicular to the ecliptic plane switched from roughly 30 km per second northward to about 20 km per second southward. Note that the outer tail segment (between *A* and *B*) did not turn.

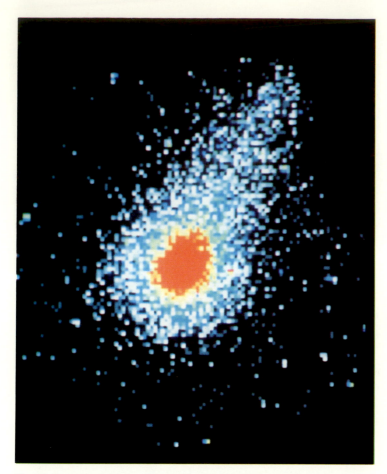

Fig. 19. A computer display of Comet Bradfield 1979 X, recorded in blue light, has been color-coded to show regions of equal brightness. The image was made by the IUE spacecraft and was supplied by Paul Feldman.

was believed possible to rendezvous with Halley's Comet at the 1985-1986 apparition. This plan required the development of a new solar-electric propulsion system called "ion drive," but approval for development of this system on a rush basis was not granted. Now being developed, this system should be ready for a rendezvous with another comet about 1990.

Interest now focuses on possible flyby missions. Although a rendezvous would have been preferred, an excellent science package can be carried in a flyby spacecraft. This would include instruments for photographing the nucleus and for measuring the properties of the neutral and ionized gas and the magnetic field. Flyby missions to Halley's Comet have been approved by the European Space Agency, the Soviet Union, and Japan.

Unable to launch its own dedicated mission due to budgetary constraints, NASA has organized a worldwide coordinated effort called the "International Halley Watch." This effort has the goal of stimulating, encouraging, and coordinating scientific observations throughout the apparition. Other facets of the organization include standardizing observing techniques, storing data, and distributing results to scientists and the public. The latter function will be particularly important during the prime time for ground-based observations and public viewing in November of 1985 and April of 1986. The International Halley Watch certainly will support the flyby missions to the comet. In addition, it encourages observations from instruments placed in Earth orbit. Candidates would be the Space Telescope and instruments on board Spacelab; these should provide valuable glimpses of Halley's Comet from above the atmosphere (Fig. 19).

Clearly, the next few years promise major advances in our understanding of comets. This progress should be centered upon the 1985-1986 apparition of Halley's Comet and the missions sent to investigate it directly, the International Halley Watch, and observations by instruments in Earth orbit. We can also hope for a cometary rendezvous mission by the decade of the 1990's.

John A. Wood

Sometime after midnight on February 8, 1969, the editor of *El Heraldo* in Chihuahua City, Mexico, was called from his office by an excited night watchman, who wanted him to witness an alarming spectacle outside. The night sky was illuminated by a brilliant pulsating blue-white light that moved across the southern horizon, leaving a glowing trail behind it. The object lit the sky so brightly that it was seen in parts of the United States, hundreds of kilometers north of Chihuahua. Soon the editor's phone began to ring. Agitated voices rushed to tell him what they had seen. Some were convinced that the end of the world was at hand.

Even greater was the fright of residents 240 km to the south, directly beneath the dazzling phenomenon, which, of course, was an exceptionally bright fireball. As it moved in a northeasterly direction, the brilliant flare broke into two major pieces. Each of these shortly exploded into a fireworks display of diverging lights. All this time a chain of tremendous detonations (sonic booms) assailed the ears of onlookers. Thousands of stones and pebbles dropped from the dark winter sky.

The first of these was found the next day, dug slightly into the ground, only a few steps from a house in the small village of Pueblito de Allende. Regional newspapers told of its discovery. These reports, together with accounts of the spectacular fireball, brought meteorite hunters to the area: scientists from Mexican and U.S. museums and universities and the NASA Lunar Receiving Laboratory in Houston (which was gearing up to process the first samples from the Moon in that same year), mineral dealers, and amateur collectors. An untold number of specimens (such as the one seen in Fig. 1) was found, which yielded an aggregate weight estimated to be roughly two tons. An even cruder estimate of the amount of unrecovered material adds another two tons.

The discovery sites of meteorite specimens were distributed in an elongated ellipse, called a *strewnfield*, approximately 50 km long by 10 km wide. Larger specimens, which had been least decelerated by the atmosphere, were found at the northeast end of the strewnfield. The largest of all (110 kg) defined its northeast tip. Progressively smaller and more easily decelerated specimens fell farther to the southwest. This is the classic fall pattern for members of a meteorite shower.

THE ALLENDE METEORITE

All stones from this particular shower are named *Allende* for the locality of first discovery. Other meteorites are similarly named for their recovery sites. Allende has come to be the most intensively studied of all meteorites, partly because it is available in such copious quantities. Also, when Allende became available, a number of research groups were preparing to study the cherished Apollo rocks, and they welcomed a new extraterrestrial sample on which to practice. But in large part, Allende is simply an exceptionally interesting meteorite about which several very important discoveries have been made.

The individual specimens of Allende are coated with thin black layers of slaggy material that was melted during atmospheric deceleration. Minerals immediately beneath the fusion crusts show no sign of thermal damage, however, because ablation peels surface material away from a mete-

Fig. 1. A search party at the Allende meteorite strewnfield, in February of 1969. Clutching a meteorite at left is G. J. Wasserburg of Caltech, who later made pioneering discoveries concerning the prehistory of the solar system by studying this material (see text). With him are Prof. Schmitter of the University of Mexico, and Sr. Martinez, one of the farmers whose land the stones dropped on. Photograph by D. P. Elston.

orite as fast as it becomes hot during flight through the atmosphere. Thus little heat has a chance to penetrate the meteorite's interior.

When broken open as in Fig. 2, an Allende stone is found to consist of dense, reasonably hard, dark gray material devoid of cavities or visible porosity. Its substance is perceptibly heavier (specific gravity, 3.67 grams per cubic centimeter) than that of an ordinary terrestrial rock of equivalent size. Close examination reveals a conglomeration of small objects, mostly about a millimeter in size, that range from spherical to highly irregular in shape and from white to dark gray in color (Fig. 3). These small stony inclusions are embedded in a matrix of very-fine-grained, dark gray, earthy-looking matter. The meteorite does not closely resemble any known type of terrestrial rock.

An obvious first thing to do with an unknown extraterrestrial sample like this is submit it to chemical analysis. When this was done to Allende, the meteorite exhibited the pattern of chemical-element abundances shown in Fig. 4 below. It is interesting to plot Allende's composition against the spectroscopically derived composition of the atmosphere of the Sun: the two are very similar. It is as if a mass of solar material had been ripped out of the Sun and allowed to cool and condense. Only a few elements depart far from the 45° line of equal concentrations. Hydrogen, carbon, nitrogen, oxygen, and the noble gases are so volatile, or form compounds so volatile, that they are incapable of condensing in the inner solar system. Consequently, we would expect them to be underrepresented in condensed solar material. Lithium, on the other hand, is destroyed by thermonuclear reactions in hot stellar interiors; thus, after 4.6 billion years of attrition in the Sun, this element has been depleted in the solar atmosphere.

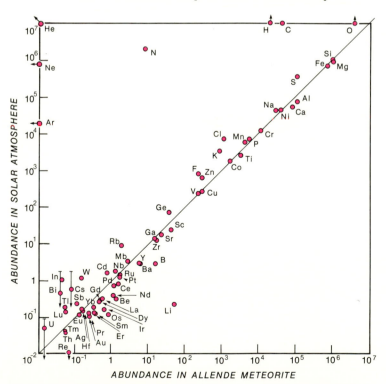

Fig. 2. (top). A large stone from the Allende shower with a 2-cm scale bar. The near end of the stone has been broken off, exposing unaltered meteoritic material. Other surfaces, rounded by ablation during entry into the atmosphere, are coated with a fusion crust and Mexican clay. Fig. 3. (above). A closer view of the broken surface on the Allende specimen, showing inclusions and chondrules (one or a few millimeters across) embedded in a dark matrix.

Another interesting measurement to make on an exotic stone is to determine its radiometric age, the time that has elapsed since it crystallized and cooled to the point where the isotopic daughters of radioactive decay in it could begin to accumulate. The amount of argon-40 found in the meteorite corresponds to that which would be generated by 4.57 billion years of decay of the potassium-40 in it. This is, to within the error of the measurement, the accepted age of the solar system.

The great ages and unfractionated chemical compositions of chondritic meteorites (of which Allende is an example) are convincing evidence that they are specimens of primitive planetary material, preserved in the same state they assumed when the planets first accreted. Rocks from geologically active planets like the Earth are highly fractionated and have younger ages (less than 0.5 billion years for most Earth rocks).

Fig. 4. Concentrations of 69 chemical elements in Allende, plotted against their concentrations in the solar atmosphere (using logarithmic axes). In both cases, abundances are relative to one million silicon atoms.

ALLENDE COMPONENTS, AND EVENTS LONG AGO

Most of the excitement Allende has generated stems from the small objects or inclusions, mentioned earlier, with which it is studded. On closer study, each of these was found to be an integral mineral system that had formed more or less independently of the meteoritic material now surrounding it (Fig. 5). It seems clear that the inclusions were dispersed in space when they formed, and subsequently they accreted, together with fine mineral dust, into planetary objects of some type. The dust, packed together, comprises the earthy matrix of the present meteorite.

Many of the inclusions consist of minerals that are rather uncommon on Earth, and that had not previously been recognized as abundant meteoritic phases: melilite, perovskite, hibonite, fassaite, spinel, and others. What these minerals have in common are high concentrations of calcium, aluminum, and titanium, relative to the mean abundances of these elements in bulk Allende material. And the significance of these elements is that they are the most refractory of the major elements in meteoritic or planetary matter. That is, if meteoritic material were heated to progressively higher temperatures in a gas of solar composition, these would be the last major elements to vaporize; if the hot vapor were cooled, they would condense first.

This is intriguing because it agrees with some of our preconceptions concerning the formation of planetary matter as a by-product of the Sun's origin. In 1962, A. G. W. Cameron of the Harvard-Smithsonian Center for Astrophysics developed a model of solar system formation, with roots extending back to Immanuel Kant and Pierre Simon Laplace, wherein the gravitational collapse of a cloud of interstellar gas and dust gave rise both to the proto-Sun and to a rotating accretion disk or nebula, also of solar composition. Compression of the nebular gas as it fell together was thought to have heated it temporarily, so that in the region of the terrestrial planets, temperatures were high enough to vaporize completely the interstellar dust incorporated in the nebula. As the hot nebula cooled by radiation, the first things we would expect to condense from it would be grains rich in the least volatile elements, namely calcium, aluminum, and titanium — just what we find in the Allende calcium- and aluminum-rich inclusions (CAI's). See Fig. 7 on the next page.

The correspondence between theory and observation is actually much more striking than that just indicated. If one assumes some not unreasonable things about the nebula — that is, elemental abundances equivalent to those presently observed in the Sun and gas pressures in the range of 0.0001 to 0.001 atmosphere (from Cameron's model) — it is possible to use the tool of chemical thermodynamics to predict which minerals would condense, and in what order, in a hypothetical nebula that cooled from the totally vaporized state. This was done in 1972 by Lawrence Grossman, then a graduate student at Yale University. A simplified depiction of the condensation sequence of minerals appears in Fig. 6 at right. The minerals at the top of the diagram (highest temperatures) closely approximate the minerals actually found in the Allende CAI's.

In addition to CAI's, Allende contains another class of inclusions. These consist largely of the minerals olivine and

Fig. 5. Allende inclusions and chondrules as viewed in the microscope. A slice of the stone has been sawed and ground to a thinness of 30 microns, and illuminated from beneath the microscope stage. The field of view is 1.5 mm wide. Chondrules tend to be ovoid in shape and contain brightly colored minerals. CAI's are more irregular and less colorful. (These colors are not the minerals' true colors, but are caused by introducing polarizing filters in the optical train.)

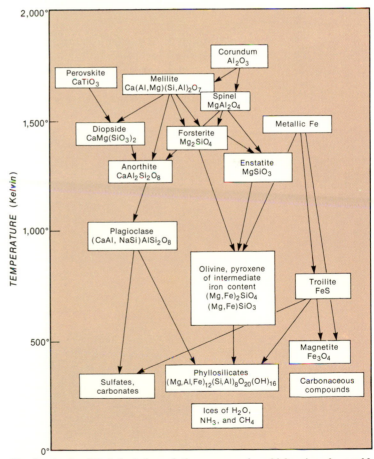

Fig. 6. A simplified depiction of the sequence in which minerals would condense from a cooling gas of solar composition. Arrows signify that continuing reaction with cooling residual gases would transform minerals from the upper boxes into those minerals beneath them. Rapid cooling would prevent complete transformations, accounting for the preservation of the spinel, melilite, perovskite, and so on in Allende CAI's. Figure from J. A. Wood, *The Solar System*, Prentice-Hall, 1979.

pyroxene, which appear lower in the condensation sequence than the CAI minerals. Their principal elements are magnesium, iron, and silicon; their compositions much more nearly reflect elemental abundances in Allende overall (and in the solar atmosphere) than CAI compositions do. Named chondrules, they are actually much more abundant in Allende than CAI's, since Mg, Fe, and Si occur in greater concentrations in the meteorite than do Ca, Al, and Ti. The textures of many chondrules show that they have been melted; they tend to be more regular in shape than CAI's, and some are nearly spherical. A widely held view is that they are frozen droplets of planetary material that were melted and splashed about by high-velocity impacts on accreting parent meteorite planets.

Now the plot thickens. A major input to the problem of understanding meteorites was made by Robert N. Clayton and co-workers at the University of Chicago in the early 1970's, by determining the isotopic composition of oxygen in both meteorites and components of meteorites. There are three isotopes of oxygen: ^{16}O (99.756 percent of oxygen atoms on Earth), ^{17}O (0.039 percent), and ^{18}O (0.205 percent). These are stable isotopes of ordinary oxygen — they have nothing to do with radioactive decay; they are what the solar system was given to work with when it came together. Proportions of these isotopes found by Clayton and co-workers in a number of planetary and meteoritic samples are shown in Fig. 8.

The oxygen isotopes tend not to enter different minerals or phases in exactly the same proportions. The vibrational energy of an oxygen atom in a crystal depends on its mass; that it, upon which isotope it is. A given mineral site slightly prefers a particular oxygen isotope because it has the lowest vibrational energy at that position. This leads to a dissimilar partitioning of oxygen isotopes between, for example, a mineral condensing from a vapor and the oxygen remaining in the vapor, or a mineral crystallizing from a melt and the remaining melt. The details of the partitioning are very complicated. There is one simple regularity, however. Wherever partitioning introduces a bias in the $^{17}O/^{16}O$ ratio (where a difference of one mass unit is involved) between two phases, it should introduce twice as large a bias in the $^{18}O/^{16}O$ ratio (a difference of two mass units). This effect is nicely demonstrated by the upper curve in Fig. 8. One expects that all terrestrial samples, having partitioned oxygen isotopes among one another in circumstances similar to those just sketched, would plot along a line where the $^{18}O/^{16}O$ variation is just twice that of $^{17}O/^{16}O$, and they do. The same trend should hold for lunar samples, though it could not have been predicted that the Earth and Moon curves would coincide.

The remarkable thing about the oxygen isotope plot is the lower curve, formed by data from CAI's and chondrules and minerals separated from Allende and other similar meteorites. Here the $^{18}O/^{16}O$ variation is not twice that of $^{17}O/^{16}O$ but almost exactly equal to it (the curve has a 1:1 rather than a 1:2 slope). This curve cannot be explained by equilibrium partitioning between various possible oxygen sites. The only explanation seems to be that it results from a mixing, in various proportions, of two fundamentally different oxygens, each with its own isotopic composition. One of these oxygen end-members seems to plot on or near the prosaic Earth-Moon partitioning line. The other is way to the lower left. It could be pure ^{16}O, which is what the mixing line points to. Or it may merely be

Fig. 7. Schematic cross-section of the hypothetical solar nebula, in which interstellar dust grains are thought to have been thermally processed to produce CAI's. At an earlier time, this material was a collapsing interstellar cloud; later, it spawned the planets before finally dissipating as gaseous residue.

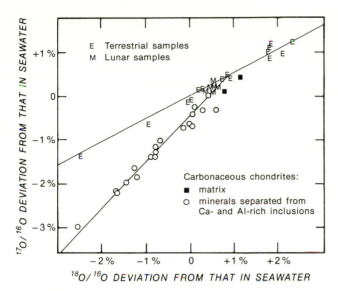

Fig. 8. Isotopic compositions of oxygen in terrestrial, lunar, and meteoritic materials. The lowermost curve, pertaining to carbonaceous chondrite components, appears to require mixing of two oxygen components which predate the solar system (see text). After R. N. Clayton *et al.*, *Science 182*, 1973, 485-488.

oxygen enriched in ^{16}O from its terrestrial content of 99.756 percent to something over 99.766 percent.

What can we conclude from this? First, the early solar system (presumably the nebula) contained two discrete components of some sort, each with oxygen of a characteristic isotopic composition. We understand fairly well by now that the various isotopes of the chemical elements were created in a variety of astrophysical settings: some are relics of the "Big Bang"; some are the products of fusion reactions in stellar interiors; some were created in supernova explosions; ordinary novas and the effect of cosmic rays on interstellar dust may have generated other isotopes; however, much is still uncertain. It is not difficult to picture several components of interstellar gas and dust contributing to the infant solar system, each of which consisted of atoms from a different mixture of astrophysical sources, and so each of which contained elements having different isotopic compositions. This would account for the two different oxygens in the primordial solar system.

However, this concept also predicts that the other chemical elements should display similar isotopic variability, and this is *not* observed. With a few exceptions, elements other than oxygen in the meteorites and meteorite components are remarkably invariant in their isotopic compositions. Apparently the two oxygen components of the early solar system were not associated with two different batches of interstellar gas and dust that never mixed completely — such a situation would have left isotopic anomalies in many of the elements. Instead we can draw a second conclusion, that the two components must be identifiable with the *gas* and the *dust* that came into the solar system; each of these components must have been reasonably well mixed.

This makes a difference because of a special circumstance connected with oxygen: it is the only element that would have been abundant in both the gas *and* the dust of the nebula. All the other elements were concentrated in one component or the other: H, He, C, N, etc. in the gas; Si,

Mg, Fe, S, Al, and so on in the dust. But about one-seventh of the solar system's oxygen would have been present as oxides in the dust, and the remainder would have been in the gas phase as water and carbon monoxide molecules. If the gas contained oxygen that was close to terrestrial oxygen in isotopic makeup and the dust was enriched in ^{16}O (a third tentative conclusion), then during high-energy events in the nebula that transformed the dust into CAI's, some incomplete degree of exchange of oxygen between the two components probably would have occurred, leading to the mixtures in various proportions of the two oxygen endmembers that are observed in CAI's. The other elements, concentrated either in the gas or dust, would have remained isotopically homogeneous.

What were these high-energy events in the nebula? Nobody knows, is the answer. However, one thing can be said (conclusion number four): apparently the nebula didn't get hot enough to vaporize and then recondense the dust wholesale. If this had happened, the two oxygens would have mingled and their isotopic compositions would have been homogenized. Fortunately, this conclusion agrees with a more recent reevaluation of the solar nebula model by Cameron, in which he concludes that compression of the nebula would not have heated the gases enough to vaporize dust particles after all. It appears instead that the nebula was relatively cold, but local high-energy events of some sort (still unspecified) must have heated dust aggregations enough to distill off their more volatile elements, leaving Ca and Al concentrated in residual masses (CAI's). At the same time partial exchange of oxygen between CAI's and the surrounding gas occurred.

The oxygen situation is exciting, because it confirms that we can find in the meteorites not just clues about events and processes in the nebula at the time when the planets were being formed, but even a faint record of the solar system's sources of raw material. The main thrust of research activity in meteoritics currently lies in this area of pre-solar system events and processes. One of the major advances on

Fig. 9. Meteorite collecting in Antarctica, during the 1978-79 season. Here the specimen is a tiny fragment of carbonaceous chondrite. Kazuyuki Suraishi of the National Institute of Polar Research, Tokyo, photographs the meteorite before moving it; Ursula Marvin of the Harvard-Smithsonian Center for Astrophysics holds a scale and numbering device that will appear in the photograph for identification. This photograph is by W. A. Cassidy.

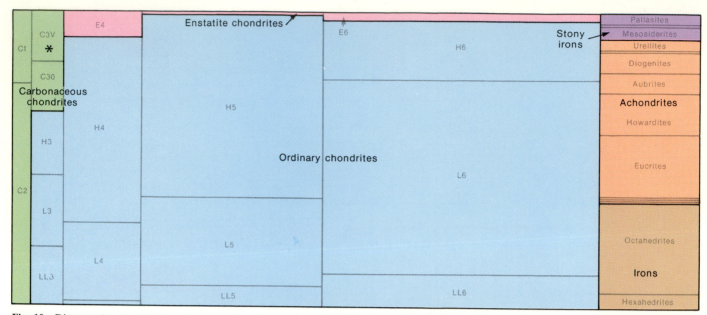

Fig. 10. Diagram showing the six major categories of meteorites. Within each category, the breakdown of subtypes is suggested. The box areas are proportional to abundances of meteorite types among observed falls. (An asterisk marks the position of Allende in the scheme.) "Ordinary chondrites" contain major elements in solar proportions (like Allende), but have been thermally metamorphosed to varying degrees. The degree of such metamorphism increases to the right.

this front was made by G. J. Wasserburg and co-workers of the California Institute of Technology in 1976, when they found unequivocal evidence of the former presence of ^{26}Al in Allende CAI's. This isotope has a very short half-life, only 720,000 years, toward its decay into ^{26}Mg. For any detectable amount of it to have been "alive" in Allende inclusions requires that it was created immediately before or during the formation of the solar system, and promptly mingled with the solar system's raw materials. It seems inescapable that a supernova (which is capable of creating ^{26}Al, among other things) occurred near enough to the nascent solar system in space and time to contribute important amounts of freshly synthesized nuclides to it. An unlikely coincidence? Cameron and another theorist, James W. Truran, have argued that it was not a coincidence at all: the supernova that contaminated a cloud of interstellar gas and dust with ^{26}Al also sent a shock wave through it which, by compressing the gas, initiated the formation of a group of stars including our own Sun.

METEORITES GENERALLY

So far this discussion has focused on a particular meteorite, Allende. Of course Allende is only one of many, albeit an uncommonly interesting and provocative meteorite. Meteorites fall to Earth all the time. Most are so small that they are consumed by ablation in the atmosphere, or if one survives, it is such a diminutive pebble as to have no chance of being found on the ground. (A rare exception is the Ras Tanura chondrite, only six grams, which dropped on the end of a petroleum-loading dock in Saudi Arabia in 1961.) But a half-dozen or so observed falls are recovered each year on the world's land areas, and another dozen or two meteorites are found that were not observed in their falls to Earth.

A recently discovered bountiful source of "finds" is Antarctica. Over the years, glacial flow carries

Fig. 11. Two schematic cross-sections suggest the evolution of interiors of meteorites' "parent" planets. These accrete (*left*) as chondritic mixtures of metal and sulfide grains (black), minerals of relatively low melting temperature (yellow), and abundant high-melting Mg and Fe silicates (green). Internal heating of the planets causes them to melt partially (*right*). Molten metal and sulfide drain to the centers, forming cores (the source of iron meteorites?). Silicates that melt at low temperatures produce a low-density magma, which tends to erupt to planetary surfaces (the source of Ca-rich achondrites?).

Fig. 12. This fragment of the Nakhla, Egypt, achondritic meteorite (an igneous rock), consists entirely of silicate minerals; no metal alloys are present. Nakhla is the only meteorite known to have struck and killed a living creature — a dog — when it fell.

Fig. 13. Polished slice of the Pavlodar, Siberia, pallasite, a stony-iron meteorite. This consists of a close-packed accumulation of rounded crystals of olivine, with metal alloys filling the spaces between them. The metal is similar in composition and structure to that in iron meteorites.

Fig. 14. A slab of the Edmonton, Kentucky, iron meteorite has been sawed, ground, polished, and chemically etched to reveal the *Widmanstätten structure*, which consists of intersecting plates of kamacite alloy with taenite in the interstices. The irregular inclusion toward the left is an iron-nickel phosphide mineral.

meteorites toward the margins of the continent, and in certain critical areas they tend to accumulate in remarkable numbers. Fig. 9 shows two researchers collecting them in such an area. Natural history museums have been collecting meteorites since the early nineteenth century, when their extraterrestrial origin was first appreciated. By now roughly three thousand discrete meteorites have been catalogued. (A meteorite shower, like Allende, counts as one meteorite in this reckoning.) Meteorites also fall on the Moon, of course. Lunar soil samples contain a component of smashed-up meteorite material amounting to as much as 4 percent by weight.

There are diverse varieties of meteorites (Fig. 10). The traditional breakdown is into stones, irons, and stony-irons. A more meaningful division, however, would be into *undifferentiated meteorites* (chondrites) and *differentiated meteorites*. The most abundant chemical elements are present in all chondritic meteorites, as they are in Allende, in the same proportions as they are in the solar atmosphere. The significance of this is that the chondrites must be samples of primitive planetary material that have never been melted, since rock has an irresistible tendency to separate (differentiate) into layers or volumes of more specialized composition if it ever melts (see Fig. 11). The subtypes of chondrites vary in their content of minor and trace elements, particularly elements that are more or less volatile. This has been interpreted to mean they are samples of material that accreted at various temperatures in the nebula, and some materials accreted at temperatures sufficiently high that the most volatile elements were still in the gas and did not join the accreting particles. (The depletion of volatile elements correlates with the degree of planetary metamorphism of chondrites, however, a topic touched on in the next section. A minority view among meteoriticists is that the volatile elements were "cooked out" of the parent meteorite planets during high-temperature metamorphism, not excluded during accretion.)

Differentiated meteorites apparently *are* the products of melting and separation of more primitive planetary material. Meteorites in this major category include irons, stony-irons, and some stones (the achondrites). Three different samples are seen above in Figs. 12, 13, and 14.

Where do meteorites come from? That is, where do they reside between the nebular and Antarctic stages of their careers? Most and very probably all of them are fragments of asteroids, though some workers have made a case for an association with comet nuclei. Trajectories of several meteorites in the atmosphere have been photographed with sufficient precision to allow the meteorites' orbits in space to be calculated, and these closely resemble the orbits of Apollo (Earth-crossing) asteroids, shown in Fig. 15. A problem still remains in understanding where Apollo asteroids come from and how they are related to belt asteroids, but the correspondence between meteorites and asteroids generally is well established. The spectra of telescopic images of sunlight reflected from asteroid surfaces are very similar to laboratory-derived reflection spectra of pulverized meteorite specimens (Fig. 16). In both cases the characteristic shape of the spectrum is caused by the selective absorption of light at particular wavelengths by the minerals present in the reflecting material.

THE PLANETARY HISTORY OF METEORITES

There is abundant evidence that the meteorites have not always wandered through interplanetary space in the form of rocks as small as they are now, but were once parts of planetary bodies. Presumably impacts among the bodies broke them up and released fragments, a few of which have been captured by Earth as meteorites (technically, the word *meteorite* is applied to an object only after it has fallen to Earth). The evidence is as follows:

First, there is the apparent correspondence between

meteorites and asteroids, many of which are hundreds of kilometers in dimension, and so qualify as small planets. Many or perhaps all of the smaller asteroids may be fragments of once-larger objects that collided.

Second, cosmic rays have the effect of generating certain isotopes (such as ^{3}He, ^{20}Ne, ^{38}Ar) in meteorites while they are at large in space as small rocks, but not when they are buried more than a few meters deep inside larger objects, where they are largely shielded from cosmic radiation. Measurement of the abundances of cosmic-ray-generated nuclides in meteorite specimens shows that they were shielded from cosmic radiation during most of the life of the solar system, typically until a few tens of millions of years ago.

Third, the great majority of meteorites show evidence of having been affected by high temperatures for relatively long periods of time. Some have been melted and chemically differentiated, as already noted. A great many more (most of the chondrites) have not been melted, but were *metamorphosed,* that is, held at temperatures of approximately 1,300° K for long enough (many years) to make their silicate textures coarsen and become more granular, and to cause the compositions of their silicate minerals to become more uniform (by solid-state diffusion of elements from one mineral grain to another). We do not have certain

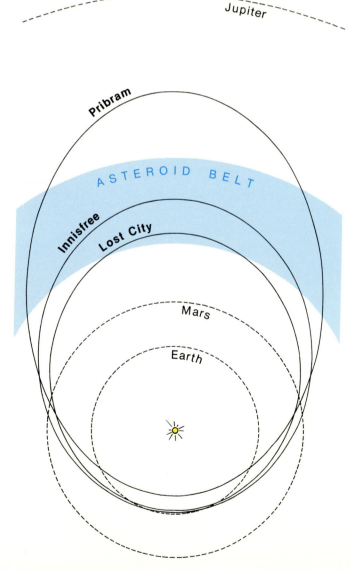

Fig. 15a. Pre-impact orbits of three meteorites that were photographed during atmospheric entry, plotted relative to the asteroid belt.

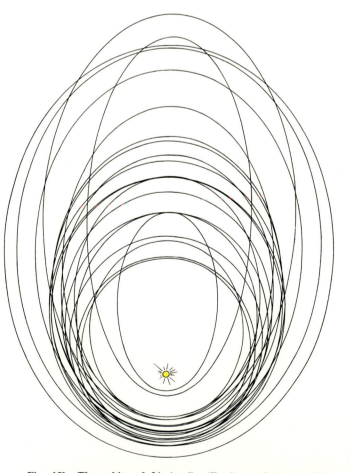

Fig. 15b. The orbits of 21 Apollo (Earth-crossing) asteroids. Perihelia of the orbits have been aligned for these plots, and the orbits themselves depressed to the plane of the paper. Inclinations in most cases vary from 0-20°. "Belt" asteroids, which have relatively small eccentricities, do not cross the Earth's orbit.

knowledge of the source of this heat, but planetary interiors are better able to retain heat, once generated, than are small dispersed rocks.

The decay of the radioactive elements uranium and thorium, and of potassium's radioactive isotope ^{40}K, is an important source of heat in large planets. But asteroids are not large enough to retain heat generated as slowly as these isotopes decay. The heat is conducted to their surfaces and radiated away into space almost as fast as it is deposited. The most likely bet for asteroids is that the early solar system contained a radioactive nuclide with a much shorter half-life than the isotopes of U, Th, and K. This would have delivered its heat in a pulse short enough to have heated small planets substantially, and would have decayed to undetectable levels by today. The ^{26}Al found in Allende CAI's by Wasserburg and co-workers may very well have done the job. If all the aluminum in the early solar system contained proportionately as much ^{26}Al as the aluminum of these CAI's did, its decay would have released enough energy to melt the centers of objects as small as 20 km in diameter.

There is a way of gauging the rate at which meteoritic material cooled from the high temperatures it experienced. Most meteorites contain metallic nickel and iron in varying (and in some cases dominant) amounts. These metals form two alloys, kamacite and taenite, the proportions and compositions of which vary with temperature if the alloys are allowed to react with one another to equilibrium (see Fig. 14). The degree to which they *do* continue reacting and equilibrating in a cooling meteoritic system depends on how fast the system cools. If it cools rapidly, alloy compositions appropriate to high temperatures are frozen in. If it cools more slowly, the alloys take up lower-temperature configurations before the diffusive motions of Ni and Fe are frozen in. There is substantial uncertainty in the method, but the cooling rates it points to (typically about 10°-20° per million years) are those that would be experienced 30-50 km deep in a small body after its heat source was "turned off" and its cooling was restrained only by the insulating effect of the rock above it.

Granted that the meteorites are fragments of planets, is it possible that they and the asteroids all came from various sites in a single large planet, the missing fifth planet of Bode's law, which once orbited between Mars and Jupiter but was somehow demolished? This concept is astonishingly popular and durable everywhere except among meteorite and asteroid researchers. There are a number of difficulties. First, we might note that the depths of burial indicated by metal-alloy cooling rates (just discussed) are not all that great. Second, rock in the interior of largish planets is subjected to high pressures, and high pressures produce certain characteristic minerals such as spinel and garnet which are not found in the meteorites that appear to have equilibrated in planetary interiors.

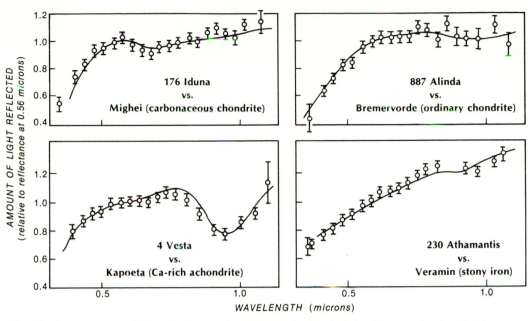

Fig. 16. A comparison of the reflection spectra of four asteroids (points with error bars) with laboratory-determined spectra of meteorites representing four major subtypes (continuous curves). After C. R. Chapman, *Geochim. Cosmochim. Acta, 40,* 1976, 701-719.

Third, there are important chemical differences among the various meteorite subclasses which make it awkward to require that they coexisted in a single body. Meteorites are much easier to understand if they came from one or two dozen parent bodies, each of which was an isolated chemical system.

Fourth, it is energetically difficult to break up a large planet. Certainly it is energetically impossible to spall fragments off the surface of a sizeable object without smashing them to smithereens. This is nicely demonstrated by the Moon. The Moon is constantly being hit by high-velocity projectiles, and fragments of lunar debris must often be knocked away at greater than the lunar escape velocity, but does any of this material ever come to Earth in a collectible and recognizable form? The answer, now that we know what lunar rock looks like, is no. Acceleration to 1.7 km per second (the lunar escape velocity) must stress lunar debris so greatly as to pulverize or melt and scatter it in particles too small to survive as meteorites.

ASTEROID REGOLITHS

Many stony meteorites, if broken or sawed so an interior section can be examined, are found to consist not of integral rocky material but of a collection of angular fragments, tightly welded together. This involves a different and coarser structure than the conglomeration of inclusions and matrix that comprises Allende, described earlier. The angular fragments can be centimeters in size, and if the meteorite in question is a chondrite, each fragment will itself contain numerous chondrules and inclusions. It appears that the chondritic system went through discrete episodes of (1) assembly of chondrules, inclusions, and matrix; (2) lithification (welding together) of this conglomeration into a hard mass; (3) shattering of the chondritic rock into fragments; (4) lithification of the fragmental debris into a new hard mass; and (5) another shattering event that broke free a piece of the rock and left it in an orbit that eventually intersected Earth's.

Such aggregations of rock formations are called *breccias* (Fig. 17), an Italian word. It is not hard to imagine how the breaking-up occurred. The asteroid belt is a crowded place, and collisions take place with calculable frequency. At least for the present distribution of asteroid orbits, the mean collision velocity among asteroids is about 5 km per second, more than adequate to smash hard rock.

Fig. 17. This fragment of the Cumberland Falls, Kentucky, meteorite is a breccia: a welded mass of angular fragments that aggregated long ago and far away. The internal components are derived from diverse types of meteorites: achondrites (white) and enstatite chondrites (black).

Our instincts tell us that the debris from such collisions would scatter into space in all directions and have no chance of being collected and lithified into a breccia. After all, asteroids are so small that their feeble gravity fields would have little chance of holding onto the flying debris fragments from a high-energy impact. Recent consideration of the asteroid-impact problem has shown that this is not entirely true, however, as explained in Clark Chapman's chapter on asteroids. Impacts produce fragments with a wide spectrum of velocities, ranging from very high (comparable to the impact velocity) down to almost zero. Though most of the debris may have velocities in excess of the tiny escape velocity of the asteroid, and be lost from it, some fraction of the fragments will have been moved at less than the escape velocity. These fall back to the surface and join a layer of residual, unconsolidated debris.

The Moon is covered with such a layer of debris or "soil," which we term the *regolith*, and the Apollo program taught us a great deal about the development and properties of regoliths. One point is that a regolith develops during a succession of events (impacts), which progressively break up and sometimes melt the regolith particles while moving them to different levels in the regolith profile ("gardening"); this process is depicted in Fig. 18. Thus a particular particle may spend millions of years buried deeply enough in the regolith to be largely shielded from cosmic rays and protected from the other rigors of the space environment. Then an impact may redeposit it right on the surface of the Moon, where it is bombarded by solar-wind ions and micrometeorites as well as cosmic rays. Then an ejecta blanket may bury it again.

Apparently the same thing happens in asteroidal regoliths, except the mean residency time of a regolith particle (before being ejected at greater than the escape velocity) and the mean steady-state thickness of the regolith layer are smaller on an asteroid than on the Moon. In any case, many of the meteoritic breccias have all the earmarks of

regolith samples; nuclides implanted in grain surfaces by an ancient solar wind, cosmic-ray generated nuclides, and melting effects attributable to micrometeorite impacts.

Many breccias consist of fragments all of the same meteoritic material, but in some cases a rich variety of meteorite types are present, occasionally types completely outside our previous experience. Evidently a complex history of impacts, with mingling of debris from both projectiles and target, gave rise to these.

The process that tends to relithify (or weld together) loose meteorite fragments into a breccia is the same type of event that broke them up in the first place: an impact. Beneath an impact crater, particles and fragments are jammed against one another at momentarily tremendous pressures. The stresses may cause some melting along fragment surfaces, and this liquid acts as a glue when it resolidifies. Breccias that are not "well-glued" may exist in space, but they would be too weak to survive passage through the Earth's atmosphere and so are not represented in our collections.

This discussion has suggested some of the things meteorites are telling us about the dramatic events that shaped the solar system. Meteorites contain a record of these events, but it is not written in plain language. Their message is not unlike the plaque showing a nude man and woman and a diagram of the planetary system carried by Pioneer 10 into interstellar space. Vital information is there, but it is susceptible to being misinterpreted in several ways or simply overlooked.

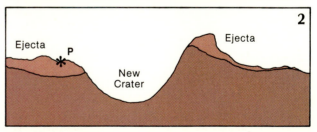

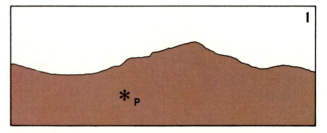

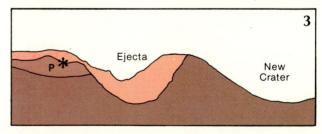

Fig. 18. Three cross-sectional sketches of a planetary regolith, evolving with time, suggest how a particular pebble or grain (*p*) moves up and down in the regolith profile.

Small Bodies and Their Origins

William K. Hartmann

Until about a century ago, the solar system seemed like an orderly place, with just a few well-defined classes of inhabitants. The planets were the large objects going around the Sun. The satellites were all intermediate-sized objects going in direct orbits around the planets (traveling counterclockwise as seen from above the solar system). By 1840, thirteen such satellites had been discovered, all sizeable and all moving in direct orbits. The solar system of the 1800's was also populated by asteroids, which seemed to be an entirely different class of objects, all small, and all neatly grouped in the planetary niche that had been predicted by "Bode's rule" between Mars and Jupiter. The fourth class of objects, the comets, were again "obviously" different, with their fuzzy comas and enormous diffuse tails.

SMALL BODIES: CONFUSION AMONG TYPES

Some problems soon appeared. In 1846, Neptune's large satellite, Triton, was discovered, and it was in a retrograde (clockwise) orbit. This seemed perhaps only an exception that proved the rule, since the next six satellites discovered were all again in direct orbits. This group, however, included asteroid-sized satellites, such as Phobos and Deimos (discovered circling Mars in 1877) and Amalthea, Jupiter's fifth satellite (discovered in 1892).

The tidy state of affairs began to break down even more dramatically around 1900, and planetary astronomers began to recognize that the solar system was quite different from what they had thought. In 1898, the small, retrograde, outermost satellite of Saturn, Phoebe, was discovered. Thus by 1900 the list of known satellites included 21 objects, most of which were large and moving prograde; but two were retrograde and two others were far below 100 km in diameter. By 1960, nine more satellites had been discovered, of which four were retrograde and six were less than 100 km across. Thus satellites became an unholy mixture of planet-sized, asteroid-sized, prograde, and retrograde bodies.

To make matters worse, asteroids began popping up in unexpected parts of the solar system. In 1898, a Berlin astronomer discovered asteroid 433 Eros, whose perihelion lies inside the orbit of Mars. Asteroid 624 Hektor, discovered in 1907, turned out to be moving in the same orbit as Jupiter but 60° ahead of the planet. Numerous other asteroids were subsequently discovered in Jupiter's orbit, 60°

ahead of and 60° behind the planet. These are called Trojan asteroids (named after Trojan war figures). In 1932, asteroid 1862 Apollo was found passing inside the Earth's orbit. Others of this type came to be called Apollo asteroids. Now it was clear that many small bodies moved through interplanetary space in different parts of the solar system.

The neat distinction between asteroids and comets also began to break down when observers realized that as comets recede from the Sun, they eventually lose their tails and become indistinguishable from asteroids when viewed with telescopes. Furthermore, dynamical studies during the 1950's showed that comets could be thrown by Jupiter's gravity into orbits indistinguishable from those of many asteroids, such as the Apollos. This process is called gravitational scattering. Such findings introduced confusion: How do we know an asteroid is "really" an asteroid and not an inactive comet?

Fig. 1. The unusual object Chiron, catalogued as asteroid 2060, crosses the orbit of Saturn and on rare occasions may pass close to the planet. Such close encounters have altered the orbits of many small interplanetary bodies over time. In this imaginary view by the author, Chiron passes directly over Saturn's pole; recent calculations suggest that Chiron may have passed through Saturn's satellite system within the last hundred thousand years.

Fig. 2. In this Voyager 2 photograph taken on July 8, 1979, the bright streak at right is a small Jovian satellite, 1979 J1. Jupiter's ring appears as a light gray smear running across the image, and the other streak is that of an 8.3-magnitude star. Orbiting only 129,000 km from Jupiter's center, the satellite skirts the outer edge of the planet's ring and may be involved with maintaining it.

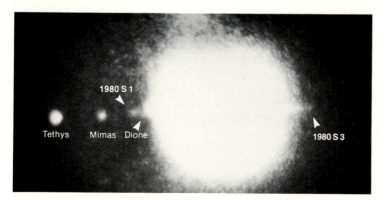

Fig. 3. Using an electronographic camera attached to one of the telescopes at Pic du Midi Observatory in France, R. Despiau, P. Lacques, and J. Lecacheux photographed two tiny satellites just beyond the glare of Saturn. When this photograph was taken, on March 1, 1980, Saturn's rings appeared edge-on to terrestrial observers and thus are not prominent here. Three larger satellites are identified.

Astronomers argued about the differences between the two groups. Asteroids, they decided, inhabit the inner solar system (out to the orbit of Jupiter) while comets drop in for visits from the outermost solar system. By the 1970's, spectroscopic observations indicated that all "true" asteroids have surfaces of stony material, whereas comets emit gases that must come from ices subliming either on their surfaces or in their interiors. Nonetheless, some asteroids with stony-looking surfaces could contain ices, and comets could contain rocky material. The surface of Ceres, for example, may contain ice, as well as minerals with chemically bound water, as L. Lebofsky and his co-workers have inferred from spectroscopic data obtained by them. And several observers have noted similarities in the colors of cometary dust and carbonaceous silicate minerals contained in certain reddish black (or *D-type;* see Chapter 10) asteroids of the outer solar system. Such data are blurring the distinction between comets and remote asteroids; both may contain icy and stony material.

Orbital distinctions between comets and remote asteroids were blurred as well when Charles Kowal discovered the enigmatic object 2060 Chiron moving in a cometlike orbit between Saturn and Uranus (Fig. 1). Several hundred kilometers across, it is larger than any known comet, though cometary diameters are imprecisely known. Recent observations by Dale Cruikshank, J. Degewij, and myself have shown that this object exhibits a color similar to dark-colored, carbonaceous asteroids rather than to a bright icy surface like Europa's.

In addition to comets and asteroids, a third class of solar system body is worth considering: the small, outer satellites of Jupiter and Saturn. American astronomer Gerard P. Kuiper, who pioneered modern astronomical studies of the solar system, pointed out as early as the 1950's that these seemed to be distinct from the large, prograde, inner satellites. Saturn's outermost satellite Phoebe and Jupiter's sixth through thirteenth moons are relatively small, have especially inclined and eccentric orbits, and in many cases move in a retrograde direction. Kuiper suggested these moons were captured interplanetary wanderers. Perhaps they originated in the interplanetary assortment of comets and asteroids.

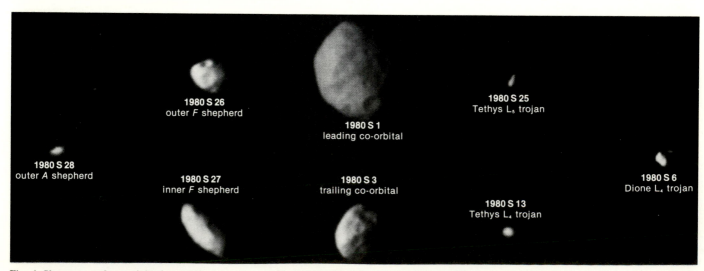

Fig. 4. Shown to scale are eight tiny satellites known to orbit Saturn. All eight either "share" orbits or "guard" rings.

By the 1960's and 1970's, other theorists like Hannes Alfvén and Gustaf Arrhenius emphasized that the planet-satellite systems were miniature versions of the Sun-planet system. Each has a few big, prograde bodies and a host of small bodies in mixed orbits. In studying the origin of the solar system, they argued that it no longer had to be viewed as a single unique case. Instead, we "aim at a general theory of the secondary bodies around a primary body," according to Alfvén and Arrhenius.

If we look at the solar system and the major satellite systems in this way, we begin to sense certain patterns that shed light on the relationships among small bodies. All these systems have a complement of large bodies. In Jupiter's case they are the Galilean satellites; in Saturn's they include giant Titan and several other intermediate-size moons (Fig. 5). Left over are the small bodies: the asteroids in the mid-solar system, the comets that lie mostly in the outer solar system, and certain groups of moons.

For example, just as the Trojan asteroids are trapped in the so-called Lagrangian points 60° behind and ahead of Jupiter in its orbit, small moons move in the corresponding Lagrangian points of Saturn's satellites Tethys and Dione, as was discovered in 1980. (Dione, however, apparently has only a leading Lagrangian companion). The small moonlets (Figs. 2-4, 9, and 10) that orbit close to Jupiter and Saturn could be called counterparts of diminutive Mercury. And just as small, irregular satellites populate the outskirts of the Jovian and Saturnian systems, the small "double planet" Pluto and Charon, Chiron, and the Oort swarm of comets reside in the outer solar system. In both the solar system and the mini-systems of the giant planets, one gets the impres-

sion that while major worlds grew in the mid-regions, the inner and outer regions became the domain of numerous small bodies.

These comparisons become more meaningful when we abandon our category-laden view of the solar system. Scientists are taught to classify things as a first step toward understanding them, but old classifications can introduce misleading "distinctions" — as may be the case with the comets, asteroids, and certain moons.

LOOKING FOR PATTERNS AMONG THE SMALL BODIES

At first glance, this new view of the solar system seems confused. Yet such confusion is common to any new stage of scientific exploration. Out of the confusion we are beginning to make new sense of the patterns observed in the solar system.

We used to look at the solar system in a certain way which I call "the nine-planets gestalt." In this view of things, the planets were regarded as the nine most important worlds. All other bodies were looked on as less important and less interesting. Planets, satellites, comets, and asteroids were regarded as obviously hierarchical categories.

Now we are experiencing the breakdown of "the nine-planets gestalt." We are looking at a system of worlds in which 25 worlds exceed 1,000 km in diameter. Today we know that some satellites are larger and more geologically active than some so-called "planets." And we are finding more of a continuum among the smaller bodies, rather than distinct categories.

To make things clearer, we can abolish the concept of putting comets, asteroids, and small moons in different classes. Let us go back to the beginning. Meteorite studies show that when the planets formed, the solar system was crowded with small bodies ranging from hundreds of kilometers across down to lesser worlds a kilometer or less in size, called *planetesimals*. An understanding of small bodies observed today in the solar system must be based on a determination of the compositions and destinies of the original planetesimals. As discussed more in John Lewis' chapter, different kinds of minerals condensed from nebular gas and formed various kinds of planetesimals at increasing distances from the Sun. In the inner solar system, the gas never got cold, and planetesimals formed from "high-temperature" minerals such as nickel-iron flecks and silicate rock grains. Earth and the terrestrial planets evolved from these minerals.

In the outer part of the asteroid belt, the temperature was lower, and certain additional minerals condensed. These include carbon-rich and water-rich compounds such as those found in the black, crumbly stone meteorites known as carbonaceous chondrites. The carbonaceous meteorites show evidence that liquid water trickled through part of their interiors as (or after) they formed, confirming that they originated in a water-rich or ice-rich part of the solar system. From the outer asteroid belt outward, low temperatures occurred and frozen water was very common in the early planetary material. Abundant hydrogen and oxygen there gave rise to huge amounts of water, making bodies from Jupiter's region and outward rich in water ice. At still colder temperatures, in the region of Uranus and beyond, methane ice could form. These ideas explain many obvious features of

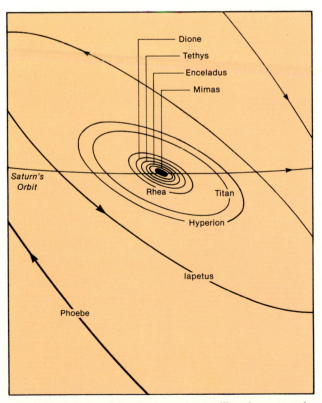

Fig. 5. The orbits of Saturn's nine major satellites shown to scale. Adapted from *The Atlas of the Universe* by Patrick Moore, Rand McNally and Company, 1970.

the solar system, such as the fact that the outer asteroid belt is rich in asteroids with "C-type" surfaces which spectroscopically resemble carbonaceous chondrites; the fact that the many satellites of Jupiter, Saturn, and Uranus have surfaces of water ice; and the fact that the comets which inhabit the outermost solar system are also made of water and other ices.

With this background, we can begin to make sense of the relationship between asteroids, small satellites, and perhaps even comets. Most planetesimals were fated to collide with planets (Fig. 6) or to get thrown out of the solar system, due to the gravitational scattering by the planets. Asteroids in the main belt are leftover rocky planetesimals of the inner-to-middle solar system, mostly formed between Mars and Jupiter and trapped there. Many icy planetesimals, in the outer solar system, had close encounters with giant planets and were thrown not quite free of the solar system, but into long, elliptical orbits. Occasionally they fall back into the inner system and are seen as comets.

Meanwhile, Jupiter's gravity perturbed many of the rocky planetesimals either inward or outward from the asteroid belt. Both the cometary and asteroidal objects may have their orbits further altered by additional close encounters with various planets. This explains why active comets, "burned-out" comets, and rocky asteroids can all end up with similar types of orbits.

One of the most remarkable recent findings about such objects is that certain ones in known cometary orbits may be temporarily captured into orbits around Jupiter and become outer satellites of Jupiter for periods ranging from months to decades or longer. This finding comes from computer projections of motions of such bodies backward or forward in time, revealing the occasional temporary captures. This may begin to explain the presence of bodies with irregular orbits on the outskirts of the Jovian and Saturnian systems.

We can also now see the asteroids in a new light. It is no longer surprising when asteroidlike objects appear outside the main belt, because planetesimals formed everywhere in the early solar system. Belt asteroids were once viewed as a swarm of objects hopelessly mixed by gravitational scattering as well as by collisions among asteroids that fragmented many and produced strange surfaces: cross-sectional exposures of the fragmented bodies interiors. Today we see that, in spite of these processes, an underlying systematic compositional trend survives in the belt.

Imagine a voyage outward through the asteroid belt. As we enter the inner edge of the belt, about 70 percent of the asteroids we encounter have stony surfaces which are essentially silicate minerals, the type of material originally formed in the inner solar system. By the time we reach the middle belt, the fraction of "C-type" carbonaceous asteroids (which resemble carbonaceous chondrite meteorites) has increased to about two thirds. As we pass into the outer belt, the fraction of C-type objects is still increasing, probably exceeding 85 percent. The percentage of ordinary stony-type asteroids, the ones that resemble most meteorites that fall on Earth, has dropped to nearly zero in the outer belt.

In and beyond the outermost belt, the few percent of non-C-type objects are mostly asteroids with unusual surfaces often unclassified. In the 1970's, the Dutch-American astronomer Johan Degewij defined a subclass among asteroids in this distant region that are as dark as carbonaceous

chondrites but are somewhat redder in color. In 1980, Jonathan Gradie and Joseph Veverka suggested that these so-called "D" (dark) asteroids are essentially C-type objects in which the carbon materials are enriched in reddish organic compounds that condense at even lower temperatures than most C-type materials. (Note that organic compounds such as amino acids have been discovered in carbonaceous chondrite meteorites.) These outermost asteroids are truly dark, reflecting only 3-5 percent of the light striking them, because they have so much black carbon disseminated throughout their material. In short, we can say that in the extreme outer asteroid belt, nearly all planetesimals are near-black, stony objects — rich in volatile materials like water and carbon.

Continuing outward, we reach Jupiter's orbit and find that all of the Trojan asteroids studied so far are also dark stony objects, classified as either C-type or D-types. Cruikshank, Degewij, and other observers have discovered that several of the outermost satellites of Jupiter are also dark stony objects and are virtually indistinguishable from Trojan asteroids. These are strange satellites, in the outermost part of Jupiter's gravitational sphere of influence. They are divided into two groups, as follows. Satellites J6, J7, J10, and J13 are all in prograde orbits, at about 11.5 million km from Jupiter, with inclinations between 26° and 29°. On the other hand, J8, J9, J11, and J12 move in *retrograde* orbits at about 22 million km from Jupiter, with inclinations between 16° and 33°. Since both groups seem physically indistinguishable from Trojans, the suggestion is very strong that they are simply captured Trojan planetesimals. Some of the dark stony objects that were forming in this re-

Fig. 6. A part of the Moon unseen before the space age. The lunar far side was first photographed by the Soviet spacecraft Luna 3 and later by Lunar Orbiter, Zond, and Apollo missions. The smooth, dark area near the left edge is Mare Crisium, with Mare Marginis and Mare Smythii to its right and lower right, respectively. Most of the Moon's far side bears innumerable impact scars, the result of aeons of meteoritic bombardment. Many of these remain from the early period in solar system history when planetesimals passed frequently among the planets.

gion of the solar system approached the outskirts of Jupiter's gravitational field and were somehow slowed down and captured by Jupiter.

This agrees with the earlier theoretical evidence for capture, but the exact mechanism is highly uncertain. Several theories have been proposed and suggestions include a three-body interaction between the Sun, Jupiter, and the planetesimal, or the slowing of an object as it passed through a mass of extended atmosphere of proto-Jupiter. Perhaps only two Trojan planetesimals were captured, one at 11.5 million and the other at 22 million kilometers; and then each of the two underwent a collision that broke it into fragments. Alternatively, the capture mechanism may be highly selective, allowing objects to end up in only the 11.5-million-km prograde orbit or the 22.5-million-km retrograde orbit. The exact origin of Jupiter's outer, Trojanlike moons remains a mystery.

Saturn's outermost moon, Phoebe (Fig. 7a), offers strong support for the capture hypothesis. Dark Phoebe lies on the outskirts of Saturn's sphere of influence, and it is in a retrograde orbit some 26 million km across and inclined 30° to Saturn's equator. Other Saturnian moons are larger, prograde, more icy, and have orbital inclinations not exceeding 15°. Once again, the suggestion is that the outermost moon, Phoebe, is an example of a captured planetesimal with a dark stony surface formed in the outer solar system.

Mars' two small satellites, Phobos and Deimos (Fig. 7b), are also dark stony bodies with C-type surface materials. With diameters of 22 and 12 km, respectively, they look much like C-type asteroids from the pool of blackish carbonaceous (and icy?) objects that formed in the outer part of the belt. This supports an earlier conjecture by some scientists that Phobos and Deimos were asteroids captured by Mars, perhaps when it was surrounded by a massive primitive atmosphere.

Fig. 7a. Saturn's dark, outermost satellite Phoebe was photographed by Voyager 2, two million kilometers away. This low-resolution picture reveals the moon's spherical form and mottled surface, although no topographic details can be discerned.

Here we can see a striking pattern. Mars, Jupiter, and Saturn all have small outer satellites circling them that are dark-surfaced, per-

Fig. 7b. The two Martian moons Phobos (left) and Deimos are reproduced at about their correct relative sizes in these images recorded by the Viking 1 orbiter in 1977. Rugged Phobos is shown from the south, with the crater Hall (about 6 km across) near the terminator. Deimos, seen south up, exhibits dark craters up to 1.3 km in diameter. These bodies reflect only a few percent of the sunlight striking them and are much darker than many other asteroids in the inner asteroid belt beyond Mars. Phobos and Deimos have spectra resembling those of certain "C-type" asteroids common to the outer asteroid belt and to Jupiter's Trojan asteroids; such material is believed to have a composition similar to carbonaceous chondrite meteorites. This suggests that perhaps they were at one time nearer Jupiter, which perturbed them into orbits that allowed their capture by Mars — a theory that is inviting but remains unproven.

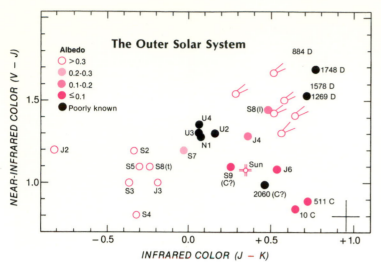

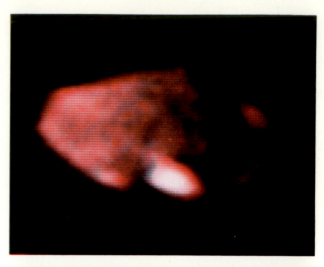

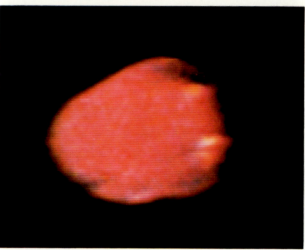

Fig. 8. This diagram characterizes small bodies associated with the outer solar system in terms of three parameters: near-infrared color (visual minus the near infrared "J" band at 1.25 microns), color further in the infrared, and albedo. All known bright icy surfaces lie in the lower-left corner; all known D-type carbonaceous asteroids plot at upper right. The diagram emphasizes the division of outer solar system objects into "black" and "white" surface types, and allows us to characterize certain remote objects that are too faint for detailed spectroscopic study. For example, 2060 Chiron has colors that suggest dark, carbonaceous-like surface materials, not clean ices. Six comets studied so far by this method have mean colors similar to D-type asteroids; these colors may be produced by a mixture of D-type dust and ice. Of special interest is the leftmost comet, Schwassmann-Wachmann 1; it ventures unusually far (6 AU) from the Sun, where it is often inactive and resembles a D asteroid. Asteroids are identified by number and type; J, S, U, and N indicate satellites of the giant outer planets. Cross (lower right) indicates typical error of measurement. Diagram adapted from data of D. Cruikshank, J. Degewij, and the author.

Fig. 9. Images of Amalthea obtained by Voyager 1 in 1979. Note the satellite's red color, believed to be the result of a coating of sulfur particles ejected from nearby Io. Amalthea's elongated shape, with the longest axis always pointed toward Jupiter, provided varying views for the spacecraft.

haps indistinguishable from carbonaceous-type Trojans and outer-belt asteroids, and are candidates for having become satellites by means of capture. If they really are bodies captured from a pool of dark planetesimals, how did the planetesimals become "capturable"? There are several possible mechanisms. Some planetesimals that formed near Jupiter may have had close encounters with Jupiter and undergone gravitational scattering, putting them into wide-ranging orbits that were frequently changed by later planetary encounters. Also, theoretical studies have revealed several ways by which Jupiter's gravitational force can kick asteroids out of various regions in the main asteroid belt and throw them onto orbits that approach other planets. Such effects explain how dark-surfaced C-type objects that formed near Jupiter may have been widely scattered in the early solar system, and why asteroids now in the inner solar system seem to be a mixture of the various stony and C-type objects, formed in different parts of the main asteroid belt.

So far we have encountered only dark objects whose surfaces are colored by stony or carbonaceous materials of various kinds. But chemical condensation theory (Chapter 20), supported by densities measured for satellites of Jupiter and Saturn, implies that from Jupiter's realm outward worlds were formed with a large percentage of ices in their interiors. Frozen water covers the satellites of Jupiter and Saturn, but recent spectroscopic observations of the frigid, outermost worlds such as Triton and Pluto indicate that frozen methane (CH_4) dominates in that distant region. Voyager pictures have made bright, ice-covered worlds familiar to us. Still, remarkably, *every* interplanetary worldlet beyond the asteroid belt studied so far has a dark, blackish surface. If the moons and interplanetary bodies of the outer solar system formed in similar cold regions from similar mixtures of ice and rock, how are we to explain the remarkable dichotomy (Fig. 8) between black worlds (mostly interplanetary bodies or outer moons) and white worlds (which tend to be inner satellites of larger size)?

SURFACE EVOLUTION OF SMALL SATELLITES AND RELATED BODIES

In an attempt to explain these surface types, let us take an imaginary trip into the inner gravitational sphere of influence in one of the satellite systems. We start outside this region among the Trojan asteroids of Jupiter. The largest one, Hektor, measures about 100 by 300 km and is reddish-black in color. The next largest ones are spheroidal and range downward in diameter from 190 km. These are probably too small to ever have been heated much by internal radioactivity or to have had volcanic eruptions on their surfaces.

If our theory is correct, these objects are likely to have formed with a substantial component of water ice as well as a component of carbon-rich silicate minerals. If they are

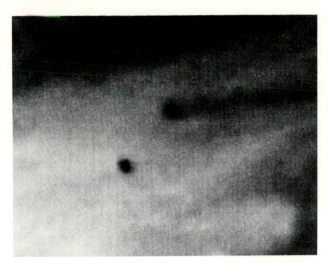

Fig. 10. Another recently discovered Jovian satellite, 1979 J 2, appears as a dark spot against Jupiter's clouds in this Voyager 1 image taken on March 4, 1979; its shadow is also visible, toward upper right at the end of a dark feature on the planet.

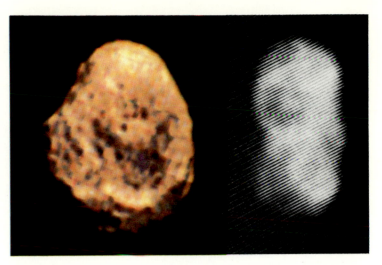

Fig. 11. Hyperion, Saturn's third-from-outermost satellite, seen face-on and side-on by Voyager 2. A major collision on this moon (measuring 205 by 130 by 100 km) may have blown away part of its surface, resulting in the observed hamburger shape.

rich in ice, why are their surfaces so dark? I have suggested elsewhere that slow sublimation and microcratering processes, having continually hammered away at the surfaces of these bodies, selectively vaporized the ice into water vapor which would escape, leaving the dark stony material concentrated in the surface layers.

As we now venture towards the gravitational sphere of influence of Jupiter, we encounter similar types of objects that were captured into satellite orbits. Passing inside these, we begin to encounter much larger satellites that were not captured but actually formed in orbit around Jupiter (see Chapter 14). Because we are at Jupiter's distance from the Sun, we can expect that these satellites also formed from a mixture of icy and stony material. Callisto, the outermost Galilean satellite, seems to confirm this. It has an ancient cratered surface that reflects 19 percent of the incident light, and its surface has been identified as a mixture of ice and dark stony material. As we travel inward, we encounter Ganymede, which has a surface that is clearly no longer primitive. Some ancient areas of Ganymede look almost like Callisto, being heavily cratered but dark. But Ganymede seems to have been heated enough for the ice to melt, causing water to erupt in some areas. This has resurfaced parts of Ganymede with bright, lightly cratered ice flows — the equivalent of lava flows on the Earth and Moon. The pattern supports two of the major ideas so far. First, there is a fair amount of ice in Callisto and Ganymede, but there is also a component of dark stony material. Second, the most heavily cratered regions are darkest while the fresh ice regions are brightest.

As we travel inward toward Jupiter, we next encounter Europa, which also supports the pattern. Europa has a very bright surface reflecting 69 percent of the light and composed of almost pure water ice. It is virtually uncratered. Heating has apparently strongly melted ice in the interior, causing eruptions of water that resurfaced the whole satellite with bright ice. But what was the source of the heat?

All of these Galilean moons are large and may have had some internal heating due to intrinsic radioactivity. However, note that the level of eruptive activity, which has

destroyed the primitive cratered surfaces and produced fresh ice, has increased as we move in toward the planet. This suggests that Jupiter itself may have something to do with heating. There are two possibilities, either or both of which may explain the observations. First, we know that the giant planets radiated a large amount of heat as they formed. This may have been enough to melt the surface layers of the inner satellites but not the outer satellites.

Second, as discovered in 1979, tidal flexing within the innermost satellites was enough to produce substantial amounts of heat. The confirmation of this is the amazing inner Galilean satellite, Io, which is so heated by tidal flexing that it has continual volcanic activity. This volcanic activity, in turn, explains the absence of water on Io. So much of the material has been heated and erupted that all water has long since vanished from the outer layers of the satellite. This explains why the surface of Io is completely anomalous, dominated neither by the water-ice component nor the dark stony component, but rather by sulfurous "lavas" that were left after the H_2O was exhausted.

Moving inward from Io, we encounter Amalthea (Fig. 9). It too has an anomalous surface — reddish, dark, and cratered, but unlike C-type or D-type asteroids or satellites. Veverka and his co-workers have suggested that Amalthea is simply coated with sulfur compounds blown off of Io and spiraling in toward Jupiter. Tiny neighboring satellites, 1979 J 1, 1979 J 2, and 1979 J 3 (Figs. 2 and 10), may also have anomalous surfaces, for the same reason.

The innermost moons 1979 J 1 and 1979 J 3 orbit near the edge of the Jovian ring. Their relationship to the ring sheds further light on the evolution of satellites in general. The two moonlets are continually sandblasted by micrometeorites (which are most concentrated and fastest moving close to the planet) falling into Jupiter's gravitational field. Particles blown off the satellites orbit Jupiter and, influenced by certain dynamical forces, spiral in toward the planet. Apparently, these minuscule dust particles are the main constituents of the ring. The Jovian ring, then, is not static, primordial debris. Rather, it contains material from these tiny moonlets flowing through it and then falling into

Jupiter. Just as with a river, the material we see in it today will not be the same material in it tomorrow.

On Saturn's satellites (and perhaps those of Uranus and Neptune) the cratered surfaces do not seem to have darkened as much as in Jupiter's system. Perhaps the proportion of ices was even higher so far from the Sun, with too little dust to darken these satellites efficiently. However, there are trends that support the ideas discussed above. The reflectivity of Saturn's moons generally increases with decreasing distance to the planet. On Enceladus, one of the interior moons, some cratered areas seem to have been resurfaced with fresh ice (see Chapter 16); its surface is one of the most reflective known in the solar system. And Saturn's outermost satellite, Phoebe, is also the darkest.

Phoebe and its inward neighbor Iapetus give striking support for our ideas of darkening as a result of cratering. The side of Iapetus facing the direction of orbital motion may be darkened by dust knocked from the surface of Phoebe (see Chapter 16). Here we are not faced with a gradual deposition process; the dust striking Iapetus does so at high speed and does not reach the bright, presumably native material on the trailing hemisphere.

We know from samples of the lunar regolith that incoming meteoritic material never constitutes more than a few percent of the soil generated on a planet because larger meteorites keep digging up fresh planetary material from below, mixing it in with the surface material. Thus we can say that the leading hemisphere of Iapetus probably has no more than a few percent of the dark stony component from Phoebe. Perhaps this is enough to cause the observed blackening of the leading side of Iapetus, but it is more likely that the sublimation of the ice is also involved, as we postulated above. In any case, this seems to be a clear example of a cratering process which has darkened the target surface.

Fig. 12. This highly detailed image of Saturn's satellite Rhea from Voyager 1 reveals surface features as small as 2 km in diameter. Rhea's heavily cratered surface has probably survived intact since soon after the moon formed some 4.5 billion years ago. Light-colored rims on several craters may be fresh exposures of ice or condensed volatiles clinging to steep slopes.

SUMMARY

We have looked at the small bodies of the solar system in a new way. For a moment, we put aside the old observational categories of asteroids, comets, moons, and so on, which were originally based on telescopic appearance and present-day orbits, and later confused by evidence about their differing surface compositions. Instead, we looked upon these bodies as "planetesimals" with different initial compositions dependent mainly on their original location in the solar system. Some formed in orbit around planets and were moons from the start, possibly altered early by heat sources associated with nearby planets. Some formed in the depths of interplanetary space and were perturbed into various final locations, somewhat diffusing the original orderly spacing of composition groups.

The planetesimals that formed in the mid-to-inner asteroid belt were primarily rock objects, like ordinary meteorites; they have rocky surfaces. Some odd surface types among them may be associated with major collisions that split them and revealed interior materials such as metallic cores. The ones formed beyond that point had two important components: a very dark carbonaceous-rich stone component and a bright icy component. Ice-rich planetesimals passing through the inner solar system, where the ice component can sublime and produce a tail, came to be known as comets; the others are catalogued as asteroids — a diverse group with varied orbits and compositions. The proportions of ice varied with location, and in the region from the outermost asteroid belt to Uranus, the icy component was mostly frozen water. If these planetesimals were not disturbed by heating, their surfaces grew dark as ice was vaporized out of surface soils by impacts, leaving a dark regolith akin to carbonaceous chondrites. This applies to most outer solar system bodies smaller than a few hundred kilometers across. If heating was important, the ice melted, turning into a watery lava that may have erupted, producing bright ice surfaces. This situation occurred on many of the larger, inner moons. Some small satellites, such as Phobos, Deimos, J6 through J13, and Phoebe may have formed in the asteroid-to-Jupiter region as C-type or D-type planetesimals and then were captured by planets. Trojan asteroids may be related survivors of this group.

In short, the processes of planetary evolution have conspired to present us with a rich variety of mini-worlds.

Putting It All Together

John S. Lewis

The most challenging aspect of our study of the solar system is to understand the extraordinary diversity of the objects in it. It is difficult to discern the similarities and relationships between bodies as different as Vesta, Earth, Io, Saturn, Halley's Comet, and the Allende meteorite. And yet, in the most fundamental sense, they are all siblings in a single family, all sharing in a common genetic inheritance.

BEFORE THE PLANETS

These bodies, and every other member of the solar system, were shaped by physical and chemical processes in a large, dense gas and dust cloud surrounding the forming Sun. This gas and dust cloud probably formed from the collapse of a very massive interstellar cloud (Fig. 1). If it was at all typical, the cloud would have had thousands of times the mass of the entire solar system. During collapse, it may have repeatedly fragmented to produce thousands of stellar and

Fig. 1. Excited by the fierce radiation from hot young stars like those in the closely spaced quartet at center (the Trapezium), dense clouds of gas and dust glow with the bright green emissions of oxygen in the heart of the Orion nebula. Here, as elsewhere in countless galaxies, we are witnessing the collapse of an interstellar cloud and the formation of new stars. This photograph was taken by George Herbig in 1961, using the 3-m reflector of Lick Observatory.

planetary systems. From our fragment of this cloud, the solar nebula, grew all the bodies we observe today. Most of the other nebular fragments probably gave rise to double, triple, and more complex star systems. Our knowledge of stellar populations suggests that anywhere from a few percent to perhaps 20 percent of all stars are *not* members of multiple systems. These, plus very close double stars and extremely separated multiple stars, are candidates for formation of planetary systems at least vaguely similar to our own. All three classes of systems share the characteristic that planets would have stable orbits over billions of years.

We have pieced together some idea of the events that preceded our solar system's formation. To reconstruct the sequence theorists believe most likely, we first assume the original interstellar cloud to be an inhomogeneous mixture of debris from various stellar explosions and mass-loss processes. Its collapse produces occasional large fragments possessing little angular momentum. These quickly collapse to form stars with masses 20 or more times that of the Sun and luminosities 40,000 times greater. Such stars might run through their entire hydrogen-fusion lifetimes very rapidly, concluding their evolutionary courses in less than one million years. In contrast, stars comparable to the Sun last some 10,000 times longer, or about 10 billion years.

A stellar nebula evolves into a stable hydrogen-burning star in only about 100,000 years — a remarkably brief time relative to the age of the universe. This is actually a fortunate circumstance, because a group of stellar objects formed together will become smeared out within their galaxy in about 10 million years. Thus enormously massive stars are born, live their entire lifetimes, and die spectacularly in supernova explosions, while other fragments of the same interstellar cloud are still collapsing. The shock waves from supernova explosions rapidly dissipate the more tenuous outer portions of the cloud complex, while at the same time violently compressing the denser inner region. This compression encourages and may even cause the collapse of individual prestellar nebulae, and hence acts as a trigger for star formation.

Radioactive debris from a supernova explosion expands outward, as in Fig. 2, riding the shock wave at a few percent of the speed of light, cooling rapidly and condensing. First to solidify are nonvolatile (refractory) minerals containing short-lived radionuclides such as aluminum-26, which collect into small grains. These radioactive grains can then strike prestellar nebulae. If these nebulae are dense enough, the mineral grains decelerate, become captured, and then find their way into growing solid planetesimals.

Because of the strong instability of a hot, low-density shock front entering a cool, dense medium, the shock will break up into narrow tongues which can penetrate deep into the cooler nebular gas. As a result, the distinctive supernova grains mix inhomogeneously within the nebula.

Further collapse of a nebula leads to establishment of a well-defined spin axis. Cooling by infrared (heat) radiation from the resulting nebular disk permits continual shrinkage, countered by mass flow into the disk that maintains the high luminosity and disk temperatures. The disk flattens further, while angular momentum hinders material from moving inward toward the spin axis.

Temperatures in such a disk range from well over 1,000° K near the center to only a few tens of degrees near the periphery (Fig. 3). Pressures in the disk plane range from about 0.1 bar near the center to a mere one-millionth of that near the edge (1 bar is atmospheric pressure on the Earth). Under these conditions, not all mineral grains inherited from the interstellar medium vaporize; however, over the innermost portion of the disk they do, permitting elements contained in the local dust and gas to mix thoroughly. The nebula thus homogenizes material near its center effectively, but cannot erase certain types of chemical and isotopic inhomogeneities which survive at any temperature short of outright vaporization. In other words, the features most easily preserved are those carried by highly refractory mineral grains.

This, then, is our scenario. And while it will certainly seem oversimplified in years to come, we can still use it to attempt explanations for the general features of solar system bodies. The goal of such *cosmogony* is to generate soundly based physical and chemical models of the evolution of the nebular disk, the composition and accretion of solid objects within the disk, and the internal evolution of these bodies to their present states. Working the problem backwards from the present, however, proves extremely difficult (and per-

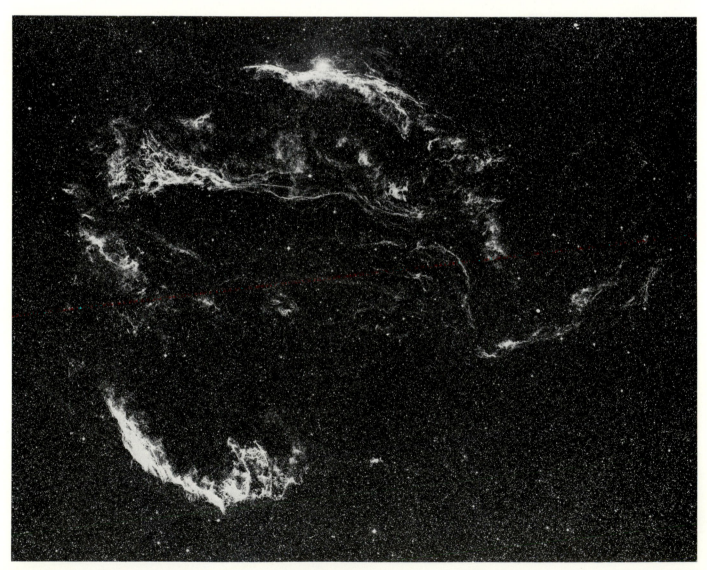

Fig. 2. The Cygnus Loop, a giant bubble of glowing gas and dust some 98 light-years (nearly one million billion kilometers) in diameter, is probably all that remains of a supernova explosion that occurred about 50,000 years ago. The delicate filaments seen in this red-light photograph from Hale Observatories consist mostly of interstellar matter swept up and compressed by the expanding blast.

haps impossible), since disruptive processes and the coupling of certain objects into resonant, repetitive motions has caused a wide variety of early logical possibilities to converge into the system we observe at present.

THE SOLAR SYSTEM TODAY

The preceding chapters provide a many-faceted discussion of the present state of our solar system. Our conclusions are based on data from the Earth, meteorites, other planets, the Sun itself, and astronomical observations of distant objects. The quality of this information varies, however. For example, we have enormously more data on Earth's structure, composition, and history than for any other body. Even so, we are limited in our knowledge of the Earth by two important factors. First, because of the difficulty of studying our planet's deep interior, we rely heavily on inferences from seismic data. Perhaps 99 percent of our geological and geochemical data on Earth pertain to the most accessible 1 percent of the mass of the planet. Second, we are too often unable to distinguish other planets' very general and even universal properties from the rare, idiosyncratic features unique to Earth.

These difficulties are alleviated somewhat by comparison of Earth data with meteorites. Most of the roughly 3,000 known meteorites, the *chondrites,* are primitive in structure (that is, they have not undergone melting and density-dependent geochemical differentiation). Historically, much of our understanding and discussion on the composition of Earth's mantle and core has been based on the study of the compositions of chondritic meteorite classes. The diversity and heterogeneity of meteorites make this a complex but rewarding process.

It is difficult to apply trends in meteoritic compositions to the planets because of our ignorance concerning where the various meteorite classes formed. But detailed chemical and mineralogical studies of a particular class reveals a great deal about physical conditions at some location in the accretion disk at a well-defined and very ancient time — about 4.6 billion years ago. Until the last few years it was commonly assumed by meteoriticists that *ordinary chondrites* came from the asteroid belt and volatile-rich *carbonaceous chondrites* from comets. We are now skeptical of these ideas, since photometric and spectroscopic data on asteroids show that the belt is dominated by carbonaceous material, and ordinary chondritic material is extremely rare and may be wholly absent from the belt. This means that the main meteoritic contribution on Earth is probably due to a small, local population of Earth-crossing asteroids with diameters of 1-20 km. These are in relatively unstable short-lived orbits, and the question of the source for replenishment of these bodies is open. Calculations reveal that most of these bodies collide with Earth and the other terrestrial planets after 10-100 million years.

Lunar data are derived mainly from synoptic studies of the Moon by spacecraft and from Earth-based analyses of a large suite of returned lunar samples from Apollo and Luna landers, all of which provide important scientific knowledge. Such data show the Moon to be a highly devolatilized and differentiated body with strong chemical similarity to the Earth and to chondritic meteorites.

Unfortunately, our sampling of the Moon is limited to nine sites, all on the near side and mostly in basalt-flooded mare basins, which comprise a very minor proportion of our satellite's total surface. Information on the Moon's deep interior is still limited because of the rarity of natural seismic events which serve as probes of the structure. The very existence of a lunar core is uncertain, yet in theory one with 20 percent of the Moon's diameter is possible. A further problem concerns the genetic relationship of the Moon to other solar system bodies. Variants of all the major pre-Apollo scenarios for lunar origin (fission from Earth, simultaneous accretion, and capture) are still tenable.

Planetary orbiters and landers have provided a limited glimpse of the abundances of radioactive elements at three points on the surface of Venus and of the major rock-forming elements at two points on Mars. Mass-spectrometric and gas-chromatographic analyses of the atmospheres of both planets have yielded considerable data on reactive-gas and rare-gas abundances. The very complex gravity field of Mars (and the simple one of Venus) have been mapped. Viking spacecraft searched for and failed to find life on Mars, but studied the local meteorological conditions for a full Martian year. The photographic mapping of Mars has revealed a wealth of surface detail, including huge volcanic constructs and obvious, widespread manifestations of great amounts of water and crustal ice. Radar mapping of Venus (through its clouds) has disclosed a topography strikingly unlike ours, with no clear evidence of global tectonics. Recent revelations on the atmospheric composition of both planets have stimulated much interesting work, but the dearth of geochemical and historical data hinders our interpretive efforts. Knowledge of the chemical, physical and thermal states of their interiors borders on calculated guesses, with no seismic data on Venus and virtually none on Mars.

Mercury resembles the Moon, but has a far higher density. We know almost nothing of the crustal

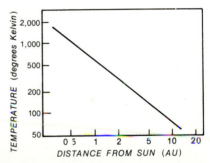

Fig. 3. As material in an accumulating nebula collapses inward toward its gravitational center, it begins to transform from a shapeless mass into a spinning, flattened disk and generates heat in the process. The small graph demonstrates how temperatures probably varied throughout the early presolar nebula.

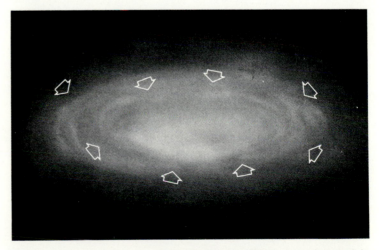

composition beyond a marginal hint of ferrous oxide (FeO) detected spectroscopically near 0.9 micron. A very large accumulation of metal in the interior is clearly indicated by Mercury's high bulk density, yet once again the very existence of a core is still a matter of inference from models of thermal history (Mariner 10 discovered a small magnetic field which *may* require a core dynamo). An intrinsic Mercurian atmosphere is virtually ruled out.

Our interplanetary wanderings have also moved outward to the Jovian planets. Voyager and Pioneer spacecraft have provided planetologists with a wealth of information about the atmospheres, magnetospheres, and satellites of both Jupiter and Saturn. Yet fundamental questions remain unanswered: What explains the colors observed (from Earth) in Jupiter's atmosphere? The global distributions of heat emanating from both giants? Their sporadic radio emissions? Perhaps atmospheric probes like that included on the Galileo mission will resolve such puzzles.

Voyager's exciting observations of Jupiter's Galilean satellites and of Saturn's Titan-dominated family of moons open new chapters in our study of the solar system. The astonishing spectacle of massive eruptions and sulfurous lava flows on Io demonstrates behavior encountered on no other known world. Likewise, revelations of tectonic activity in the rock-hard crusts of icy satellites challenge our theories and broaden our concepts of satellite formation. Yet, despite all this, we are no closer to understanding the gas giants and their satellites than we were in the study of Venus and Mars 15 years ago. Nothing is known experimentally about the interior structures of these bodies, and we have only inferences to guide us in understanding their thermal states and chemical compositions.

Uranus, Neptune, Pluto, and their satellites have not yet been visited by spacecraft and each confronts us with fundamental problems. Awaiting explanations, for example, are the thermal balance of Uranus, which alone among the Jovian planets appears to have no internal heat source; the odd

satellite system of Neptune, with massive Triton in a retrograde orbit; and the likelihood of a Pluto-Charon "double-planet" in a Neptune-crossing orbit. Little more than density and crude compositional data exist for Pluto, while the faint, distant satellites are not even *that* well characterized.

Comets remain the object of remote observations from Earth, largely by infrared and ultraviolet spectroscopy. Their evaporation, photolysis, and solar-wind interaction processes, while very interesting, are complex and very poorly understood.

GENERALIZATIONS AND GENESIS

The silicate surfaces of inner solar system bodies and the water-ice surfaces of many of the satellites surrounding Jupiter and Saturn give way to methane farther from the Sun. Similarly, the oxidized atmospheres of Venus, Earth, Mars, and Io give way to the reduced gases (such as hydrogen, methane, and ammonia) of the Jovian planets and the large, most distant satellites. The densest planet, Mercury, gives way to less dense Venus and Earth, then the even less dense Moon, Mars, and Io. Ice-rich bodies abundantly populate the outer solar system, while the Jovian planets, rich in hydrogen and helium, are least dense of all.

Harrison Brown suggested 30 years ago that the main compositional components of the solar system could be conveniently described as gases (H_2, He, and Ne), ices (H_2O, NH_3, CH_4), and rock (Fe, FeS, $(Fe,Mg)_2SiO_4$, and so on). Seeking trends for these compounds, we realize that objects in the solar system reflect differences in the vapor pressures of the rocky and icy components; in other words, far from the infant, still-condensing Sun, nebular temperatures were low enough for the condensation of ices.

We can now highlight ideas that try to explain the rich range of data described in this book. In reviewing these theories, we should keep in mind that all of them probably contain large elements of truth, but that no single model could reasonably be expected to explain all we presently

Fig. 4. One model of the solar system's formation envisions the accretion of asteroid-sized bodies, from which the planets ultimately accumulate. Residual material in the nebula dissipates quickly once the embryonic Sun ignites and begins radiating energy; planets then accrete in the absence of gas.

know. These genetic models have been proposed in the spirit that intentionally simplified descriptions with few variables can readily generate specific and testable predictions.

Virtually all models for the origin of the solar system begin with the accretion of primitive solar nebula gas and dust into larger bodies. But theorists have suggested two radically different ways of doing this, both of which may have been very important in different parts of the solar system. In the first method (Fig. 4), dust settles rapidly onto the symmetry plane of the nebula, forming grains by some combination of adhesive forces (electrostatic, magnetostatic, and so on) and gravitational instablity. This leads to the buildup of asteroid-sized bodies, which then accumulate in a few million years to a swarm of roughly lunar-sized objects in nearly circular orbits. "Stirring" of these orbits by Jupiter's gravity and by their mutual perturbations permits them to intersect, and the bodies slowly accrete to form the terrestrial planets. The entire process takes about 100 million years to form Earth, a little less for Mercury and Venus, and a little longer for Mars. By comparison, the nebular disk itself lasts only 100,000 years.

The second model (Fig. 5) works best far from the disturbing tidal pull of the Sun. Dynamic instabilities in the nebular disk shed rings of material from its outer edge. These rings then subdivide into several large, gravitationally bound, gaseous protoplanets, which eventually accumulate into planets with about the same elemental composition as we now find in the Sun.

In the dust-accretion model, the material available for each planet's formation is dominated by local dust, which is (to some degree) in chemical equilibrium with gases in the nebula around it. This equilibrium is very sensitive to temperature: brief exposure to intense heat permits the dust-gas mixture to react more completely than if this all happens over longer periods of time at temperatures that are several hundred degrees lower. Thus we presume the resulting compositions reflect the maximum temperature achieved in each locality during the nebula's condensation (Fig. 6).

What actually happens is obviously more complex. Very small grains may continue to react with the surrounding gas as it cools, long after the formation of larger grains rich in refractory elements (those with high evaporation temperatures). This would infuse the small grains with relatively more volatile compounds like carbon, water, and the halogens. Meanwhile, turbulence can preferentially carry these smaller grains far from the nebula's central plane into regions of lower density, where radiative cooling is more effective and the temperature lower. Turbulence can also lead to the mixing of small grains formed at various distances from the Sun, so that both volatile-rich and refractory-rich dust populations may be found in any location (Fig. 7).

In these preplanetary solids, the volatiles reside in minerals like hydroxyl silicates, sulfides, halides, and reactive forms of iron oxide. We find this general explanation works well for primitive chondritic meteorites, which show a higher concentration of all volatile elements in their fine-grained ("matrix") material than in their larger crystals of high-temperature compounds.

Since volatile-rich minerals are but a small fraction of the total preplanetary mass, large solid objects would have compositions dominated by the material condensing locally. This means that the chemistry and major-element mineralogy of a planet reflects, rather closely, the equilibrium composition of solid particles formed nearby in the solar nebula, at the highest temperature experienced by the dust-gas mixture. A planet's entire volatile inventory may be concentrated in as little as 1 percent of its mass — a quantity too small to influence the bulk density of the body detectably.

So far in this scenario, we have discussed only the consequences of chemical equilibration between dust and gas. There are very strong theoretical reasons to believe that, at the temperatures prevalent throughout much of the nebula, certain chemical reactions *cannot* reach equilibrium over its 100,000-year lifetime. For example, close to the Sun, CO is

Fig. 5. Another formation scenario involves the detachment of unstable rings of material from the contracting nebula's outer edge. These eventually coalesce into the planets and other bodies.

the chemically stable occurrence of carbon and N_2 the most stable nitrogen species. But in regions cooler than 680° K, methane (CH_4) is more stable than CO; and below 330° K, ammonia (NH_3) is more stable than N_2. So as the temperature drops, we should observe carbon and nitrogen converting to their more reduced forms. However, at such low temperatures the reactions involved are extremely sluggish, such that no more than a few percent methane or ammonia evolves even over the entire lifetime of the nebula. Intermediate products of the reduction reactions, species that ordinarily would be unstable, accumulate instead. These compounds include solid polymorphs of carbon, such as carbynes (crystalline long-chain polyacetylene solids) and a rich variety of organic matter.

Low-temperature chondritic meteorites do contain these species, but the *most* volatile-rich meteorites (the C1 and C2 carbonaceous chondrites) also incorporate much larger amounts of magnetite, carbonates, sulfates, hydroxyl silicates, and elemental sulfur. Yet no satisfactory scenario for making any of these species in the nebula is known! Moreover, strong evidence argues that these meteorites formed from the infusion of liquid water, carbon dioxide, oxidizing agents, and other volatiles into some more conventional type of chondritic material. For example, we believe that the oxidation of certain reduced forms of sulfur (primarily sulfides) has occurred in the presence of liquid water near its freezing point. Such processes are conceivable on the surface of a parent body, but are improbable in a cold, low-density nebula. So we must question whether any C1- or C2-type material was available at the time of planetary accretion in the inner solar system.

THE FORMATION OF THE PLANETS

Our concepts of chemical behavior within the nebula and the laboratory data on meteorite composition are used in different ways by those who model the formation of the terrestrial planets. Theorists agree with the idea of heterogeneous solids accreting to form each planet, but still debate the various details found among their individual models. At one extreme, large amounts of C1 material (some 10-20 percent of each planet's mass) could have accumulated together in a homogeneous mixture with refractory (volatile-poor) material, followed by the massive escape of all the volatiles in excess of the amounts now found on the terrestrial planets. This concept, the work of A. E. Ringwood, is an adaptation of his earlier theory in which each planet began as entirely C1 material. Taking a different approach, Edward Anders suggests that the distribution of material outward from the Sun allowed planets to accrete in layers or "zones" that differ in composition outward from their centers. Karl Turekian favors a scenario in which high-temperature condensates are accumulated first, followed by successive layers of ever more volatile-rich material which accretes as it condenses in the solar nebula. Here the final "veneering" of the Earth by C1 material brings in most of the volatiles.

My co-workers and I find no convincing reason to make C1 compounds an important component of the forming Earth. Instead, as it accreted the Earth may have captured solid material with a "tail" that extended out beyond its orbit; most of its volatiles would then come from C3 chondritic material near the orbit of Mars. Other accretion

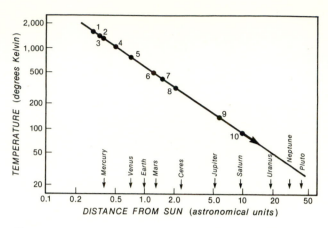

Fig. 6. Indicated in this diagram are the temperatures and locations at which major planetary constituents would be expected to condense from the primordial solar nebula (at a more evolved and cooler stage than in Fig. 3): *1*, refractory minerals like the oxides of calcium, aluminum, and titanium, and rare metals like tungsten and osmium; *2*, common metals like iron, nickel, cobalt, and their alloys; *3*, magnesium-rich silicates; *4*, alkali feldspars (silicates rich in sodium and potassium); *5*, iron sulfide; *6*, the lowest temperature at which unoxidized iron metal can exist; *7*, hydrated minerals rich in calcium; *8*, hydrated minerals rich in iron and magnesium; *9*, water ice; and *10*, other ices.

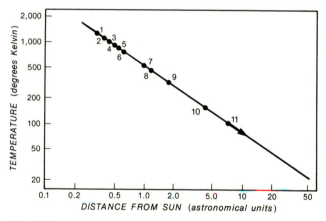

Fig. 7. A second diagram identifies the condensation temperatures of those minerals expected to be carriers of volatile elements: *1*, uranium and thorium (the source of helium through radioactive decay); *2*, iron phosphide; *3*, potassium compounds (the source of argon-40 through the decay of potassium-40); *4*, sodalite (rich in chlorine); fluorapatite (rich in fluorine); *6*, iron sulfide; *7*, the lowest temperature at which unoxidized iron metal can exist; *8*, calcium-rich hydrated silicates; *9*, iron- and magnesium-rich hydrated silicates; *10*, water ice and simple ammonium and carbonate salts; and *11*, other ices.

studies show that this method of collection should work, but the composition of these hypothetical captured solids is open to debate. As a point in its favor, this model correctly predicted the asteroid belt's domination by volatile-rich carbonaceous chondrites — a conclusion later confirmed by astronomical observations.

Furthermore, since the planets almost certainly took much longer to accrete than the nebula's 100,000-year existence, it seems unlikely that the accretion could have taken place in step with the cooling nebula, and the idea of strong compositional layering would be ruled out. We would still expect the terrestrial planets to have collected volatile-en-

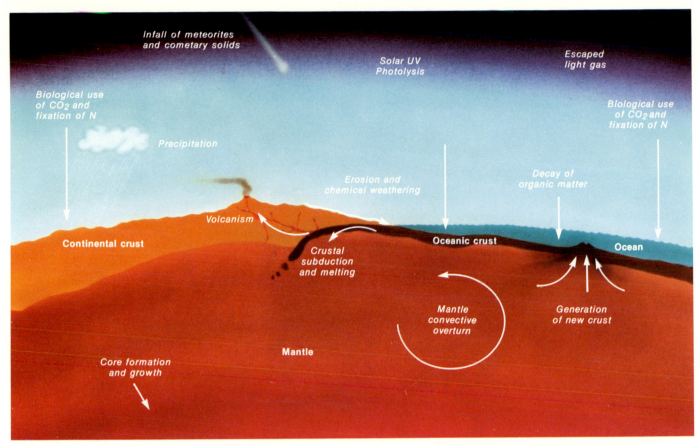

Fig. 8. **Are the atmospheres that now surround the planets the same ones present just after the formation of the solar system? Almost certainly not, since all of the processes illustrated here have been modifying atmospheric compositions since that time. The two most important mechanisms for supplying the atmospheres of the terrestrial planets are internal differentiation (followed by outgassing) and the accumulation of volatile-rich objects like comets.**

riched material near the end of their accretion, but in a different way. Small bodies near the fringes of the early solar system, composed largely of icy substances, in time would have been perturbed inward by gravitational tugs from the nearly completed planets. Migrating into the vicinity of the terrestrial planets, they would soon be swept up, supplying the Earth and its neighbors with a late surge of volatiles.

We can discuss briefly another consideration in solar system modeling. The temperature and composition trends that accompany increasing distance from the Sun are mimicked to some extent in the pre-planetary nebula from which the Jovian planets and their satellites condensed. For example, temperatures near the center of Jupiter's own nebula were apparently too high for ices to condense, and rocky material containing only a small proportion of water (in hydroxyl silicates) probably dominated. If the amount of metallic iron in such material is low enough, it can be completely oxidized by the water, with some water left over. In the thermal evolution of such a moon-sized body, heating from radionuclide decay will result in early outgassing of water vapor and carbon dioxide, which will readily escape from the body's gravity field. Once all hydrogen compounds are lost as well, sulfur and sulfur dioxide remain as the principal volatiles. In this way, a body with a roughly Mars-like composition could evolve to resemble Io if it is small enough — but not *so* small that the sulfur gases can also escape!

LOOKING TO THE FUTURE

The study of the origin and evolution of the solar system is still in its infancy. Our present sketches of (qualitatively) plausible histories are generally consistent with observation (Fig. 8). While encouraging, they are far from the last word. New observations may, at any moment, either shake or support these ideas. As we have seen, the realms of our present ignorance about the solar system are vast indeed.

Any satisfactory description of the early solar system will almost certainly draw upon many of the competing theories we have discussed. New elements will surely be introduced by both experiment and theory. Perhaps in the lifetimes of many of the readers of this volume we shall know in detail why Mercury has such a high metallic content, whether Venus ever had oceans, where the Moon came from, whether there was once a benign climate on Mars, what materials color the clouds of Jupiter, why planetary ring systems are so diverse, whether massive deposits of organic matter exist on Titan, why Uranus' spin axis is tipped at right angles to the axis of its orbit, what Pluto is made of, where comets were born, and which asteroids spawn meteorites. For now, all these questions are debated. Even where clear answers are known, not all can be sensibly integrated into the overall picture of solar system origin and evolution. There is a vast amount yet to find out, and the rewards of the search will be proportional to our effort.

Chapter Authors and Suggested Readings

Introduction

CARL SAGAN is the director of the Laboratory for Planetary Studies and David Duncan Professor of Astronomy and Space Sciences at Cornell University. He was a member of the science teams for the Mariner 9 and Viking missions to Mars and is currently on the imaging team for the Voyager missions to the outer planets. Dr. Sagan has served as chairman of the American Astronomical Society's Division of Planetary Sciences, chairman of the Astronomy Section of the American Association for the Advancement of Science, and president of the Planetary Section of the American Geophysical Union. For 12 years, he was editor-in-chief of *Icarus,* the international journal of solar system studies. He is also the president of The Planetary Society (P. O. Box 3599, Pasadena, California), a non-profit organization devoted to promoting planetary exploration. In addition to 400 published scientific and popular articles, Dr. Sagan has authored, coauthored, or edited more than a dozen books, including *Intelligent Life in the Universe, The Cosmic Connection, The Dragons of Eden, Murmurs of Earth* and *Broca's Brain.* In 1978 he received the Pulitzer Prize for literature. Dr. Sagan was the creator and host of the enormously successful television series *Cosmos,* first shown on the Public Broadcasting Service network in late 1980.

(The following books and articles involve general discussions of planetary science and solar system exploration.)

Chapman, C. R., *Planets of Rock and Ice* (Scribner's, 1982)

Cook, A. H., *Interiors of the Planets* (Cambridge, 1980)

Francis, P., *The Planets* (Penguin, 1981)

Hartmann, W. K., *Moons and Planets: An Introduction to Planetary Science* (Wadsworth, 1972)

Kaufmann, W. J., *Exploration of the Solar System* (Macmillian, 1978)

Kuiper, G. P., and B. M. Middlehurst, *Planets and Satellites* (Univ. of Chicago, 1961)

Kuiper, G. P., and B. M. Middlehurst, *The Earth as a Planet* (Univ. of Chicago, 1954)

Kuiper, G. P., and B. M. Middlehurst, *The Moon, Meteorites, and Comets* (Univ. of Chicago, 1963)

Kuiper, G. P., and B. M. Middlehurst, *The Sun* (Univ. of Chicago, 1953)

McAlester, A. L., ed., *The Prentice-Hall Foundations of Earth Science Series* (13 volumes of general interest)

Mitton, S., ed., *The Cambridge Encyclopaedia of Astronomy* (Crown, 1977)

Schultz, P., *Moon Morphology* (Univ. of Texas, 1976)

Short, N. M., *Planetary Geology* (Prentice-Hall, 1975)

The Solar System, a bound collection of articles appearing in the September, 1975, issue of *Scientific American* (W. H. Freeman, 1975)

Whipple, F. L., *Orbiting the Sun* (Harvard Univ., 1981)

Wood, J. A., *The Solar System* (Prentice-Hall, 1979)

Chapter 1: The Golden Age of Solar System Exploration

NOEL W. HINNERS is the director of NASA's Goddard Space Flight Center in Greenbelt, Maryland. After receiving a B.S. in Geology and Soil Chemistry, he studied at the California Institute of Technology and then at Princeton University, where he earned a PhD in geochemistry and geology in 1963. Dr. Hinners served as the head of the lunar exploration department at Bellcomm, Inc., where he evaluated scientific tasks for the first manned lunar landing. He was also involved in the selection of Apollo landing sites, development of the Apollo Lunar Surface Experiment Packages, and planning the scientific investigations to be carried out on the lunar surface and from orbit. In 1972, Dr. Hinners joined the staff of NASA and in 1974 was named Associate Administrator for Space Science. In this position he oversaw NASA's lunar and planetary programs, as well as those in physics, astronomy, and life sciences. Dr. Hinners was the founding editor of *Geophysical Research Letters,* a journal of the American Geophysical Union, and for three years was director of the Smithsonian Institution's National Air and Space Museum in Washington, D. C.

Beatty, J. K., "NASA and the Selling of Space Science" (*Sky and Telescope, 63,* 243-245, March 1982)

Beatty, J. K., "The Origin of NASA: A 20-Year Perspective" (*Sky and Telescope, 56,* 276-278, Oct. 1978)

Hall, C., *Lunar Impact: The Story of Project Ranger* (U. S. National Aeronautics and Space Administration, SP-4210, 1977)

Logsdon, J. M., "An Apollo Perspective" (*Aeronautics and Astronautics, 17,* Dec. 1979)

Newell, H. E., *Beyond the Atmosphere* (U. S. National Aeronautics and Space Administration, SP-4211, 1981)

Press, F., "Science and Technology in the White House, 1977-1980, Part 2" (*Science, 211,* 249-256, Jan. 16, 1981)

Report on Space Science 1975 (U. S. National Academy of Sciences, 1976)

Strategy for Exploration of Primitive Solar-System Bodies — Asteroids, Comets, and Meteoroids: 1980-1990 (U. S. National Academy of Sciences, 1978)

Chapter 2: The Sun

JOHN A. EDDY is a senior scientist at the High Altitude Observatory of the National Center for Atmospheric Research in Boulder, Colorado. He began his professional career as a line officer for the U. S. Navy, following his graduation from the U. S. Naval Academy in 1953. In 1958 he began working at the High Altitude Observatory and in 1962 he received a PhD from the University of Colorado. Affiliated with both these institutions since that time, Dr. Eddy also became a research associate of the Harvard-Smithsonian Center for Astrophysics in Cambridge, Massachusetts in 1977. His research interests include solar physics, the influence of the Sun on climate, the history of astronomy, astronomy of the American Indians, and archaeoastronomy. Dr. Eddy has written many journal articles on these subjects, and authored the NASA Special Publication *A New Sun,* published in 1979.

Eddy, J. A., *A New Sun: The Solar Results from Skylab* (U. S. National Aeronautics and Space Administration, SP-402, 1979)

Gamow, G., *A Star Called the Sun* (Viking, 1964)

Gibson, E. G., *The Quiet Sun* (U. S. National Aeronautics and Space Administration, SP-303, 1973)

Kiepenheuer, K. O., *The Sun* (Univ. of Michigan, 1959)

Maag, R. C., *et al.*, eds., *Observe and Understand the Sun* (Astronomical League, 1977)

Meadows, A. J., *Early Solar Physics* (Pergamon, 1970)

Menzel, D. H., *Our Sun* (Harvard Univ., 1959)

O'Leary, B., "The Stormy Sun" (*Sky and Telescope, 60,* 199-201, Sept. 1980)

Washburn, Mark, *In the Light of the Sun* (Harcourt Brace Jovanovich, 1981)

Chapter 3: Magnetospheres and the Interplanetary Medium

JAMES A. VAN ALLEN is one of the pioneers in space physics, having initiated scientific work with high-altitude rockets in late 1945. Since that time he has conducted experiments with rocketborne equipment at locations ranging from 82° north latitude to 77° south latitude and has been a principal investigator on 24 Earth-orbiting, interplanetary, and planetary spacecraft. In early 1958, during the International Geophysical Year, he and his students George H. Ludwig, Carl. E. McIlwain, and Ernest C. Ray discovered the radiation belts of the Earth with their instrument on the first successful American satellite, Explorer 1. Dr. Van Allen's research since that time has emphasized planetary magnetospheric physics, most recently that of Saturn. He has also done extensive work on solar X-rays, the propagation and acceleration of energetic solar electrons and protons in interplanetary space, and the intensity of the primary cosmic radiation at great heliocentric distances. He has been Professor of Physics and head of the Department of Physics and Astronomy at the University of Iowa since 1951.

Akasofu, S.-I., and S. Chapman, *Solar Terrestrial Physics* (Oxford Univ., 1972)

Brandt, J. C., *Introduction to the Solar Wind* (W. H. Freeman, 1970)

Fimmel, R. O., J. Van Allen, and E. Burgess, *Pioneer — First to Jupiter, Saturn, and Beyond* (U. S. National Aeronautics and Space Administration, SP-446, 1980)

Hundhausen, A. J., *Coronal Expansion and Solar Wind* (Springer-Verlag, 1972)

Krimigis, S. M., *et al.*, "Low-Energy Charged Particles in Saturn's Magnetosphere: Results from Voyager 1" (*Science, 212,* 225-230, 1981)

Stern, D. P., and N. Ness, *Planetary Magnetospheres* (U. S. National Aeronautics and Space Administration, TM-83841, Oct. 1981)

Chapter 4: The Collision of Solid Bodies

EUGENE M. SHOEMAKER is a geologist at the U. S. Geological Survey in Flagstaff, Arizona, and a Professor of Geology at the California Institute of Technology. Dr. Shoemaker conducted the first detailed geologic investigation of Meteor Crater in Arizona, the results of which helped provide a foundation for cratering research on the Moon and planets. He has helped develop methods of lunar geologic mapping and established a lunar geologic time scale. He was a coinvestigator on the television experiments for the Ranger and Surveyor unmanned lunar spacecraft. Dr. Shoemaker also participated in planning the geologic investigations for the Apollo landings. At present he is a member of the imaging team for the Voyager outer planets missions. He received a B.S. from CIT in 1947, Master's degrees from CIT and Princeton University, and a PhD from Princeton in 1960. Dr. Shoemaker's current research concerns paleomagnetism, the satellites of Jupiter and Saturn, and impact processes in the solar system.

Alvarez, L. W., *et al.*, "Extraterrestrial Cause for the Cretaceous-Tertiary Extinction" (*Science, 208,* 1095-1108, 1980)

Grieve, R. A. F., and M. R. Dence, "The Terrestrial Cratering Record II. The Crater Production Rate" (*Icarus, 38,* 230-242, 1979)

Hartmann, W. K., "Cratering in the Solar System" (*Scientific American, 236,* 84-89, Jan. 1977)

Oort, J. H., "The Structure of the Cloud of Comets Surrounding the Solar System, and a Hypothesis Concerning Its Origin" (*Bull. Astron. Inst. Netherlands, 11,* 91-110, 1950)

Öpik, E. J., *Interplanetary Encounters: Close-Range Gravitational Interactions* (Elsevier, 1976)

Strom, R. G., "Mercury: A Post-Mariner 10 Assessment" (*Space Science Reviews, 24,* 3-70, 1979)

Wetherill, G. W., "Late Heavy Bombardment of the Moon and Terrestrial Plants" (*Proc. 6th Lunar Sci. Conf.*, Pergamon, 1975, 1539-1561)

Wetherill, G. W., "Steady State Populations of Apollo-Amor Objects" (*Icarus, 37,* 96-112, 1979)

Chapter 5: Surfaces of the Terrestrial Planets

JAMES W. HEAD, III, is a Professor of Geological Sciences at Brown University. He studied as an undergraduate at Washington and Lee University (B.S. 1964) and received his PhD from Brown in 1969. From 1968 to 1972 he participated in selecting the landing sites for the Apollo program, while serving at Bellcomm, Inc., in Washington, D.C. During this time Dr. Head also helped plan and evaluate the package of experiments to be deployed on the Moon. In 1973-74 he was interim director of the Lunar Science Institute in Houston, Texas, which directs the program of lunar sample analysis. Dr. Head's research centers on the study of the processes that form and modify planets' surfaces and crusts, and how these processes vary with time and interact to produce the historical record preserved on the planet. In addition to his participation in unmanned planetary exploration, Dr. Head has helped to train astronauts who will fly on the Space Shuttle. He is an editor of *The Moon and the Planets,* an international journal, and directs the NASA-Brown University Regional Planetary Data Center.

Chapman, C. R., *The Inner Planets* (Scribner's, 1977)

Continents Adrift and Continents Aground: readings from *Scientific American* (W. H. Freeman, 1976)

Guest, J., *et al.*, *Planetary Geology* (Wiley, 1979)

Hartmann, W. K., "The Significance of the Planet Mercury" (*Sky and Telescope, 51,* 307-310, May 1976)

Head, J. W., *et al.*, "Geological Evolution of the Terrestrial Planets" (*American Scientist, 65,* 21-29, Jan.-Feb. 1977)

Murray, B. C., *et al., Earthlike Planets* (W. H. Freeman, 1981)

Mutch, T. A., *Geology of the Moon — A Stratigraphic View* (Princeton Univ., 1970)

Nozette, S., and P. Ford, "A World Revealed: Venus by Radar" (*Astronomy, 9,* 6-15, March 1981)

Ozima, M., *The Earth, Its Birth and Growth* (Cambridge, 1981)

Siever, R., "The Earth" (*Scientific American, 233,* 83-90, Sept. 1975)

Toksoz, M. N., "The Subduction of the Lithosphere" (*Scientific American, 233,* 88-98. Nov. 1975)

Chapter 6: Atmospheres of the Terrestrial Planets

JAMES B. POLLACK is a senior scientist in the space-science division of NASA's Ames Research Center in Moffett Field, California. He participated in the Mariner 9 mission to Mars as the imaging team member in charge of photographing the Martian moons Phobos and Deimos. Dr. Pollack served on the lander imaging team for the Viking Mars mission, where he was responsible for studies of the Martian atmosphere, and was an interdisciplinary scientist on the Pioneer Venus mission. He is currently an imaging team member for the Voyager outer-planet encounters. Dr. Pollack received an A.B. in physics from Princeton University in 1960, an M.A. in physics from the University of California at Berkeley in 1962, and a PhD in astronomy from Harvard University in 1965. From 1965 to 1968 he was a research physicist at the Smithsonian Astrophysical Observatory in Cambridge, Massachusetts. From 1968 to 1970 Dr. Pollack was Senior Research Associate at Cornell University's Center for Radiophysics and Space Research, in Ithaca, New York. Dr. Pollack's research interests center on planetary atmospheres and on understanding climatic change on Earth and other planets. He is a member of the editorial board for *Icarus*.

Barbato, J. P., and E. A. Ayer, *Atmospheres* (Pergamon Press, 1981)

Goody, R. H., and J. C. G. Walker, *Atmospheres* (Prentice-Hall, 1972)

Hoffer, W., "Looking Up To Know Ourselves" (*Mosaic, 11,* 39-44, Sept.-Oct. 1980)

Leovy, C. B., "The Atmosphere of Mars" (*Scientific American, 237,* 34-43, July 1977)

Sagan, C., "The Planet Venus" (*Science, 133,* 849-858, March 24, 1961)

Chapter 7: The Moon

BEVAN M. FRENCH has been NASA's discipline scientist for planetary materials research since 1975. In this position he administers the agency's research on lunar samples, meteorites, and cosmic dust. He received an A.B. in geology from Dartmouth College in 1958, an M.S. from California Institute of Technology in 1960, and a PhD from Johns Hopkins University in 1964. That same year he joined the NASA-Goddard Space Flight Center and remained there for eight years, studying ancient terrestrial meteorite impact structures. Beginning in 1969, Dr. French studied the lunar rocks returned from the Apollo 11, 12, and 14 missions. He was also selected as one of a small group of scientists to study material returned by the Russian Luna 16 unmanned probe, and in 1971 and 1972 he participated in astronaut training trips with the Apollo 16 and 17 crews. Dr. French has published several books and more than 35 technical papers about natural iron minerals, chemical reactions in rocks, terrestrial meteorite craters, and lunar samples. He has also written extensively about space science discoveries for the non-scientist. In 1977, Penguin Books published *The Moon Book*, his nontechnical description of recent lunar discoveries for the general reader.

Cadogan, P., *The Moon, Our Sister Planet* (Cambridge, 1981)

Collins, M., *Carrying the Fire: An Astronaut's Journeys* (Ballantine, 1974)

Cooper, H. S. F., Jr., *Moon Rocks* (Dial, 1970)

Cortwright, E. M., ed., *Apollo Expeditions to the Moon* (U. S. National Aeronautics and Space Administration, SP-350, 1975)

French, B. M., *The Moon Book* (Penguin, 1977)

Hartmann, W. K., "The Moon's Early History" (*Astronomy, 4,* 6-16, Sept. 1976)

Masursky, H., *et al.,* eds., *Apollo Over the Moon: A View From Orbit* (U. S. National Aeronautics and Space Administration, SP-362, 1978)

Shoemaker, E. M., "Geology of the Moon" (*Scientific American, 211,* 39-47, Dec. 1964)

Taylor, S. R., *Lunar Science: A Post-Apollo View* (Pergamon, 1975)

Taylor, G. J., *A Close Look at the Moon* (Dodd-Mead, 1980)

Wood, J. A., "The Moon" (*Scientific American, 233,* 93-102, Sept. 1975)

Chapter 8: Mars

HAROLD MASURSKY has been involved in nearly every U. S. program of lunar and planetary exploration. He received a B.S. in geology from Yale University in 1943 and his M.S. in 1951. Since that time he has been a member of the scientific personnel for the Ranger, Surveyor, and Lunar Orbiter unmanned lunar missions, the Mariner 9 and Viking missions to Mars, the Pioneer Venus mission, and the Voyager missions to the outer planets. For the Apollo program, Mr. Masursky was involved in landing site selection, geologic training of the astronauts, and analysis of the orbital and surface data brought back. In addition to his participation in lunar and planetary missions, Mr. Masursky is involved in the planning of future missions, including a Venus radar mapper and the Galileo Jupiter orbiter. He has written numerous publications dealing with the geology of the Moon and the terrestrial planets, and coauthored (along with other Mariner 9 television team members) the NASA publication *Mars as Viewed by Mariner 9*. Mr. Masursky is a senior scientist at the Branch of Astrogeologic Studies at the U. S. Geological Survey in Flagstaff, Arizona.

Arvidson, R. E., *et al.,* "The Surface of Mars" (*Scientific American, 238,* 76-89, March 1978)

Carr, M. H., *The Surface of Mars* (Yale Univ., 1981)

Mars as Viewed by Mariner 9 (U. S. National Aeronautics and Space Administration, SP-329, 1974)

Mutch, T. A., *et al., The Geology of Mars* (Princeton Univ., 1976)

Scientific Results of the Viking Project (*Journal of Geophysical Research, 82,* 3959-4681, 1977)

Viking Orbiter Views of Mars (U. S. National Aeronautics and Space Administration, SP-441, 1980)

Chapter 9: Life on Mars?

GERALD A. SOFFEN is the Director for Life Sciences, National Aeronautics and Space Administration in Washington, D. C. He is responsible for all medical and biological research within that agency. Prior to his present position, Dr. Soffen was the chief environmental scientist at NASA's Langley Research Center in Hampton, Virginia, where he oversaw scientific work in the atmospheric sciences, air and water environmental quality, and the Earth's climate. Dr. Soffen served as the Project Scientist for the Viking Project which successfully landed two spacecraft on Mars in 1976. From the start of the project in 1969 to the completion of its major objectives, he was responsible for all the spacecraft's scientific investigations and the activities of the science teams. Prior to Viking, Dr. Soffen managed biological instrument development at the Jet Propulsion Laboratory in Pasadena, California. He received a B.A. in zoology from the University of California at Los Angeles and a PhD in biophysics from Princeton University in 1960. Dr. Soffen is the author of many papers on exobiology, space biology, and planetology.

Cooper, H. S. F., Jr., *The Search for Life on Mars* (Harper and Row, 1980)

Dickerson, R. E., "Chemical Evolution and the Origin of Life" (*Scientific American, 239,* 70-86, Sept. 1978)

Goldsmith, D., and T. Owen, *The Search for Life in the Universe* (Benjamin-Cummings, 1980)

Horowitz, N. H., "The Search for Life on Mars" (*Scientific American, 237,* 52-61, Nov. 1977)

Klein, H. P., "The Viking Mission and the Search for Life on Mars" (*Reviews of Geophysics and Space Physics, 17,* 1655, Oct. 1979)

Masursky, H., *et al.,* "Martian Channels and the Search for Extraterrestrial Life" (*Journal of Molecular Evolution, 14,* 39-55, 1979)

Chapter 10: Asteroids

CLARK R. CHAPMAN is research scientist at the Planetary Science Institute, a division of Science Applications, Inc., in Tucson, Arizona. He is the author of about 100 articles on terrestrial planets, asteroids, Jupiter, and other topics relating to the origin and evolution of planets. He has written two popular books (*The Inner Planets* and *Planets of Rock and Ice*), has been a member of the National Academy of Sciences Space Science Board's planetary advisory committee, and is on the imaging team of the Galileo mission to Jupiter. Dr. Chapman did his undergraduate work at Harvard University and received his PhD from Massachusetts Institute of Technology.

Chapman, C. R., "The Nature of Asteroids" (*Scientific American, 232,* 24-33, Jan. 1975)

Gehrels, T., ed., *Asteroids* (Univ. of Arizona, 1979)

Morrison, D., and W. C. Wells, *Asteroids: An Exploration Assessment* (U. S. National Aeronautics and Space Administration, CP-2053, 1978)

Special Issue on Asteroids (*Icarus, 40,* no. 3, Dec. 1979)

Chapter 11: The Voyager Encounters

BRADFORD A. SMITH is the team leader of the Voyager imaging-science experiment and an Associate Professor of Planetary Sciences at the University of Arizona in Tucson. He received a B.S. in chemical engineering from Northeastern University in 1954. In 1958 he joined the faculty of New Mexico State University, where he developed a program of planetary photography. Later he became the director of planetary research and an Associate Professor of Astronomy. In 1972 he received a Doctorate in Astronomy from New Mexico State University. Dr. Smith served on the imaging teams for the Mariner 6, 7, and 9 missions to Mars and is a member of the Wide-Field/Planetary Camera team for the Space Telescope. At the University of Arizona, he heads a group involved in telescopic imaging using charge-coupled devices.

Beatty, J. K., "Voyager at Saturn, Act II" (*Sky and Telescope, 62,* 430-444, Nov. 1981)

Beatty, J. K., "Voyager's Encore Performance" (*Sky and Telescope, 58,* 206-216, Sept. 1979)

Berry, R., "Return to Jupiter" (*Astronomy, 7,* 6-23, Sept. 1979)

Eberhart, J., "Jupiter and Family" (*Science News, 115,* 164-173, March 17, 1979)

Gore, R., "Voyager Views Jupiter's Dazzling Realm" (*National Geographic, 157,* 2-29, Jan. 1980)

Morrison, D., *Voyages to Saturn* (U. S. National Aeronautics and Space Administration, SP-451, 1982)

Morrison, D., and J. Samz, *Voyage to Jupiter* (U. S. National Aeronautics and Space Administration, SP-439, 1980)

Science, special Voyager 1 at Saturn issue, *212,* no. 4491, 1981

Science, special Voyager 2 at Saturn issue, *215,* no. 4532, 1982

Washburn, M., *Distant Encounters* (Harcourt Brace Jovanovich, 1982)

Chapter 12: Jupiter and Saturn

ANDREW P. INGERSOLL majored in physics as an undergraduate at Amherst College, spending his summers at the Woods Hole Oceanographic Institution, where he was introduced to the subject of geophysical fluid dynamics. His interest in oceans and atmospheres was nurtured at Harvard, where he was awarded his PhD in 1966. That year Dr. Ingersoll joined the Division of Geological (and, more recently, Planetary) Sciences at the California Institute of Technology in Pasadena, California, where he has been a Professor of Planetary Science since 1976. Dr. Ingersoll is still interested in how planetary atmospheres and oceans work — how they redistribute their uneven heat inputs, why the winds and currents flow as they do, and how their climates change. He has participated as a member of several spacecraft experiment teams, including the Pioneer Venus infrared radiometer, the Nimbus 7 experiment to monitor the Earth's radiation budget, the Voyager imaging team, and the Pioneer Jupiter and Pioneer Saturn infrared radiometer experiments, the latter as principal investigator. He divides his professional time among teaching, analysis of spacecraft data, and developing dynamical models of what goes on.

Berry, R., "Voyager: Science at Saturn" (*Astronomy, 9,* 6-22, Feb. 1981)

Ingersoll, A. P., "The Meteorology of Jupiter" (*Scientific American, 234,* 46-56, March 1976)

Kerr, R. A., "Jovian Weather: Like Earth's or a Star's?" (*Science, 209,* 1219-1220, Sept. 12, 1980)

Schwartzenburg, D., "The Great Red Spot" (*Astronomy, 8,* 6-13, July 1980)

Science, special Voyager 1 issue, *204,* 945-1008, June 1, 1979

Science, special Voyager 2 issue, *206,* 925-996, Nov. 23, 1979

Wolfe, J., "Jupiter," (*Scientific American, 233,* 119-126, Sept. 1975)

Chapter 13: Planetary Rings

JOSEPH A. BURNS is a Professor of Theoretical and Applied Mechanics at Cornell University, where he received his PhD in 1966; his B.S. in naval architecture was from Webb Institute. In addition to his activities in Ithaca, he held a postdoctoral fellowship at NASA's Goddard Space Flight Center in 1967-68 and was a senior investigator during 1975-76 at NASA's Ames Research Center. Dr. Burns spent seven months in 1973 as an exchange fellow in Moscow and Prague, and worked at the Observatoire de Paris in 1979. His early publications include work on turbulence stimulation, charged-particle dynamics, kinetic art and heuristic studies in classical mechanics. For nearly a decade Dr. Burns's research interests have centered on planetary problems. His current work concerns planetary rings, the small bodies of the solar system (satellites, asteroids, and interplanetary debris), orbital evolution and tides, in addition to the rotational dynamics and strength of planets, satellites and asteroids. He edited *Planetary Satellites* (University of Arizona Press, 1977) and is Editor of *Icarus.*

Alexander, A. F. O'D., *The Planet Saturn* (Faber and Faber, 1962; Dover, 1980)

Burns, J. A., *et al.,* "Physical Processes in Jupiter's Ring: Clues to its Origin By Jove!" (*Icarus, 44,* 339-360, Nov. 1980)

Dermott, S. F., "The Origin of Planetary Rings" (*Phil. Trans. of Royal Society of London, A303,* 261-279, Dec. 1981)

Elliot, J. L., *et al.,* "Discovering the Rings of Uranus" (*Sky and Telescope, 53,* 412-416, June 1977)

Goldreich, P., and S. D. Tremaine, "Dynamics of Planetary Rings" (*Annual Reviews of Astronomy and Astrophysics,* 1982)

Owen, T., *et al.,* "Jupiter's Ring" (*Nature, 781,* 442-446, Oct. 11, 1979)

Pollack, J. B., and J. N. Cuzzi, "Rings in the Solar System" (*Scientific American, 245,* 105-129, Nov. 1981)

Chapter 14: The Galilean Satellites

TORRENCE V. JOHNSON, a member of the Voyager imaging team, has had a long-standing interest in the Galilean satellites of Jupiter. His research has centered on remote sensing of these bodies along with the Moon, the terrestrial planets, large planetary satellites, and asteroids. In the course of this research Dr. Johnson has used the 5-m (200-inch) Hale reflector, as well as the 2.5- and 1.5-m telescopes at Mount Wilson Observatory. He is now the supervisor of the optical astronomy group in the planetary atmospheres section of the Jet Propulsion Laboratory in Pasadena, California. In 1977 Dr. Johnson was named Project Scientist for the Galileo mission to Jupiter, planned for the late 1980's. Following his undergraduate work at Washington University (B.S., physics), he earned a PhD in planetary science at the California Institute of Technology in 1970.

Beatty, J. K., "The Far-Out Worlds of Voyager 1" (*Sky and Telescope, 57,* 423-427 and 516-520, May-June 1979)

Morrison, D., "Four New Worlds" (*Astronomy, 8,* 6-22, Sept. 1980)

Morrison, D., ed., *Satellites of Jupiter* (Univ. of Arizona, 1982)

Nature, special Voyager 1 issue, *280,* 725-806, Aug. 30, 1979)

Parmentier, E. M., *et al.,* "The Tectonics of Ganymede" (*Nature, 295,* 290-293, Jan. 28, 1982)

Peale, S. J., *et al.,* "Melting of Io by Tidal Dissipation" (*Science, 203,* 892-894, March 2, 1979)

Soderblom, L. A., "The Galilean Moons of Jupiter" (*Scientific American, 242,* 88-100, Jan. 1980)

Chapter 15: Titan

JAMES B. POLLACK is also the author of Chapter 6, "Atmospheres of the Terrestrial Planets." His biographical sketch is listed under that chapter.

Hunten, D. M., ed., *The Atmosphere of Titan* (U. S. National Aeronautics and Space Administration, SP-340, 1974)

Owen, T., "Titan" (*Scientific American, 246,* 98-109, Feb. 1982)

Smith, B. A., *et al.,* "Encounter with Saturn: Voyager 1 Imaging Science Results" (*Science, 212,* 163-191, 1981)

Tyler, G. L., *et al.,* "Radio Science Investigations of the Saturnian System with Voyager 1" (*Science, 212, 201-206, 1981)*

Chapter 16: The Outer Solar System

DAVID MORRISON is a planetary astronomer at the University of Hawaii, where he is Professor and Chairman of the Astronomy Graduate Program. His research interests are concentrated on the asteroids and the satellites of the outer planets and have utilized observations with optical, infrared, and radio telescopes, as well as data from spacecraft observations. Dr. Morrison was a coinvestigator on the Mariner 10 mission to Venus and Mercury, is associated with the Voyager imaging team, and serves as an interdisciplinary scientist on Project Galileo. In addition to his research activities, Dr. Morrison served for two years at NASA headquarters in Washington working with the planetary exploration program; he has also held offices in a number of scientific and professional societies, including chairman of the Division for Planetary Sciences of the American Astronomical Society. He is the author of more than 70 technical papers as well as many popular articles published in *Sky and Telescope, Astronomy, Mercury,* and *Scientific American.* His semi-popular books *Voyager to Jupiter* and *Voyages to Saturn* have recently been published by NASA, and he is the editor of a technical volume entitled *Satellites of Jupiter.*

DALE P. CRUIKSHANK spent the years following his undergraduate studies at Iowa State University working for the late G. P. Kuiper, first at Yerkes Observatory, then at the University of Arizona, where he received a PhD in 1968. A year of study at the U.S.S.R.'s Akademiya Nauk followed. In 1970 Dr. Cruikshank joined the staff of the Institute for Astronomy at the University of Hawaii in Honolulu, where today he pursues his major research interests of infrared and photometric observation of the planets (both their surfaces and atmospheres), satellites, asteroids, and comets, in support of NASA missions to these bodies. He gives particular attention to Io, Iapetus, Pluto, Triton, and numerous unusual asteroids. Dr. Cruikshank serves on several NASA committees and working groups devoted to planning future solar system exploration, and is a member of the infrared spectrometer team on the Voyager outer-planet missions. He is the author or coauthor of some 100 articles published in scientific journals and books and serves on the editorial board of *Icarus.*

Alexander, A. F. O'D., *The Planet Uranus* (Faber and Faber, 1965)

Asimov, I., *Saturn and Beyond* (Lothrop, 1979)

Berry, R., "Mysterious Pluto" (*Astronomy, 8,* 14-22, July, 1980)

Grosser, M., *The Discovery of Neptune* (Dover, 1979)

Harrington, R. S., and B. J. Harrington, "Pluto: Still an Enigma After 50 Years" (*Sky and Telescope, 59,* 452-454, June, 1980)

Hunt, G., ed., *Uranus and the Outer Planets* (Cambridge Univ. Press, 1982)

Tombaugh, C. W., and P. Moore, *Out of the Darkness: The Planet Pluto* (Stackpole, 1980)

White, A., *The Planet Pluto* (Pergamon, 1980)

Chapter 17: Comets

JOHN C. BRANDT is the chief scientist at the Laboratory for Astronomy and Solar Physics, part of NASA's Goddard Space Flight Center in Greenbelt, Maryland. He has been involved in such space projects as the Solar Maximum Mission and the International Ultraviolet Explorer mission, still in progress; he is also principal investigator on the high resolution spectrograph to be included with NASA's Space Telescope. Dr. Brandt received a B.A. in mathematics from Washington University in St. Louis in 1956 and a PhD in astronomy and astrophysics from the University of Chicago in 1960. Subsequently, he served at Mount Wilson and Palomar observatories, Kitt Peak National Observatory, and taught astronomy and physics at several major universities. Dr. Brandt has authored or coauthored more than 180 publications on such topics as astrophysics, solar physics, planetary atmospheres, and comets.

Brandt, J., and C. R. Chapman, *Introduction to Comets* (Cambridge, 1981)

Donn, B., *et al.,* eds., *The Study of Comets, Parts 1 and 2* (U.S. National Aeronautics and Space Administration, SP-393, 1976)

Gary, G. A., *Comet Kohoutek* (U.S. National Aeronautics and Space Administration, SP-335, 1975)

Moore, P., *Comets* (Scribner's, 1978)

Sekanina, Z., "Disintegration Phenomena in Comet West" (*Sky and Telescope, 51,* 386-393, June 1976)

Whipple, F. L., "The Spin of Comets," (*Scientific American, 242,* 124-134, March 1980)

Wilkening, L. L., ed., *Comets* (Univ. of Arizona, 1982)

Chapter 18: Meteorites

JOHN A. WOOD studied geology at Virginia Polytechnic Institute (B.S., 1954) and the Massachusetts Institute of Technology (PhD, 1958). While at MIT he became interested in the geologic properties of meteorites, about which little was known at that time. Since then Dr. Wood has pursued this subject at the Smithsonian Astrophysical Observatory in Cambridge, Massachusetts, where he is a member of its research staff. He has also worked at the Enrico Fermi Institute, University of Chicago, on meteoritic research. Dr. Wood participated in the analysis of lunar samples brought back by the Apollo astronauts. He is the author of *Meteorites and the Origin of Planets* (McGraw-Hill, 1968) and *The Solar System* (Prentice-Hall, 1979). Since 1973 Dr. Wood has taught solar system studies part-time in the Department of Geological Sciences at Harvard University.

Grossman, L., "The Most Primitive Objects in the Solar System" (*Scientific American, 232,* 30-38, Feb. 1975)

Sears, D. W., *The Nature and Origin of Meteorites* (Oxford Univ., 1978)

Wasson, J. T., *Meteorites* (Springer-Verlag, 1974)

Wood, J. A., *Meteorites and the Origin of Planets* (McGraw-Hill, 1968)

Chapter 19: Small Bodies and Their Origins

WILLIAM K. HARTMANN is a senior scientist at the Planetary Science Institute, a division of Science Applications, Inc., in Tucson, Arizona. He was an investigator on NASA's Mariner 9 project, which in 1971-72 mapped Mars from orbit for the first time. His research involves the origin and evolution of planetary surfaces. He has published dozens of technical and popular articles, and is author of two textbooks: *Astronomy: The Cosmic Journey,* and *Moons and Planets.* In addition to his research activities, Dr. Hartmann is an artist, specializing in astronomical subjects; he recently coauthored *The Grand Tour* with fellow artist Ron Miller (Workman, 1981). Dr. Hartmann received his B.S. at Pennsylvania State University in 1961, and advanced degrees in geology and astronomy at the University of Arizona, Tucson, in 1965.

Burns, J. A., ed., *Planetary Satellites* (Univ. of Arizona, 1977)

Hartmann, W. K., "Smaller Bodies of the Solar System," (*Scientific American, 233,* 143-159, Sept. 1975)

Larson, S., and J. W. Fountain, "Saturn's 'New' Satellites: A Perspective" (*Sky and Telescope, 60,* 356-360, Nov. 1980)

Morrison, D., and D. Cruikshank, "Physical Properties of the Natural Satellites" (*Space Science Reviews, 15,* 641-739, March 1974)

Veverka, J., "Phobos and Deimos" (*Scientific American, 236,* 30-37, Feb. 1977)

Chapter 20: Putting It All Together

JOHN S. LEWIS is a professor of planetary sciences in the Department of Earth and Planetary Sciences at the Massachusetts Institute of Technology in Cambridge, Massachusetts. Dr. Lewis has been deeply involved in planning the exploration of the outer planets, working with NASA, the Jet Propulsion Laboratory, the U.S. National Academy of Sciences, and the National Research Council. In addition, he has investigated interstellar communications, and has worked with the Committee on Planetary and Lunar Exploration (COMPLEX), a group which has developed the strategy for solar system exploration. Dr. Lewis received a PhD in geochemistry and physical chemistry from the University of California at San Diego in 1968, following his undergraduate studies at Princeton University and a Master's program at Dartmouth University, Dr. Lewis is an associate editor of *Icarus.* His research interests span inorganic and physical chemistry, meteoritics, planetary astronomy, geochemistry, and atmospheric thermodynamics.

Alfvén, H., and G. Arrhenius, *Evolution of the Solar System* (U. S. National Aeronautics and Space Administration, SP-345, 1976)

Cameron, A. G. W., "The Origin and Evolution of the Solar System" (*Scientific American, 233,* 66-75, Sept. 1975)

Cameron, A. G. W., and M. R. Pine, "Numerical Models of the Primitive Solar Nebula" (*Icarus, 18,* 377-406, Mar. 1973)

Falk, S. W., and D. N. Schramm, "Did the Solar System Start with a Bang?" (*Sky and Telescope, 58,* 18-22, July 1979)

Grossman, L., and J. W. Larimer, "Early Chemical History of the Solar System" (*Reviews of Geophysics and Space Physics, 12,* 71-101, Feb. 1974)

Lewis, J. S., "The Chemistry of the Solar System" (*Scientific American, 230,* 51-65, March 1974)

Planetary and Satellite Characteristics

Characteristics of the inner planets

	MERCURY	VENUS	EARTH	MOON	MARS
Reciprocal mass[1]	6,023,600	408,524	328,900	27,069,000	3,098,710
Mass[2] (Earth = 1)	0.0558	0.8150	1.0000	0.01230	0.1074
Mass[2] (kg)	3.302×10^{23}	4.871×10^{24}	5.975×10^{24}	7.350×10^{22}	6.421×10^{23}
Equatorial radius (Earth = 1)	0.382	0.949	1.000	0.2725	0.532
Equatorial radius (km)	2,439	6,052	6,378	1,738	3,398
Ellipticity[3]	0.0	0.0	0.0034	0.002	0.0059
Mean density (g/cm³)	5.42	5.25	5.52	3.34	3.94
Equatorial surface gravity (m/s²)	3.78	8.60	9.78	1.62	3.72
Equatorial escape velocity (km/s)	4.3	10.3	11.2	2.38	5.0
Sidereal rotation period	58.65 days	243.01 days (retrograde)	23.9345 hours	27.322 days	24.6229 hours
Inclination of equator to orbit	0°	2°	23°.44	6°.68	23°.98

Characteristics of the outer planets

	JUPITER	SATURN	URANUS[4]	NEPTUNE	PLUTO
Reciprocal mass	1,047.355	3,498.5	22,869	19,314	130,000,000
Mass (Earth = 1)	317.893	95.147	14.54	17.23	0.0022
Mass (kg)	1.900×10^{27}	5.688×10^{26}	8.70×10^{25}	1.030×10^{26}	1.31×10^{22}
Equatorial radius (Earth = 1)	11.27	9.44	4.10	3.88	(0.24)
Equatorial radius (km)	71,398	60,330	25,400	24,300	(1,500)
Ellipticity	0.0637	0.102	(0.024)	0.0266	?
Mean density (g/cm³)	1.314	0.69	(1.19)	1.66	(0.9)
Equatorial surface gravity (m/s²)	22.88	9.05	7.77	11.00	(0.4)
Equatorial escape velocity (km/s)	59.5	35.6	21.22	23.6	(1.1)
Sidereal rotation period at equator	9.841 hours[5]	10.233 hours[6]	15.5 hours	(15.8 hours)	6.3874 days
Inclination of equator to orbit	3°.08	29°	97°.92	28°.8	(≥ 50°)

Characteristics of planetary orbits

	(AU)	(10⁶ km)	Sidereal period (years)	Sidereal period (days)	Synodic period (days)	Mean orbital velocity (km/s)	Eccentricity	Inclination to the ecliptic (degrees)
MERCURY	0.387099	57.9	0.24085	87.969	115.88	47.89	0.2056	7.00
VENUS	0.723332	108.2	0.61521	224.701	583.92	35.03	0.0068	3.39
EARTH	1.000000	149.6	1.00004	365.256	—	29.79	0.0167	—
MARS	1.523688	227.9	1.88089	686.980	779.94	24.13	0.0934	1.85
JUPITER	5.202561	778.3	11.8623	4,332.71	398.88	13.06	0.0485	1.30
SATURN	9.554747	1,429.4	29.458	10,759.5	378.09	9.64	0.0556	2.49
URANUS	19.21814	2,875.0	84.01	30,685	369.66	6.81	0.0472	0.77
NEPTUNE	30.10957	4,504.3	164.79	60,190	367.49	5.43	0.0086	1.77
PLUTO	39.44	5,900.1	248.5	90,800	366.73	4.74	0.250	17.2

Satellites of Mars

Name	Discoverer	Year of discovery	Mean distance from Mars (km)	Sidereal period (days)	Orbital inclination (degrees)	Orbital eccentricity	Radius (km)	Mass (kg)	Mean density (g/cm³)
Phobos	A. Hall	1877	9,380	0.31891	1.0	0.018	14×10	9.6×10^{15}	(1.9)
Deimos	A. Hall	1877	23,500	1.26244	2.0	0.002	8×6	2.0×10^{15}	(2.1)

Satellites of Jupiter

Name	Discoverer	Year of discovery	Mean distance from Jupiter (km)	Sidereal period (days)	Orbital inclination (degrees)	Orbital eccentricity	Radius (km)	Mass (kg)	Mean density (g/cm³)
Metis	S. Synnott	1979	128,200	0.294	(0)	(0)	(20)	?	?
Adrastea	D. Jewitt, E. Danielson	1979	128,500	0.297	(0)	(0)	(15)	?	?
Amalthea	E. Barnard	1892	181,300	0.489	0.455	0.003	120	?	?
Thebe	S. Synnott	1979	223,000	0.675	(0)	(0)	(40)	?	?
Io	S. Marius, Galileo	1610	412,600	1.769	0.027	0.000	1,816	8.916×10^{22}	3.55
Europa	S. Marius, Galileo	1610	670,900	3.551	0.468	0.000	1,563	4.873×10^{22}	3.04
Ganymede	S. Marius, Galileo	1610	1,070,000	7.155	0.183	0.001	2,638	1.490×10^{23}	1.93
Callisto	S. Marius, Galileo	1610	1,880,000	16.689	0.253	0.007	2,410	1.064×10^{23}	1.81
Leda	C. Kowal	1974	11,110,000	240	27	0.147	1-7	?	?
Himalia	C. D. Perrine	1904	11,470,000	250.6	28	0.158	85	?	?
Lysithea	S. B. Nicholson	1938	11,710,000	260	29	0.12	3-16	?	?
Elara	C. D. Perrine	1905	11,740,000	260.1	26	0.207	40	?	?
Ananke	S. B. Nicholson	1951	20,700,000	617	147	0.169	3-14	?	?
Carme	S. B. Nicholson	1938	22,350,000	692	163	0.207	4-20	?	?
Pasiphae	P. Mellote	1908	23,300,000	735	147	0.40	4-23	?	?
Sinope	S. B. Nicholson	1914	23,700,000	758	156	0.275	3-18	?	?

Satellites of Saturn

Name	Discoverer	Year of discovery	Mean distance from Saturn (km)	Sidereal period (days)	Orbital inclination (degrees)	Orbital eccentricity	Radius (km)	Mass (kg)	Mean density (g/cm³)
1980 S 28 [7]	Voyager 1	1980	137,670	0.602	(0)	(0)	20×10	?	?
1980 S 27	Voyager 1	1980	139,350	0.613	(0)	(0)	70×40	?	?
1980 S 26	Voyager 1	1980	141,700	0.629	(0)	(0)	55×35	?	?
1980 S 3 [8]			151,422	0.694	(0)	(0)	70×50	?	?
1980 S 1 [8]			151,472	0.695	(0)	(0)	110×80	?	?
Mimas	W. Herschel	1789	185,540	0.942	1.517	0.020	196	(4.5×10^{19})	(1.4)
Enceladus	W. Herschel	1789	238,040	1.370	0.023	0.004	250	8.4×10^{19}	(1.2)
Tethys	G. D. Cassini	1684	294,670	1.888	1.093	0.000	530	7.55×10^{20}	1.21
1980 S 13	B. Smith et al. [9]	1980	294,670	1.888	(1.0)	(0)	17×13	?	?
1980 S 25	B. Smith et al.	1980	294,670	1.888	(1.0)	(0)	17×11	?	?
Dione	G. D. Cassini	1684	377,420	2.737	0.023	0.002	560	1.05×10^{21}	1.43
1980 S 6	P. Laques, J. Lecacheux	1980	378,060	2.739	(0)	(0)	18×15	?	?
Rhea	G. D. Cassini	1672	527,100	4.518	0.35	0.001	765	2.49×10^{21}	1.33
Titan	C. Huygens	1655	1,221,860	15.945	0.33	0.029	2,575	1.35×10^{23}	1.88
Hyperion	W. Bond	1848	1,481,000	21.277	0.4	0.104	205×110	?	?
Iapetus	G. D. Cassini	1671	3,560,800	79.331	14.7	0.028	730	1.88×10^{21}	1.16
Phoebe	W. Pickering	1898	12,954,000	550.4	150	0.163	110	?	?

Satellites of Uranus

Name	Discoverer	Year of discovery	Mean distance from Uranus (km)	Sidereal period (days)	Orbital inclination (degrees)	Orbital eccentricity	Radius (km)	Mass [10] (kg)	Mean density (g/cm³)
Miranda	G. Kuiper	1948	130,000	1,414	3.4	0.000	(160)	(2.2×10^{19})	(1.3)
Ariel	W. Lassell	1851	192,000	2.520	0	0.003	665	(1.6×10^{21})	(1.3)
Umbriel	W. Lassell	1851	267,000	4.144	0	0.004	555	(9.3×10^{20})	(1.3)
Titania	W. Herschel	1787	438,000	8.706	0	0.002	800	(2.8×10^{21})	(1.3)
Oberon	W. Herschel	1787	587,000	13.463	0	0.001	815	(2.9×10^{21})	(1.3)

Satellites of Neptune

Name	Discoverer	Year of discovery	Mean distance from Neptune (km)	Sidereal period (days)	Orbital inclination (degrees)	Orbital eccentricity	Radius (km)	Mass (kg)	Mean density (g/cm³)
Triton	W. Lassell	1846	355,000	5.877	160	0.00	(1,600)	(3.4 × 10²²)	(2.0)
Nereid	G. Kuiper	1949	5,562,000	365.21	28	0.75	(150)	(2.8 × 10¹⁹)	(2.0)

Satellite of Pluto

Name	Discoverer	Year of discovery	Mean distance from Pluto (km)	Sidereal period (days)	Orbital inclination (degrees)	Orbital eccentricity	Radius (km)	Mass (kg)	Mean density (g/cm³)
Charon	J. Christy	1978	19,700	6.387	94°	0.0	(750)	(1.6 × 10²¹)	(0.9)

Selected asteroids

Number and name	Discoverer	Year of discovery	Radius (km)	Spectral class[11]	Rotation period (hours)	Orbital period (years)	Mean distance from Sun (AU)	Mean distance from Sun (10⁶ km)	Orbital eccentricity	Orbital inclination (degrees)
1 Ceres	G. Piazzi	1801	487	C	9.078	4.60	2.766	413.8	0.079	10.6
2 Pallas	H. Olbers	1802	269	U	7.811	4.61	2.768	414.1	0.235	34.8
3 Juno	K. Harding	1804	134	S	7.21	4.36	2.668	399.1	0.256	13.0
4 Vesta	H. Olbers	1807	263	U	5.342	3.63	2.362	353.4	0.088	7.1
6 Hebe	K. Hencke	1847	98	S	7.275	3.78	2.426	362.9	0.203	14.8
7 Iris	J. Hind	1847	105	S	7.135	3.68	2.386	356.9	0.230	5.5
10 Hygiea	A. de Gasparis	1849	211	C	(18)	5.59	3.151	471.4	0.099	3.8
15 Eunomia	A. de Gasparis	1851	124	S	6.082	4.30	2.643	395.4	0.185	11.7
16 Psyche	A. de Gasparis	1852	119	M	4.303	5.00	2.923	437.3	0.135	3.1
51 Nemausa	Laurent	1858	74	U	7.78	3.64	2.366	354.0	0.065	9.9
433 Eros	G. Witt	1898	20 × 8	S	5.270	1.76	1.458	218.1	0.223	10.8
511 Davida	R. Dugan	1903	159	C	5.167	5.67	3.190	477.2	0.177	15.7
1566 Icarus	W. Baade	1949	0.7	U	2.273	1.12	1.078	161.3	0.827	23.0
1620 Geographos	A. Wilson, R. Minkowski	1951	1.0	S	5.225	1.39	1.244	186.1	0.335	13.3
1862 Apollo	K. Reinmuth	1932	0.8	S	3.063	1.81	1.486	222.3	0.566	6.4

Rings of Jupiter

	Radius (km)
Planet (at equator)	71,398
Secondary ring, inner edge	(71,398)
outer edge	122,800
Primary ring, inner edge	122,800
outer edge	129,200

Rings of Saturn

		Radius (km)
Planet (at equator)		60,330
D ring,	inner edge	67,000
C ring,	inner edge	73,200
	outer edge	91,700
B ring,	inner edge	91,700
	outer edge	117,500
Cassini division,	center	119,000
A ring,	inner edge	121,000
Encke division[12]		133,500
A ring,	outer edge	136,200
F ring (average)		140,600
G ring (average)		170,000
E ring,	inner edge	181,000
	outer edge	(480,000)

Rings of Uranus

	Radius (km)
Planet (at equator)	25,400
Ring 6	41,900
Ring 5	42,300
Ring 4	42,600
Ring α	44,800
Ring β	45,700
Ring η	47,200
Ring γ	47,700
Ring δ	48,300
Ring ε	51,200

NOTES AND EXPLANATIONS

[1]The mass of the Sun divided by the mass of the planet (including its atmosphere and satellites). [2]Satellite masses not included. [3]The ellipticity is $(R_e - R_p)/R_e$, where R_e and R_p are the planet's equatorial and polar radii. [4]Values in parentheses are uncertain by more than 10 percent. [5]Jupiter's internal (System III) rotation period is 9.925 hours. [6]Saturn's internal (System III) rotation period is 10.675 hours. [7]Alphanumeric designations denote the order within a given year that a satellite was first sighted; as with Jupiter's Metis, Adrastea, and Thebe, and Pluto's Charon, such objects have been sighted too few times to determine accurate orbital elements. [8]These co-orbital satellites were probably first seen in 1966. They were mistaken for a single object (Janus) by A. Dollfus and (1966 S 2) by J. Fountain and S. Larson. [9]B. Smith, H. Reitsema, J. Fountain, and S. Larson are members of the Space Telescope Wide-Field/Planetary Camera Team. [10]These values presume a mean density of 1.3, although a wider range of densities are possible. [11]See Chapter 10 for discussion of these classifications. [12]A gap in Saturn's Ring A often ascribed to J. Encke, although J. Keeler was likely the first to observe it.

Maps

The following 14 pages contain pictorial maps of most of the solid bodies in the solar system visited by spacecraft. Usually, the portrayal is a Mercator projection; as a result, the cited scales decrease away from the equator and polar features appear disproportionately large. Maps of the Galilean and Saturnian satellites are preliminary versions that will be upgraded with time.

In general, these illustrations were prepared using the cartographic and airbrush techniques of the U. S. Geological Survey in Flagstaff, Arizona. Exceptions are (1) the Earth map, prepared by Bruce C. Heezen and Marie Tharp (available from the latter at 1 Washington Avenue, South Nyack, N. Y. 10960), and (2) the lunar near-side map, prepared by the Defense Mapping Agency in St. Louis, Mo. The U.S.G.S. maps can be obtained through the agency's distribution centers at the Denver Federal Center, Denver, Colo. 80225; and at 1200 S. Eads St., Arlington, Va. 22202.

Mercury

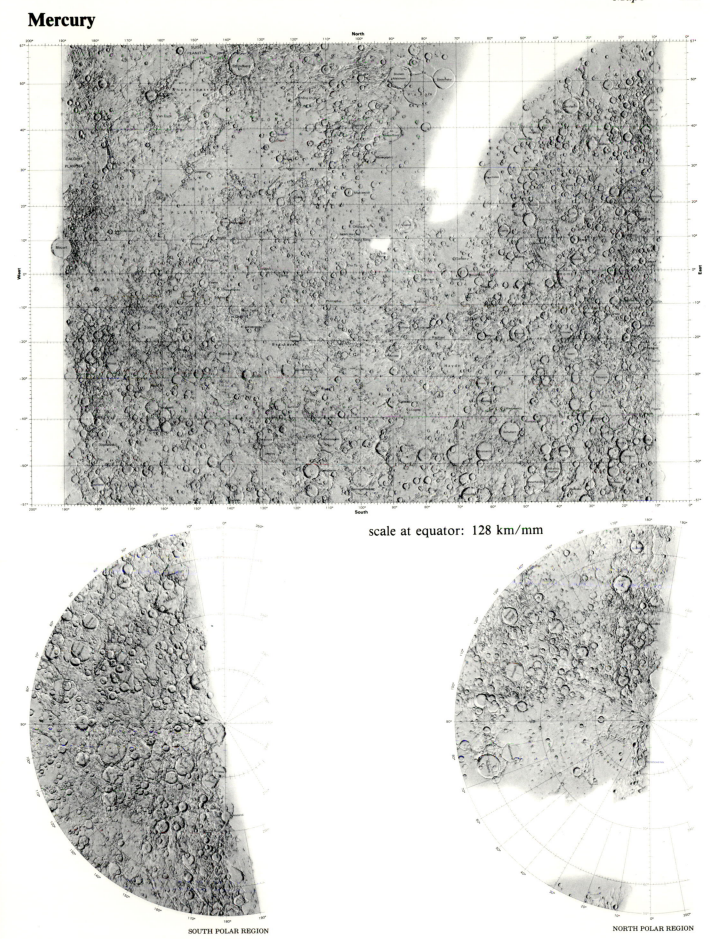

scale at equator: 128 km/mm

SOUTH POLAR REGION

NORTH POLAR REGION

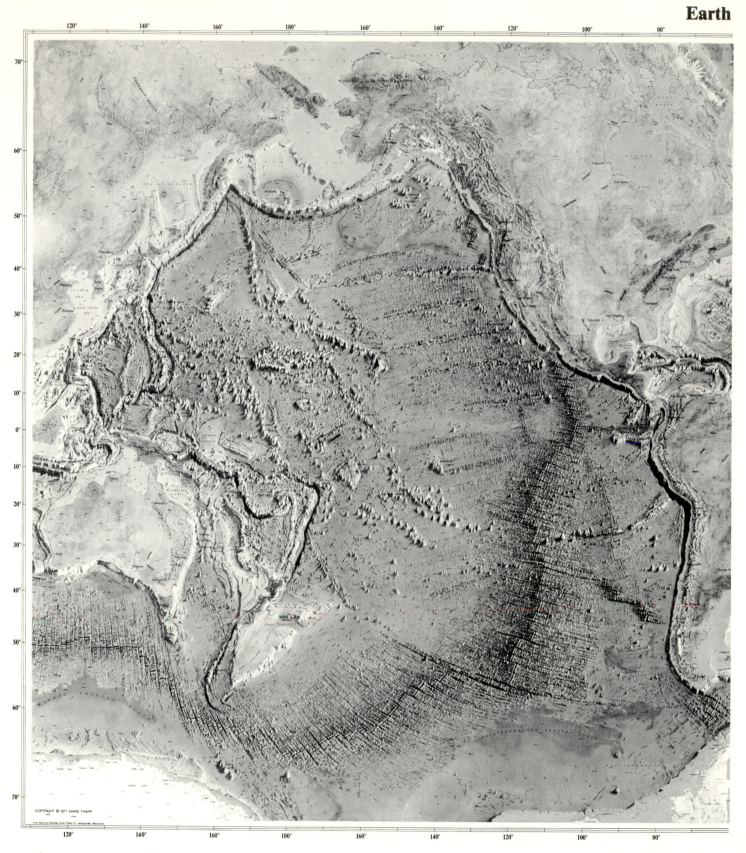

scale at equator: 115 km/mm

WORLD OCEAN FLOOR

BY BRUCE C. HEEZEN AND MARIE THARP

Based on Research and Exploration Initiated and Supported by the

UNITED STATES NAVY · OFFICE OF NAVAL RESEARCH

1977

Moon (near side)

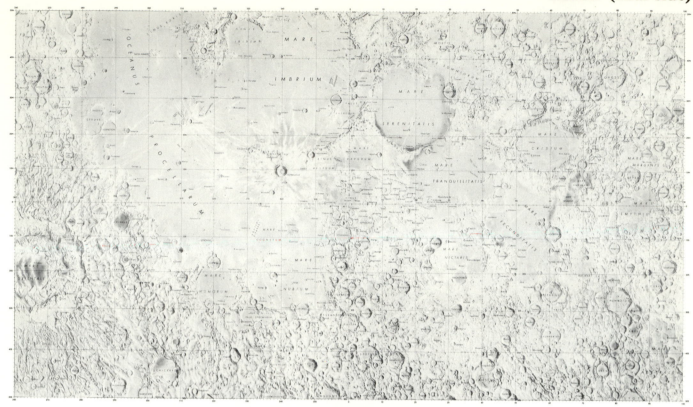

scale at equator: 35 km/mm

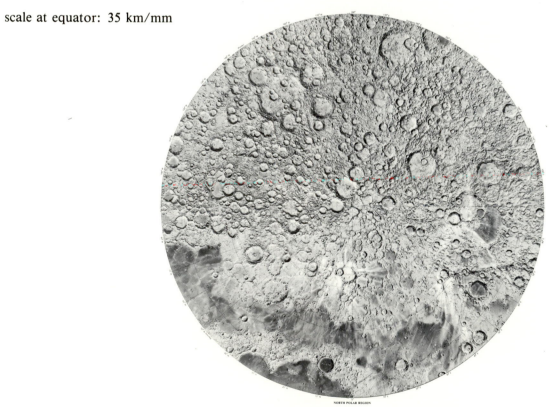

NORTH POLAR REGION

Moon (far side)

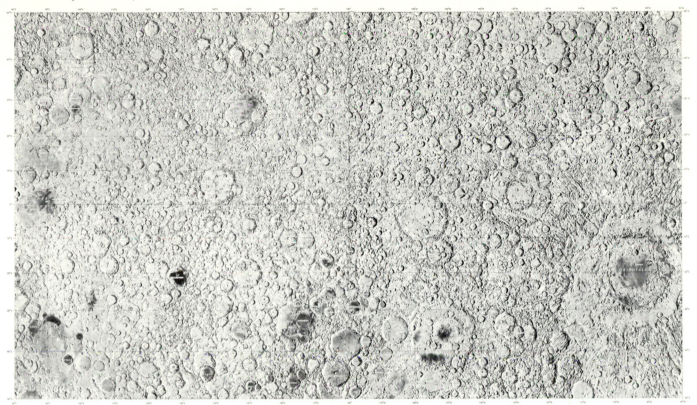

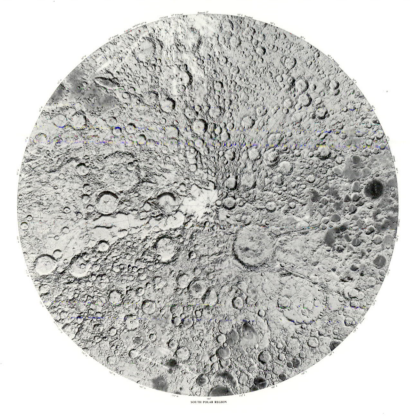

SOUTH POLAR REGION

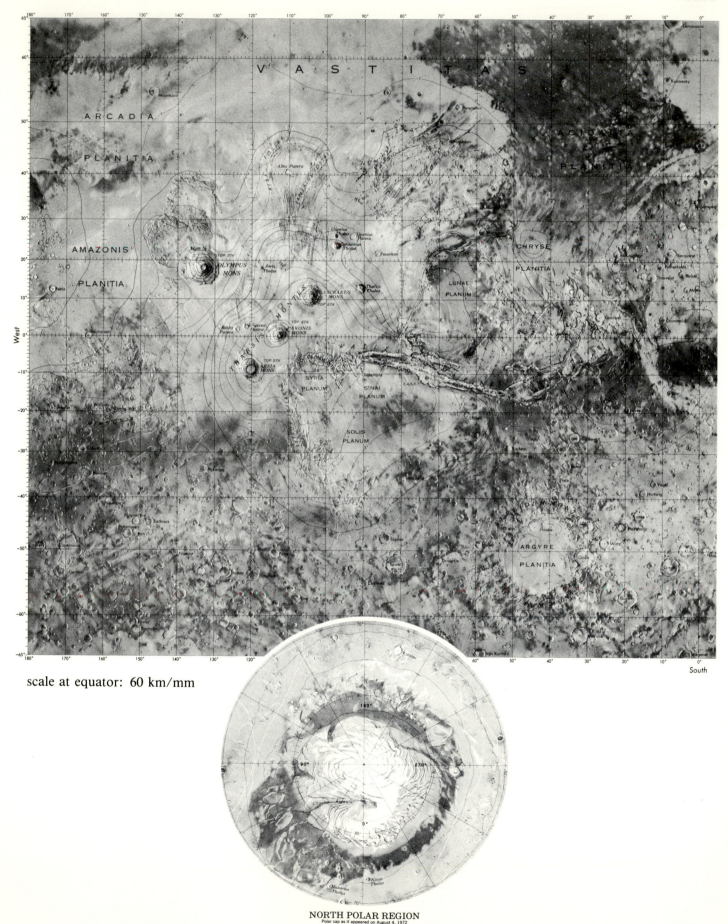

scale at equator: 60 km/mm

NORTH POLAR REGION
Polar cap as it appeared on August 4, 1972

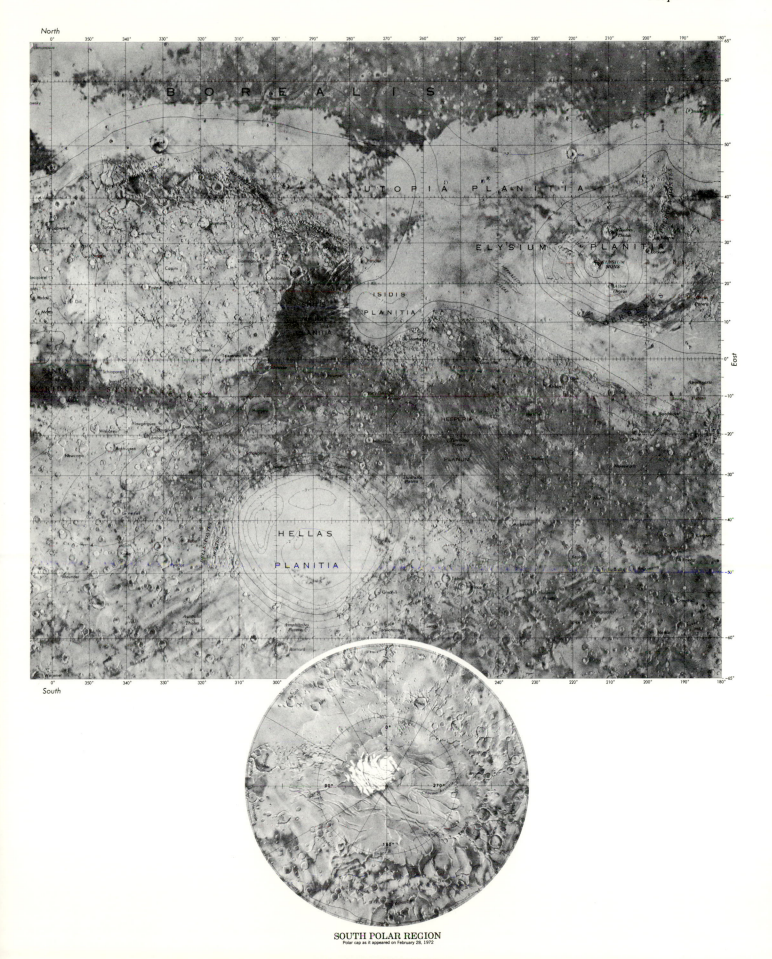

SOUTH POLAR REGION
Polar cap as it appeared on February 28, 1972

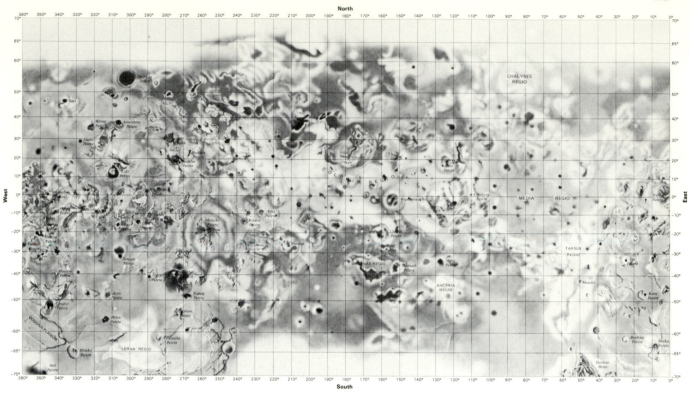

scale at equator: 67 km/mm

Europa

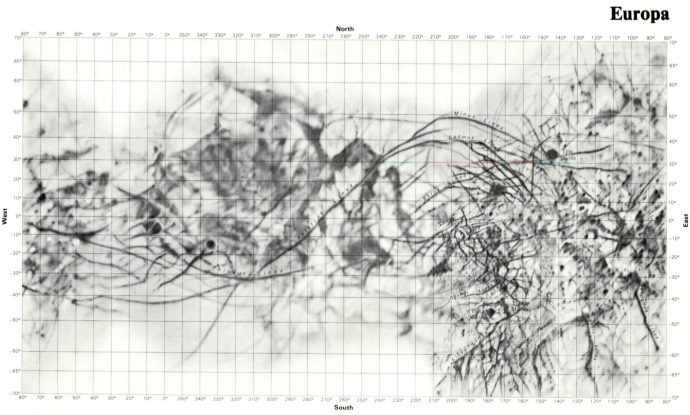

scale at equator: 58 km/mm

Ganymede

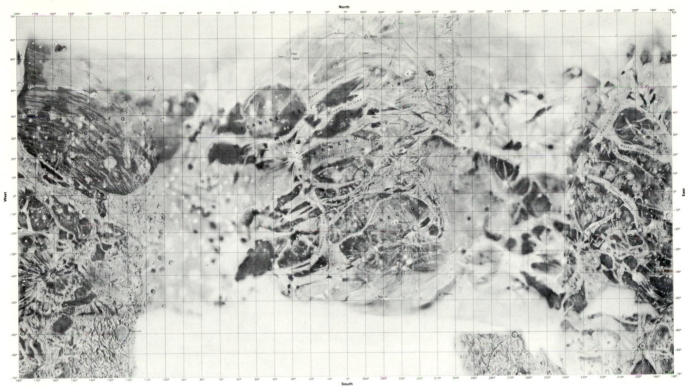

scale at equator: 96 km/mm

Callisto

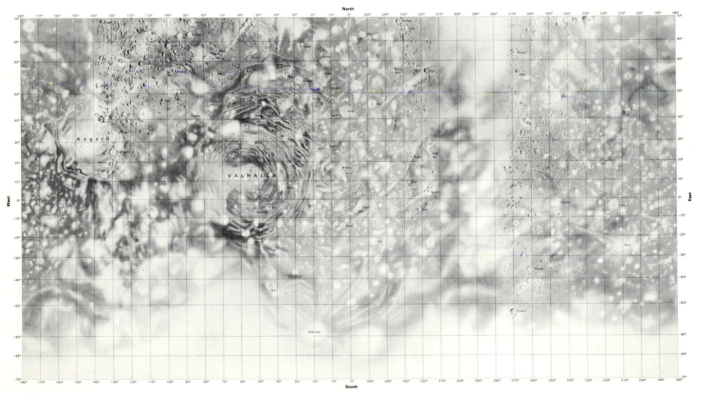

scale at equator: 88 km/mm

Mimas

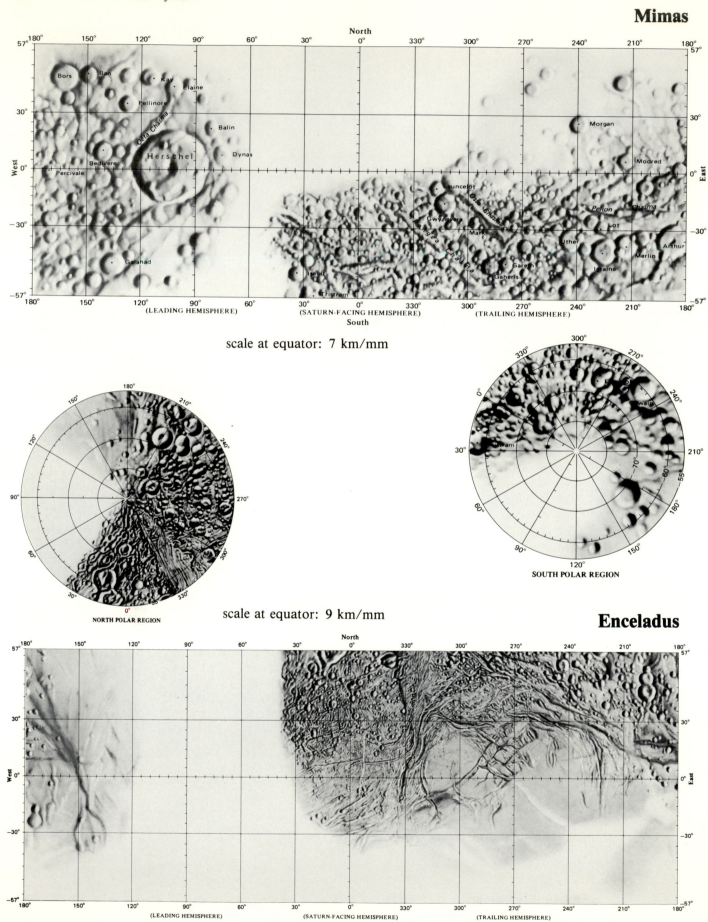

scale at equator: 7 km/mm

NORTH POLAR REGION

SOUTH POLAR REGION

scale at equator: 9 km/mm

Enceladus

Tethys

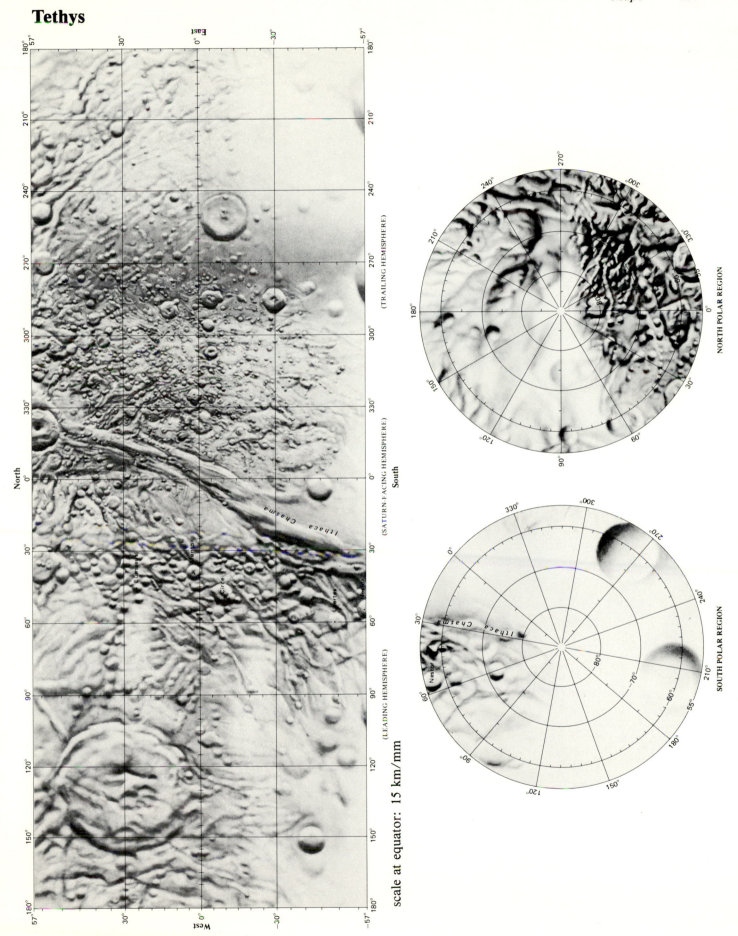

scale at equator: 15 km/mm

North

South

East

West

(TRAILING HEMISPHERE)

(SATURN-FACING HEMISPHERE)

(LEADING HEMISPHERE)

Ithaca Chasma

NORTH POLAR REGION

SOUTH POLAR REGION

Ithaca Chasma

Nestor

Dione

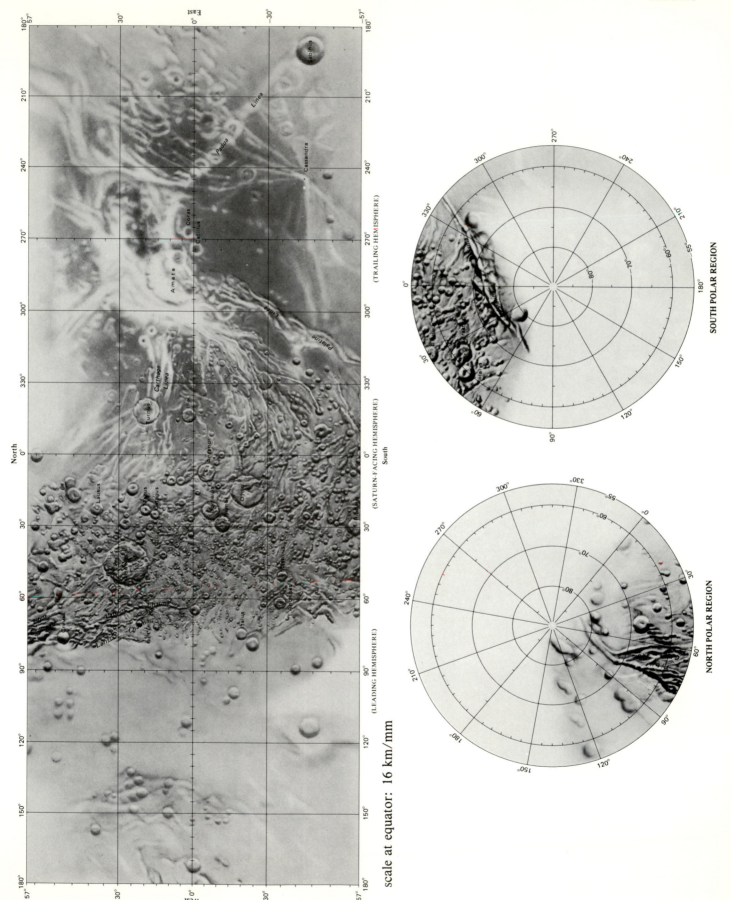

scale at equator: 16 km/mm

SOUTH POLAR REGION

NORTH POLAR REGION

Rhea

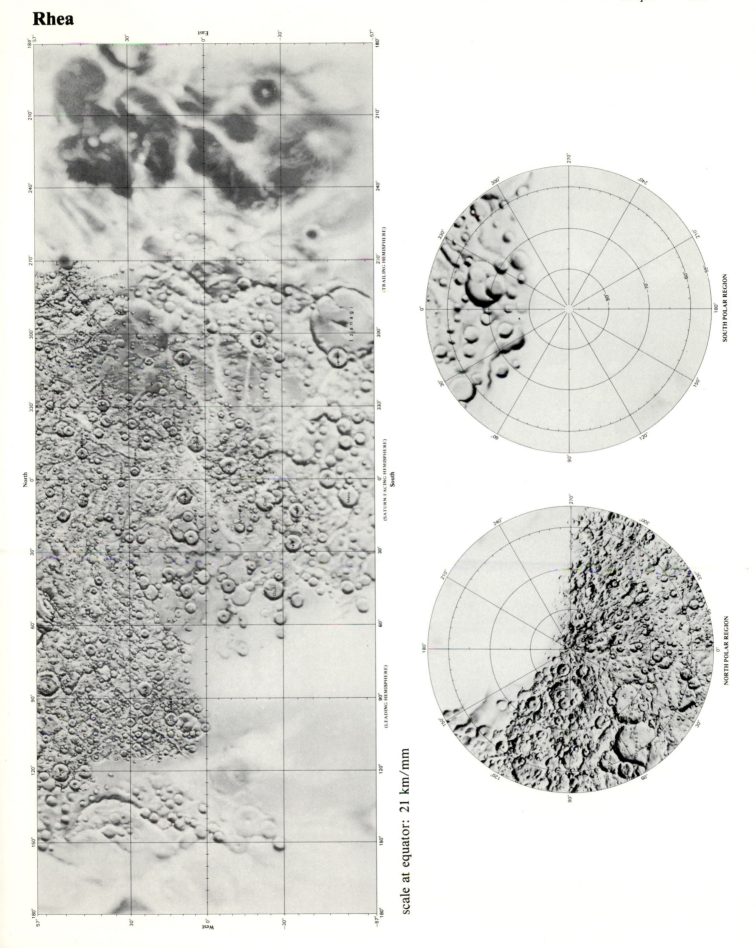

scale at equator: 21 km/mm

NORTH POLAR REGION

SOUTH POLAR REGION

Iapetus

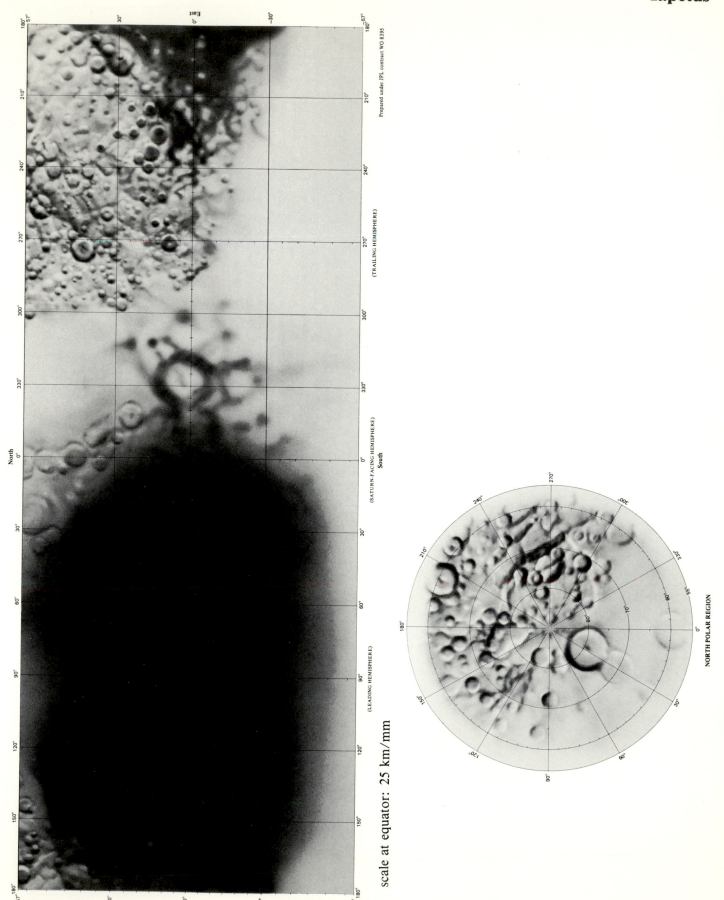

Prepared under JPL contract WO 8395

scale at equator: 25 km/mm

NORTH POLAR REGION

Index